Computer Applications in the Earth Sciences
AN UPDATE OF THE 70s

COMPUTER APPLICATIONS IN THE EARTH SCIENCES
A series edited by Daniel F. Merriam

1969 - Computer Applications in the Earth Sciences

1970 - Geostatistics

1972 - Mathematical Models of Sedimentary Processes

1981 - Computer Applications in the Earth Sciences: An Update of the 70s

Computer Applications in the Earth Sciences

AN UPDATE OF THE 70s

Edited by

Daniel F. Merriam

Jessie Page Heroy Professor of Geology
Department of Geology, Syracuse University
Syracuse, New York

PLENUM PRESS • NEW YORK AND LONDON

Library of Congress Cataloging in Publication Data

Geochautauqua (8th : 1979 : Syracuse University). Computer applications in the earth sciences, an update of the 70s.

(Computer applications in the earth sciences)
Proceedings of the 8th Geochautauqua, which was held Oct. 26-27, 1979, at Syracuse University, and which was sponsored by the Dept. of Geology, Syracuse University; the Division of Marine Geology and Geophysics, University of Miami; and the International Association for Mathematical Geology.
Includes bibliographies and index.
1. Earth sciences—Data processing—Congresses. I. Merriam, Daniel Francis. II. Syracuse University. Dept. of Geology. III. University of Miami. Division of Marine Geology and Geophysics. IV. International Association for Mathematical Geology. V. Title. VI. Series.
QE48.8.G44 1979   550'.28'54   81-10707
ISBN 0-306-40409-0   AACR2

Proceedings of the 8th Geochautauqua, held 26-27 October 1979,
at Syracuse University. The meeting was sponsored by the
Department of Geology at Syracuse University, the Division of
Marine Geology and Geophysics at the University of Miami, and
the International Association for Mathematical Geology.

© 1981 Plenum Press, New York
A Division of Plenum Publishing Corporation
233 Spring Street, New York, N.Y. 10013

All rights reserved

No part of this book may be reproduced, stored in a retrieval system, or transmitted, in any form or by any means, electronic, mechanical, photocopying, microfilming, recording, or otherwise, without written permission from the Publisher

Printed in the United States of America

in particular to our friends and colleagues

Lou Briggs
Milt Dobrin
Bill Krumbein

who contributed so much

LIST OF CONTRIBUTORS

F.P. Agterberg, Geological Survey of Canada, 601 Booth Street, Ottawa 4, Ontario, Canada KIA OE8

J.C. Brower, Department of Geology, Syracuse University, Syracuse, New York 13210

K.L. Burns, Department of Geology, Syracuse University, Syracuse, New York 13210

J.C. Davis, Kansas Geological Survey, 1930 Avenue "A", Campus West, University of Kansas, Lawrence, Kansas 66044

M.B. Dobrin, Department of Geology, University of Houston, Houston, Texas 77004 (deceased: 22 May 1980)

J.C. Griffiths, Department of Geosciences, Pennsylvania State University, University Park, Pennsylvania 16802

J.W. Harbaugh, Department of Applied Earth Sciences, Stanford University, Stanford, California 94305

W.W. Hay, Joint Oceanographic Institutions Incorporated, 2600 Virginia Avenue, N.W., Suite 512, Washington, D.C. 20037 (formerly of: Division of Marine Geology and Geophysics, University of Miami, Coral Gables, Florida 33124)

G.S. Koch, Jr., Department of Geology, University of Georgia, Athens, Georgia 30602

LIST OF CONTRIBUTORS

R.W. Le Maitre, Department of Geology, University of Melbourne, Parkville, Victoria 3052, Australia

C.J. Mann, Department of Geology, University of Illinois, Urbana, Illinois 61801

D.B. McIntyre, Department of Geology, Seaver Laboratory, Pomona College, Claremont, California 91711

D.F. Merriam, Department of Geology, Syracuse University, Syracuse, New York 13210

A.T. Miesch, U.S. Geological Survey, Denver Federal Center, Mail Stop 925, Denver, Colorado 80225

D.M. Raup, Field Museum of Natural History, Roosevelt Road at Lake Shore Drive, Chicago, Illinois 60605

R.A. Reyment, Paleontological Institute, Uppsala University, Box 558, S-751, 22 Uppsala 1, Sweden

P.G. Sutterlin, Canada Centre for Mineral & Energy Technology, 555 Booth Street, Ottawa, Canada KIA OG1

E.H.T. Whitten, Department of Geological Sciences, Northwestern University, Evanston, Illinois 60201

PREFACE

In looking back at the 1970's, the decade may prove to be a crucial one in the development of quantitative geology. After quantification had lain fallow and essentially undeveloped for 120 years, introduction of the computer in the 1950's revived interest and fostered advances in the subject. Developments continued through the 1960's at a rapid pace and the state-of-the-art was reported on in the proceedings of an international symposium held at the University of Kansas in June 1969 (Merriam, 1969).

The proceedings of the Kansas meeting, published as the first contribution in this series on "Computer Applications in the Earth Sciences" was one of 8 colloquia sponsored by the Kansas Geological Survey and the International Association for Mathematical Geology in the late 1960's. In a sense those international symposia were continued in the 1970's at Syracuse University as a series of Geochautauquas sponsored by the Department of Geology at Syracuse University and the International Association for Mathematical Geology. These proceedings report the results of the 8th Geochautauqua held in Syracuse on 26-27 October 1979.

The concept of the original meeting was simple - those working in each subdiscipline of geology would report on developments and activities in that field up to the end of 1969 and project ahead. The rationale for the first meeting was outlined in the preface by the editor

> Papers by leading experts in their field stress the "status-of-the-art." Speakers will discuss the use of computers in the earth sciences, past, present, and future. The meeting is planned for those not acquainted with the

tremendous advancements made in quantitative methods in recent years and those who are interested in future possibilities.

The concept behind the followup meeting also was simple - those who had reported on their field at the first symposium would do so again - an update of the 70's so to speak - at the second meeting. Each contributor, an expert in his area of interest, would recount developments and activity in that area for the 10-year period. The idea of having the same "expert" report on his field was excellent but in practice turned out to be not completely feasible. Some of the practitioners had moved into other areas, some new areas had come on the scene, and two of the original contributors unfortunately had died in the intervening interval. The loss of Lou Briggs and Bill Krumbein was sorely felt at the recent meeting (Milt Dobrin died after the meeting in 1980).

Most of the objectives of the meeting - a review of the decade and a look ahead for each field - were met. It is hoped that those in attendance (and those who read this volume) will benefit from exposure to the "experts". The presentations by geologic subject creates some duplication in discussion of methods and techniques but the duplication is minor.

Subjects covered by the speakers were biostratigraphy (J.C. Brower), geochemistry (A.T. Miesch), geophysics (M.B. Dobrin), information systems (P.G. Sutterlin), map analysis (J.E. Robinson), mineral- and fuel-resource forecasting (J.W. Harbaugh), mineral-resource evaluation (F.P. Agterberg), mining geology (G.S. Koch, Jr.), oceanography (W.W. Hay), paleoecology (R.A. Reyment), paleontology (D.M. Raup), petroleum exploration (J.C. Davis), petrology (R.W. LeMaitre), photointerpretation (K.L. Burns), stratigraphic analysis (C.J. Mann), and structural geology (E.H.T. Whitten).

An introduction to the subject of geology and computing was given by D.B. McIntyre with his paper entitled "Developments at the Man-Machine Interface". The stage was set for the meeting by the Dean of the Geomathematicians, J.C. Griffiths, with his presentation on "Systems Behavior and Geoscience Problem-Solving." A summation, based partly on the proceedings, is given here in the written communication by the editor (but was not presented orally at the meeting).

PREFACE

In summary - there were 7 authors who gave presentations at both meetings, 10 subjects were repeated and 3 were not (hydrology, petroleum engineering, and sedimentology), and 1 paper given orally is not presented here, 1 paper printed here was not presented, and 1 paper was available only as an abstract. So, some of the papers present only developments for the 10-year period, others give developments from the beginning, some project into the future, but all give an idea of what is in involved in that field as of the end of the decade of the 70s.

Participants in the international symposium were from many parts of the world. Countries represented were Australia, Canada, Great Britain, India, Sweden, United States, and West Germany.

About 17 percent of the participants were from industry, 56 percent from universities and colleges, 24 percent from government agencies, and 3 percent were independent consultants. A change in background of participants (e.g. contrast 60, 27, 12 and 1 % in those categories in 1969) is certainly evident. That change may reflect change in interest, differences in attitudes by the profession, location or promotion of the meetings, other considerations, or a combination of changes.

Many people helped with preparations for the meeting. In particular Ms. B.H. O'Brien, Janice Johnson, and Deborah Blose, assisted with arrangements. Several of the faculty and many of the students from the Department of Geology at SU helped with physical arrangements. The University of Miami and IAMG kindly cosponsored the meeting. Partial support for the meeting was given by V.P. Volker Weiss of the Office of Research & Graduate Affairs and was much appreciated. Janice Johnson typed the proceedings and Jim Busis of Plenum Press arranged for publication. Appreciation is expressed to all who helped make the meeting a success.

It is hoped that in a small way the efforts of the contributors may help stimulate interest and involvement of the readers of this volume in computer applications in the earth sciences. The truism that, as stated in John Griffiths' paper, the computer "...can serve as an intelligence amplifier..." seems to offer unlimited possibilities to geologists. Only the future will tell.

REFERENCE

Merriam, D.F., ed., 1969, Computer Applications in the Earth Sciences: Plenum Press, New York, 281 p.

Fontainebleau, France
August 1980

D.F. Merriam

Editor's Note: diacritical marks have been omitted

CONTENTS

List of Contributors vii

Preface ix

Systems Behavior and Geoscience Problem-
    Solving, by J.C. Griffiths 1

Developments at the Man-Machine
    Interface, by D.B. McIntyre 23

Computers as an Aid in Mineral-Resource
    Evaluation, by F.P. Agterberg 43

Quantitative Biostratigraphy, 1830-1980,
    by J.C. Brower 63

Computers in Geological Photointerpretation,
    by K.L. Burns 105

Looking Harder and Finding Less -- Use of
    the Computer in Petroleum Exploration,
    by J.C. Davis 125

Use of Computers in Seismic Reflection
    Prospecting, by M.B. Dobrin 145

Regional Mineral and Fuel Resource Fore-
    casting -- a Major Challenge and
    Opportunity for Mathematical Geologists,
    by J.W. Harbaugh 169

Computers in Oceanography (abstract),
    by W.W. Hay 179

Computer Applications in Exploration and
    Mining Geology: Ten Years of Progress,
    by G.S. Koch, Jr.   181

Some Developments in Computer Applications
    in Petrology, by R.W. Le Maitre   199

Stratigraphic Analysis: Decades of
    Revolution (1970-1979) and Refinement
    (1980-1989), by C.J. Mann 211

Computer Methods for Geochemical and
    Petrologic Mixing Problems, by
    A.T. Miesch   243

Computer as a Research Tool in
    Paleontology, by D.M. Raup 267

The Computer in Paleoecology, by
    R.A. Reyment   283

The Future of Information Systems in the
    Earth Sciences, by P.G. Sutterlin 305

Trends in Computer Applications in
    Structural Geology: 1969-1979, by
    E.H.T. Whitten 323

A Forecast for Use of Computers by
    Geologists in the Coming Decade
    of the 80s, by D.F. Merriam   369

Index   381

SYSTEMS BEHAVIOR AND GEOSCIENCE PROBLEM-SOLVING

John C. Griffiths

Pennsylvania State University

"The power of reason must be sought not in the rules that reason dictates to our imagination, but in the ability to free ourselves from any kind of rules to which we have been conditioned through experience and tradition."
R. Reichenbach, 1963
(from Weinberg, 1975).

ABSTRACT

Pure and applied science can be considered as two aspects of the same system of accumulating knowledge, a process termed the scientific method. The scientific method and its development may be viewed as a multitrack activity which includes cybernetics, operations research, and a systems approach, all of which are necessary to analyze and solve extremely complex problems. Advent of the computer, which can serve as an intelligence amplifier, was timely in that it is available now to assist in solving complex geological problems.

INTRODUCTION

It is proposed frequently that there are two types of science, pure or basic and applied, but any hypothesis built up by pure science is tested ultimately by its applications; I, therefore, consider that there is one type of science and pure and applied merely are different aspects of the same system of accumulating knowledge (see also Boulding, 1980).

Now the process of accumulating knowledge is termed "the scientific method" although it is hardly a one-track activity (Boulding, 1980, p. 533); it is more similar to a business cycle and it possesses the usual circadian rhythms. From time to time practitioners have endeavored to write the rules of the game (e.g., Chamberlin, 1897; Fisher, 1942; Weinberg, 1975) but scientific activity tends, in the long run, to evade these rules, a feature epitomized in the quote from Reichenbach.

Sometime ago I endeavored to summarize the evolution of the scientific method in a V-shaped figure (Griffiths, 1968a, fig. 1); in this view the trajectories, much simplified, converge upon General Systems Theory (von Bertalanffy, 1968), a "covering" which includes cybernetics, computers, operations research and many other recently developed activities. An alternative way of representing the same feature was included in an article on "Current Trends in Geomathematics" (Griffiths, 1970, fig. 1) based on a "classification of systems" by Beer (1964, p. 18) and again the convergence towards O.R., cybernetics, and systems approaches seem to be required to analyze, and possibly to solve, the exceedingly complex problems which now seem about to overwhelm us.

Fortunately the advent of the computer and its relatively recent application to geological problem-solving (Loudon, 1979) has served to offer us an intelligence amplifier (Ashby, 1956); the impact of the computer is captured in this masterpiece of understatement..."computers are almost certain to influence education deeply. Mathematical and logical skills will be revolutionized by it, in much the same way as the printing press led to general literacy" (Rt. Hon. Lord Robens, 1969). Comparison with the printing press suggests the following analog (Fig. 1).

Of course, for this level of development to be achieved successfully, it is necessary to accept the following recommendation; "The modern digital computer, properly used as prosthetic intelligence, rather than as a big adding machine doing faster what we always did before, enables us to extend our ability to use algebraic languages to describe and analyze the multidimensional complexity of the world we live in." (Gould, 1979). Then indeed the computer would become (a part of) a true intelligence amplifier system.

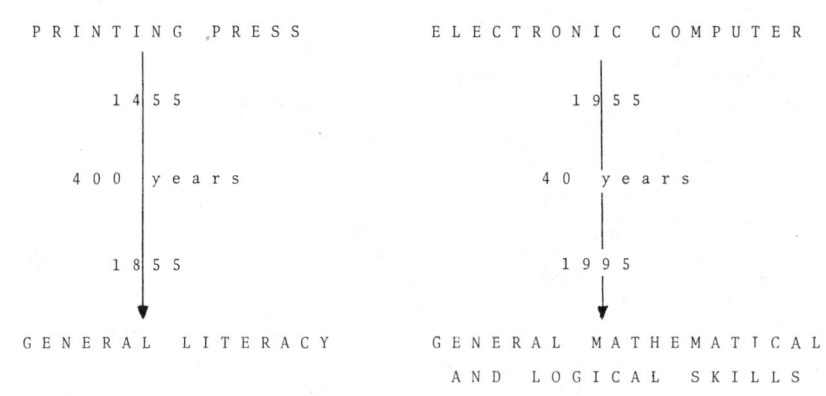

Figure 1. Analogous outcomes from effects of printing press and computer.

SYSTEMS ANALYSIS

Because systems analysis seems to be central to the procedure, it is as well to examine the concepts and some of its paradigms to understand their implications. Following Churchman (1968) a system may be defined as "a set of parts co-ordinated to accomplish a set of goals". The boundaries of a system, are identified to differentiate it from its environment and this requires the definition of those features which affect the system but which the system cannot affect. The system's environment leads to its "fixed constraints" (Churchman, 1968).

A system possesses objectives and it is necessary to postulate some manner of measuring the performance of the whole system in terms of its progress, through time, towards its objectives. In order to perform this activity the system must obtain certain resources which it usually "metabolizes". System components may be identified and then it becomes necessary to specify their activities, their goals and their measures of performance; finally there comes the "management of the system".

Such a system is easy to identify with individual organisms and with corporate institutions of various types and sizes; it is, however, less obvious that an inanimate object may be looked upon as a system in this sense. Nevertheless, because any object must be observed (by someone?) to be identified at all, then the coupling of the observer and observed achieves the level of a system. Furthermore if the idea of an inanimate object having goals is inclined to be confusing think first of the quote at the beginning and second because water "runs down hill" it evidently may be considered to possess an objective just as many processes which "strive to achieve equilibrium" do! Some profound questions in teleology and artificial intelligence arise at this juncture and it is necessary to keep an open mind until these questions have been adequately discussed. To ventilate some of these questions I have used such games as HEXAPAWN (Gardner, 1962) in which one plays a trival game against an "inanimate object" termed HER (Hexapawn Educable Robot). The game consists of a board of 3 by 3 squares and each of two players starts with three pawns; the objective of the game is to move one of the pawns from one's own to the opponent's base line. HER is composed of a set of 24 matchboxes organized so that they represent each of the decision points in the game; each matchbox contains a number of beads of two different colors, each of which corresponds to a possible move which HER may make at an appropriate point in the game. The two options are, move forward one square or take diagonally an opponent's pawn. When a position in the game requires a move by HER the matchbox corresponding to the current decision point is selected and shaken and the bead in the right-hand corner of the box determines HER's move. Each matchbox used in the game remains open until the outcome, win or lose, is attained. If HER wins, the moves which led to the successful outcome are reinforced by adding a bead of like color to each box; if HER loses then the beads in the right-hand corner of the boxes are removed. Thus appropriate moves are reinforced and "wrong" moves are removed (aversive conditioning of Skinner, 1971). In this manner HER "learns" and, after some 20 to 30 games, cannot be beaten. The question arises does HER have a purpose? If you answer yes what is the difference between HER and you? If you deny HER a purpose then can you demonstrate that you have a purpose? Before answering the question consider that "purposeful behavior requires that the acting object (i.e. system in our situation) be coupled with the goal, that is that the object registers messages from its surroundings" (i.e. be sensitive to its environment as HER is because of the built-in rewards and penalties). "If a goal is to be attained some signals from the goal are necessary at some time to direct the behavior, i.e. there must be a goal" (quotes from Rosenbleuth and Wiener, 1950; and Rosenbleuth, Wiener, and Bigelow, 1943; words in parentheses mine). These aspects of teology and artificial intelligence (such as learning) are described entertainly by Michie (1961); and then the abstruse question is raised "can a computer, or any machine, think"? This question is unresolved yet although there are strong opinions expressed in either direction (e.g. Lucas, 1961) and it is as well to realize that although the question seems to be simple enough the definitions required for its ultimate solution prove elusive and the question as usually stated may indeed be itself an example of Goedelian Incompleteness (Hofstedter, 1979).

Characteristic behavior of complex systems has been described by Forrester (1968, 1970) and may be paraphrased as follows: Complex systems tend

(1) to be counter-intuitive
(2) to be insensitive to changes in many of the (important) system parameters
(3) to be resistant to policy changes
(4) to possess influential pressure points usually in unanticipated places
(5) to compenate for external applied effects (homeostasis)
(6) to react to policy change in the short run opposite to the manner they react in the long run
(7) towards low performance.

In summary, they tend to degenerate towards, what I like to think of, as the Juggernaut Model, wherein the juggernaut is a six-wheeled flatbedded vehicle upon which is mounted a ten-ton idol surrounded by ceremonial curtains around which the "leadership" parades. The objective is to encourage the "faithful" to drag this monstrosity through loose wet sand during the monsoon season for some five miles from the jungle to the temple. There is no steering mechanism apart from persuading the faithful to hurl themsleves under the wheels for the future promise of great reward. This monster (i.e., model) should be recognizable easily, at least to organization theorists in academia, government, and industry. The sequence of events is portrayed elegantly by Argyris (1971). "In the social universe, where presumably there is no mandatory state of entropy, man can claim the dubious distinction of creating, organizations that generate entropy, that is slow but certain processes leading towards system deterioration". Expressed somewhat more vehemently "the ever accelerating conversion of resources into garbage, is, indeed, the chief characteristic of our culture" (Caster, 1970) which reflects some of the roots of the growing problem of "waste disposal". Apparently our overall system seems to be running down!

SOME CHARACTERISTICS OF EXCEEDINGLY COMPLEX PROBABILISTIC SYSTEMS

The deterioration of many of the subsystems in which we live emphasizes the feature that these systems are not simple closed systems but, in the spirit defined by Beer (1964), they are exceedingly complex probabilistic systems and they possess characteristics which are important in geological problem-solving.

One of the features characteristic of this type of system is it's memory; in geological terms the Principle of Uniformitarianism implies, in some sense, a long, almost infinite memory and many aspects of geological (and other related sciences rely on this feature to justify their procedure. Yet it is considered difficult to almost impossible to predict forwards in time for short periods of from 5 to 10 years. Memory is discussed by de Bono (1971) and from his description I have constructed the examples in Figure 2. In this situation, faced with the distribution of marbles, one has to reconstruct the procedure and starting position used in obtaining their disposition; in Figure 2A the substrate is loose (wet) sand and the positions of the marbles show that they were dropped at equal intervals on a straight line. In Figure 2B the substrate is a corrugated surface and the disposition of the marbles now is more complex; given that the investigator knows that the surface "was" corrugated he may reconstruct the manner in which the pebbles were dropped. But, if the form of the substrate is not known, then there is no unique solution to the problem. In this situation the observations solely of the disposition of the marbles is insufficient to reconstruct the events. Finally in the third example the marbles were dropped on a concrete surface and the marbles end up randomly distributed; in this situation it is not possible to reconstruct the preceeding events. Obviously memory becomes more and more limited from Figures 2A to 2C. How many geological processes possess these characteristics?

A second feature which is related closely to this limited memory is the system's property known as

GEOSCIENCE PROBLEM-SOLVING

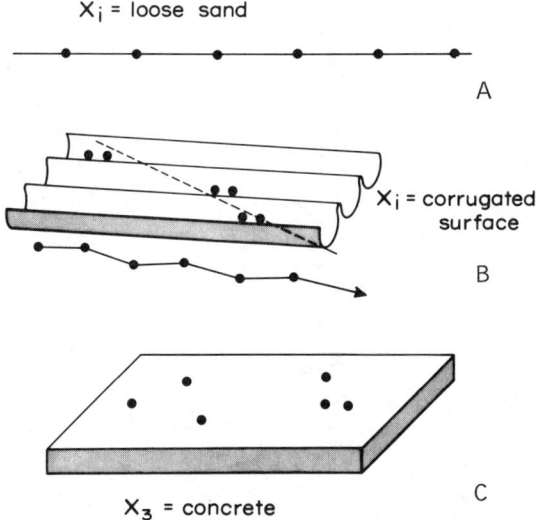

Figure 2. Exogenous versus endogenous variability or interaction between heredity and environment represented as disappearing memory trace (after de Bono, 1969).

equifinality (von Bertalanffy, 1968) in which the same outcome may be obtained from a series of events, or processes, by different routes or from different starting states. The distribution of marbles dropped on concrete is an example; the random disposition of the marbles may be obtained by several different combinations of starting states and procedures. Many processes with Markovian memories possess these characteristics.

Another characteristic of some importance is mentioned by Simon (1969, p. 23ff.) where he describes the path of an ant to a goal (see Fig. 3) and deduces that "an (ant) viewed as a behaving system, is quite simple, the apparent complexity of its behavior over time is largely a reflection of the complexity of the environment in which it finds itself" (Simon, 1969, p. 25). He then goes on to suggest that the word in parentheses may be replaced by man, skier, sloop, or the search for a problem solution and the trajectory yet would be appropriate. He states that "Only human pride argues that the apparent intricacies of our path stem from a quite different

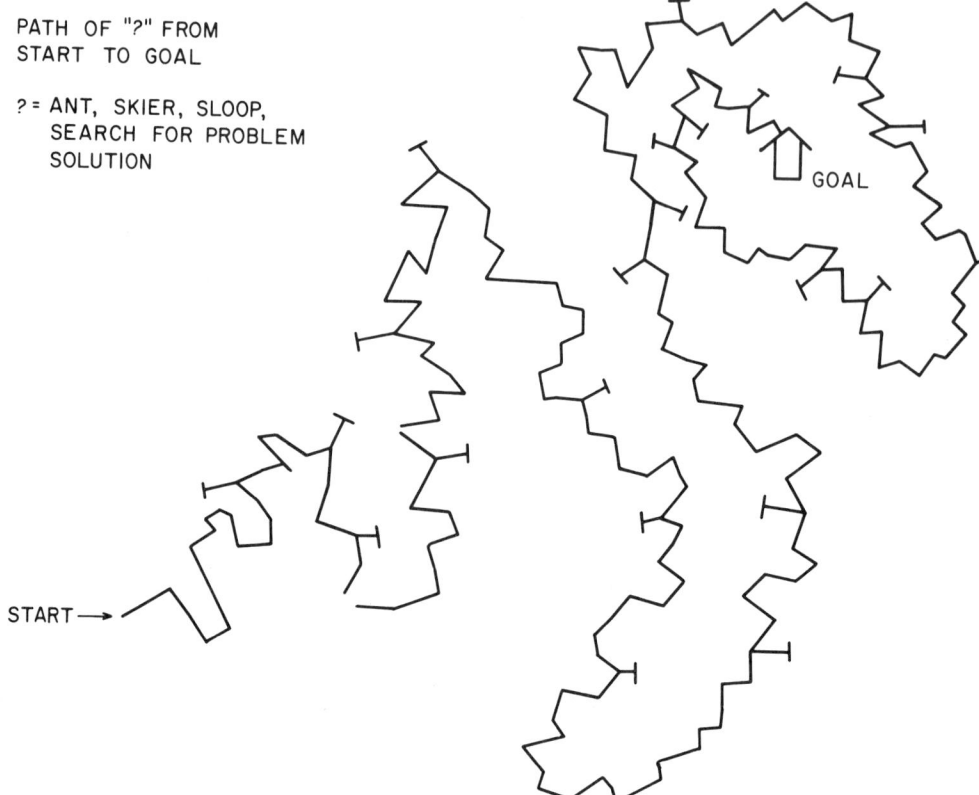

Figure 3. Path of ant to goal (after Simon, 1969).

source than the intricacy of the ant's path" (Simon, 1969, p. 53). At an earlier stage he concludes that "the possibility of building a mathematical theory of a system or of simulating that system does not depend on having an adequate microtheory of the natural laws that govern the system components. Such a microtheory might indeed be simply irrelevant" (Simon, 1969, p. 19), and this leads me to consider the approach of using black-box technology as an appropriate method of problem-solving in exceedingly complex probabilistic geologic systems.

Consider for a moment the simple input-black box-output model (Fig. 4); here the throughputs are linked together in the normal manner from "cause" to "effect" but there is a feedback loop which leads from the output back to the black box or to the input or both. Therefore

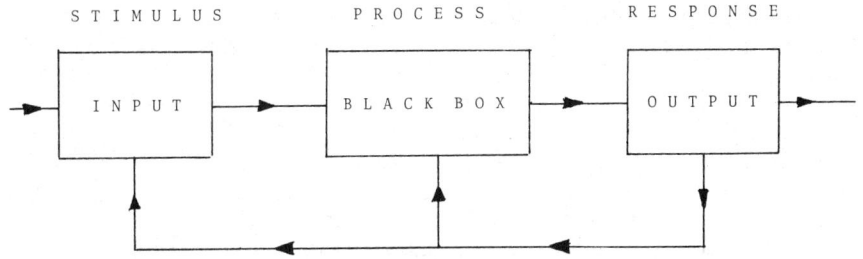

Figure 4. Simple cybernetic model of input-black box-out (I-BB-O) system.

a simple unidirectional cause-effect relationship is inadequate to represent this system; here, again, a number of geological processes may be modeled isomorphically onto this black box model where they possess a source material (input), process, and product (output).

Most scientific investigations are aimed at evaluating what is in the black box and this pursuit may be labeled a search for "explanation". However it is not obvious that this is the most appropriate procedure or even that it is feasible in many situations (cf. Boulding, 1980). It is much simpler and may be more effective to study the association between input and output in a stochastic or statistical relationship and let the box remain black. This then leads back to Simon's claim that a microtheory of what is inside the black box may be irrelevant.

As an example of a system which is likely to remain black-box consider Ashby's discussion of the complexity of a system (Ashby, 1964); his example consists of a board of 20 by 20 lamps each of which may be on or off leading to 2^{400} possible patterns. This number is very large and so Ashby discusses what "very large" means in this context and concludes with Ashby's Law: "Everything material stops at 10^{100}" (Ashby, 1964, p. 163). He then shows that 2^{400} is about 10^{120}, already exceeding the limit of Ashby's Law. Suppose now we wish to subdivide this set of patterns into those which possess some property P or not (i.e. suppose some are red) then we are led to $2^{10^{120}}$ which is approximately $10^{10^{120}}$; in other words in this rather simple system we are led to a number the *exponent* of which is beyond Ashby's Law. Such systems are likely to prove intractable to our present epistemological approaches (Ashby, 1964, p. 168); and again there are many possible geological examples.

An alternate procedure for solving this type of problem consists of relating the input to the output through a transfer function (Box and Jenkins, 1970). It is convenient to use matrix notation in this situation as in Griffiths (1966a) and it is there shown that two matrices should be specified to determine a unqiue third. When the input and output matrices are specified appropriately, the black box (process) can be determined in the form of a set of matrix equations, and the coefficients may be identifiable. On the other hand, whether or not they are identifiable the relationships may be useful in prediction, that is the "microtheory" may be relevant, and the box may be left "black".

EXAMPLES OF SYSTEMS BEHAVIOR IN A GEOLOGICAL CONTEXT

As an initial example of these system properties consider Vistelius' introduction of Markov models into geological problems (1949); if the matrices of lithological changes are powered to follow their development through various stages they may reach a final state in a finite, relatively short, number of stages, (Griffiths, 1966b), that is they behave similar to regular ergodic chains (Kemeny and others, 1959, p. 391 ff.). This final state is composed of a constant probability vector and no further powering will change it. For example, in the Khev-grdzeli section the constant vector is reached after 16 steps and consists of 33.8 percent sand, 41.5 percent silt, and 24.7 percent clay. Given this final state it is not possible to retrace the steps followed to achieve it, that is memory of the starting state and specific process stages is lost; the outcome is equifinal as in Figure 2C.

It is possible to represent the formation of detrital sediments as an input-black box-output model (see Figure 5; and Griffiths and Ondrick, 1969, p. 75); the input is equivalent to material from the source area which passes through several processes, termed erosion, transportation, deposition, and diagenesis, to emerge as a detrital sediment (the output). Much attention has been given to these processes but for convenience we may look upon them as a black box or, if they are to be treated individually, as a series of black boxes.

It is interesting to note that, as the source area is eroded, if the output is allowed to accumulate it gradually blankets the source area and inhibits further

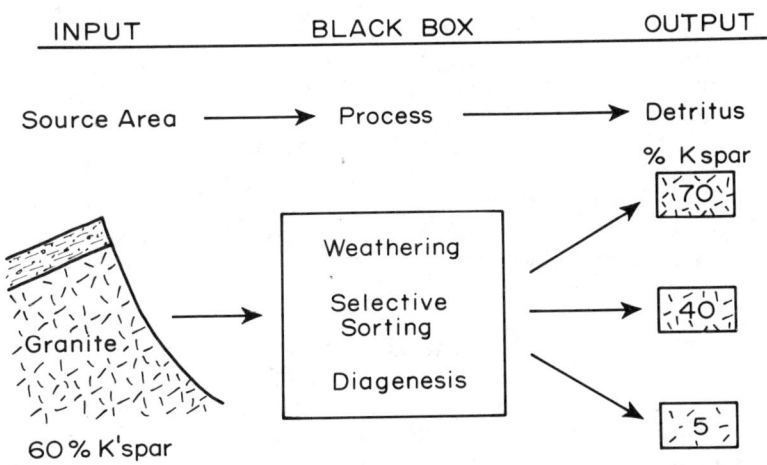

Figure 5. Petrogenetic system of arkosic detrital sediments represented as I-BB-O system.

erosion; in other words there is a negative feedback loop which gradually tends to reduce the effect of the continued process. Of course, if, as is usual, the eroded material is removed, transported, and fresh source material continually is exposed, the process continues. In practice, erosion is coupled with transportation and subsequently deposition and these, conceptually different stages in the process, are difficult to separate because they are obviously not independent. It is possible to describe the performance in the black box as a process of selective sorting on the basis of density (= type of mineral), grain size, and shape, and it is possible to study variation in this process in terms of the average grain size and size sorting of the mineral constituents in the detrital output.

The petrogenetic system is represented in this manner for the arkose rock type in Figure 5, following Krynine's description (Krynine, 1948, 1950). The input is represented as fresh and weathered granitic material containing 60 percent K-spar. This material passes through the black box labeled weathering, selective sorting, and diagenesis and yields three types of output, which differ in proportion of feldspar from much less, to much more, than that in the source material; each of these outputs may be achieved by changing the process in the black box. The point to be emphasized is that studies of variation in the quantitative composition of the output, a typical arkosic sediment, is unlikely to be unique to a specific source area unless the exact nature of the process can be specified in *quantitative* terms; this situation is equivalent to the example illustrated in Figure 2B. It also shows that to identify uniquely a component of a system of the I-BB-O type, it is necessary to specify at least two of these three components.

An alternative situation is represented in Figure 6 in which the source material differs in proportion of feldspar, a fairly reasonable expectation under "real world" conditions, but the product contains no feldspar and is equifinal. The fact that a detrital sediment contains no feldspar does not permit us to infer that there is no feldspar in the source area, another example of the phenomenon of Figure 2C.

When attempts are made to analyze variation in the properties of detrital sediments by components analysis (Griffiths, 1966a; Griffiths and Ondrick, 1969) and attempts are made to identify the components, it is determined that about 5 to 7 components account for the major proportion (70 to 90 percent) of the total variation. Two of these, usually but not always, the dominant ones, are identifiable as grain size and size sorting and two more, the proportion of silica and carbonate cements[1], usually are attributable to diagenesis. The remaining components are weak, accounting for less than 10 to 15 percent of the variation, and are not readily identifiable. In other words, the conceptual processes represented by the operational agency of selective sorting plus the effects of diagenesis overprint the effects of previous phenomena and destroy the memory of source area and associated, weathering, processes.

Because all detrital sediments follow a similar geological history, that is there is always a source area, a complex process or sequence of processes and a

---

[1] The cementitious properties may be dominant over selective sorting but nearly always these four components represent the major proportion of the variation in all the properties studied.

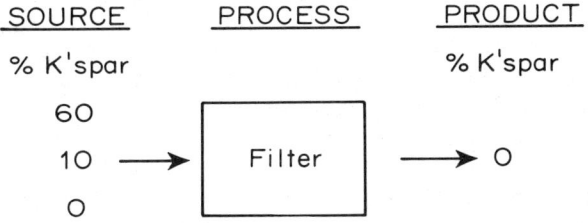

Figure 6. Example of equifinality in petrogenesis of detrital sediments.

product; and because all the processes, physical and chemical, generally reduce the complexity in composition of the source material the end-product is usually a quartzite as envisaged by Krynine (1948). This end-product, a quartzite, is equifinal; it has no memory of its source or the trajectory by which it developed; it is an example of Figure 2C.

On the basis of these several system characteristics I have proposed to represent the classification of detrital sediments, using Krynine's philosophy and nomenclature with some modification in definition, in the form of a cone (Griffiths, 1968b, 1972); this geometrical symbolism permits the representation of the detrital sediments in a dynamic relationship as contrasted with the conventional static arrangement in a square or rectangle. It purports to show that all detritus irrespective of original source area degenerates ultimately towards an equifinal quartzite. Each of the major tectonic environments of Krynine, which results in a specific type of detrital sediment, such as an arkose, high-rank and low-rank graywacke, is represented by a band from base to apex along the cone.

To identify where a rock-specimen falls in the cone it is necessary to recognize two features: first, irrespective of composition (i.e. type of source area) if a specimen is from near the base of the cone (i.e. is near its source area) the diversity in composition of *neighboring* grains will be large. On the other hand, if the specimen is from near the apex, the diversity in composition is small. Because at the apex of the cone the product is equifinal, it may not be possible to assign the specimen to its proper tectonic environment; hence all such assignments are made by rock associations, that is quartzites always are interbedded with associated rocks which are more representative, that is possess better "memories", of their tectonic environment.

To differentiate tectonic environments it is necessary to examine a specimen from the associated rock sequence which is from as near to the base of the cone as possible; then the types of feldspar, if any, and rock fragments usually may be used to assign the specimen to one of the three clans. The critical features are:

(1) The sampling arrangement of selecting (a few) neighboring grains and observing their compositional diversity is the essential procedure.
(2) Presence or absence of a specific type of constituent is more important than its proportions. Indeed quantitative proportions may be misleading for this purpose.
(3) Variation (diversity) in grains sizes of a specific detrital constituent, such as quartz for example, also will reflect the position of the specimen in the cone - if, and only if, the grains taken for measurement are contiguous.

The geometrical display as a cone also may be used to illustrate that the progress from near to distant from the source area in terms of time or space or both is best measured by the arrangement of grains of different types, sizes, and shapes. Near source, diversity among *neighboring* grains is high whereas distant from source, diversity among *neighboring* grains is low. Progress is measured essentially by the degree of differentiation from undifferentiated near source to layered strongly and laminated far from source, an example of a gradual increase in negative entropy (Fig. 7). It seems worth emphasizing that if a channel[1] sampling procedure is used, there will be little difference in grain size between the material which is near or distance from the source area.

---

[1] In other words without carefully controlled stratigraphic sampling procedure (i.e. sedimentation unit sampling; Otto, 1938) the structure, and therefore degree of intensity of the process, is lost.

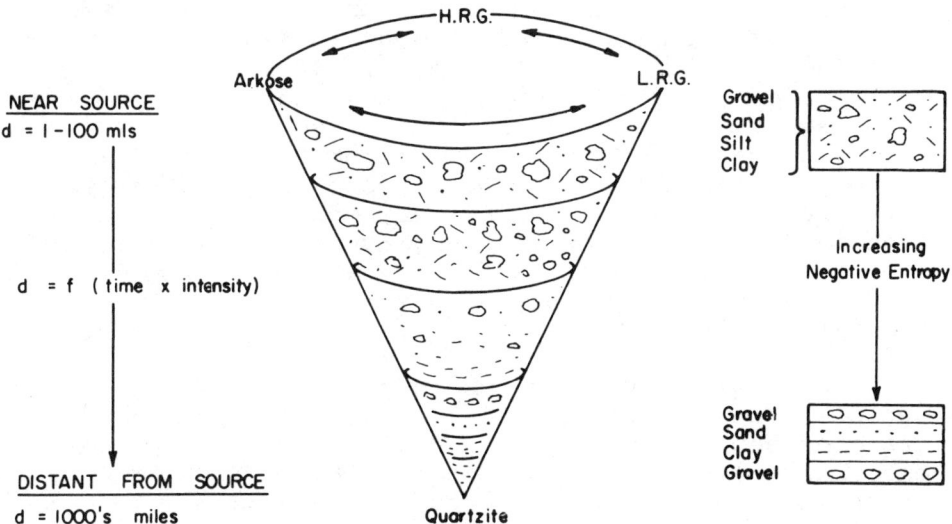

Figure 7. Classification of detrital sediments represented in cone.

This "conical" model which includes several different starting states, trajectories which converge, and a common equifinal end-product may be applied to the formation of igneous rocks as expressed for example, in Bowen's Reaction Series (Bowen, 1928). The more modern concept of "petrogeny's residua system" is the equifinal end-product. Attempts to confine the origin of granite to a single unidirectional process and a single starting state led to the granite controversy (Gilluly, 1948); this type of argument possibly reflects the fact that rival hypotheses are represented in a form which insures that they are examples of Goedel Incompleteness (Hofstedter, 1979). The solution requires the invention of a metalanguage reflected in the change of question from "How is granite formed?" to "How much granite is formed in this manner and how much that?" - a feature emphasizing that the answerable question needs a language which comprehends a quantitative answer.

A similar model may be used to represent the origin of clay minerals and the finer grained sediments of which they are largely composed; there may be several starting states (source materials) and the trajectories converge towards the equifinal end-product, laterite (Griffiths, 1952).

As a final example consider the procedure of mineral-resource assessment represented as an I-BB-O model (Fig. 8; and Griffiths, 1978, fig. 1, p. 448); here the input consists of two parts, one herited, the geological characteristics, and the other, the degree of commitment, an acquired characteristic derived from the current socio-economic infrastructure. Geological factors determine the upper and lower bound of the amount of resource present in a region (Menzie, Labovitz, and Griffiths, 1977; Labovitz, Menzie, and Griffiths, 1977); whereas the amount produced is mainly a function of the degree of commitment (Griffiths, 1978).

The geological factors are concepts and the actual objects examined are the rocks which represent the memory trace of geological events - "all we know of the past is the record of it..." (Boulding, 1980, p. 834). The types of rocks were standardized, by a transducing table (Fig. 2; Missan and others, 1978) based upon the classification in Figure 7, into 13 different rock types and five ages; this yields 65 different possible outcomes so that the input may be characterized as chosen from 2^{65} possibilities. Similarly, the output is standardized into 77 commodities (Labovitz, Menzie, and Griffiths, 1977) and any specific output therefore is chosen from 2^{77} possibilities. The black box contains therefore $2^{65} \cdot 2^{77}$ combinations or approximately $10^{3 \times 10^{24}}$ which comfortably exceeds Ashby's Law. Geological reasoning and observation of the frequency of association is aimed at reducing this enormous complexity by, for example, observing that tin usually occurs in granite, a massive reduction to a 1:1 mapping of rock type to commodity. This reduction is an oversimplification because some 70 percent of the world's tin comes from "alluvium", actually from detrital sediments derived from granite, or in our terms, from arkoses. Tin also is associated with quartz porphyry in Bolivia and with rhyolite in Mexico (Sainsbury, 1969). The major source of tin in North America is as a by-product from the lead-zinc deposits at the Sullivan Mine, Kimberley, British Columbia. It is clear therefore that the input is at least 2^5 and because the outputs frequently contain more than one other commodity the output, let us say, also is composed of at least 2^5 possible alternatives leading to a set of $2^5 \cdot 2^5 = 2^{160} = 1.46 \times 10^{48}$ possible combinations.

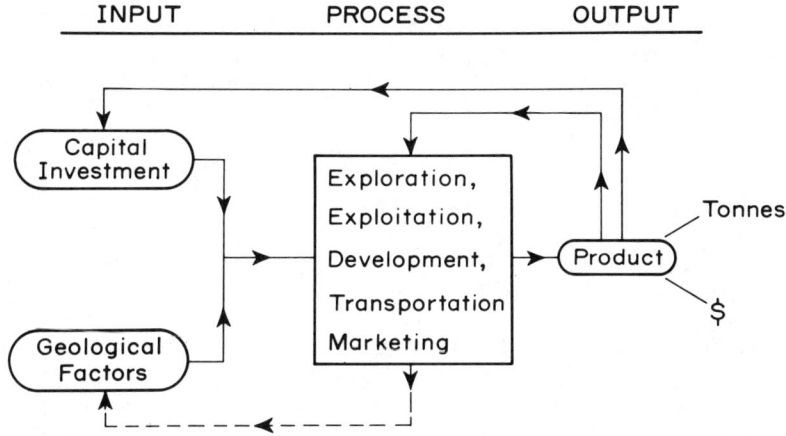

Figure 8. Mineral-resource assessment as I-BB-O model.

It is possible that some ore deposits also may be examples of equifinal products so that a black-box approach may be appropriate for studying the associations of geological events (= rocks) with commodities.

CONCLUDING REMARKS

Elsewhere I have described several additional aspects of problem-solving in geoscience (Griffiths, 1969, 1974) and, in particular, the possible challenge arising from the existence of Goedel Incompleteness (Beer, 1964) in the formulation of geological questions. The equivalence of structure, characteristic of the systems view of the epistemology of problem-solving, is displayed delightfully, by Hofstedter (1979) in tracing the correspondence between the logic of Goedel, the art of Escher, and the music of Bach. It would be surprising indeed if the equivalence cannot be extended into various other subject matter areas including geoscience.

REFERENCES

Argyris, C., 1971, Management and organizational development: McGraw-Hill Book Co., New York, 211 p.

Ashby, W.R., 1956, Design for an intelligence amplifier, in Automata studies: Ann. Math. Studies, no. 34, p. 215-234.

Ashby, W.R., 1964, Introductory remarks at panel discussion, *in* Mesarovic, M.D., ed., Views on general systems theory: John Wiley & Sons, Inc., New York, p. 165-169.

Beer, S., 1964, Management and cybernetics: John Wiley & Sons, New York, 211 p.

Boulding, K.E., 1980, Science: our common heritage: Science, v. 207, no. 4433, p. 831-836.

Bowen, N.L., 1928, The evolution of the igneous rocks: Princeton Univ. Press, Princeton, New Jersey, 334 p.

Box, G.E.P., and Jenkins, G.M., 1970, Time series analysis: Holden-Day, San Francisco, 553 p.

Caster, J.H., 1970, Etzioni's view of the environment: Science, v. 169, no. 3945, p. 529.

Chamberlin, T.C., 1897, The method of multiple working hypotheses: Jour. Geology, v. 5, no. 8, p. 837-848.

Churchman, C.W., 1968, The systems approach: Dell Publ. Co., Inc., New York, 243 p.

DeBono, E., 1971, The mechanism of mind: Penguin Books Ltd., London, 281 p.

Fisher, R.A., 1942, The design of experiments (3rd ed.): Oliver and Boyd Ltd., Edinburgh, 263 p.

Forrester, J.W., 1968, Planning under the dynamic influences of complex social systems: OECD Working Symposium on Long-range Forecasting and Planning (Bellagio, Italy) Mimeo. d-1123, 22 p.

Forrester, J.W., 1970, Counterintuitive behavior of social systems: Testimony for the Subcommittee on Urban Growth of the Committee on Banking and Currency, House of Representatives, Washington, D.C., d-1383-1, 27 p.

Gardner, M., 1962, Mathematical games: Scientific American, v. 206, no. 3, p. 138-144.

Gilluly, J., 1948, Origin of granite: Geol. Soc. America Mem. 28, 139 p.

Gould, P., 1979, Polyhedral dynamics: an introduction for social scientists, geographers and planners: Center de Etudes Geographicos, Inst. Nac. de Investigacao Cientifica (Lisboa, Portugal), 51 p.

Griffiths, J.C., 1952, Reaction relation in the finer grained rocks: Clay Mins. Bull., v. 1, no. 8, p. 251-257.

Griffiths, J.C., 1966a, A genetic model for the interpretive petrology of detrital sediments: Jour. Geology, v. 74, no. 5, pt. 2, p. 653-672.

Griffiths, J.C., 1966b, Future trends in geomathematics: Min. Ind., Pennsylvania State Univ., v. 29, no. 1, p. 1-8.

Griffiths, J.C., 1968a, Operations research in the mineral industry: Proc. Symp. on Decision-Making in Mineral Exploration, Vancouver Research Council, p. 5-9, Western Miner., v. 41, p. 22-26.

Griffiths, J.C., 1968b, Classification of geologic data: Proc. Symp. on Decision-making in Mineral Exploration, Vancouver, British Columbia Res. Council, p. 37-42; Western Miner., v. 41, p. 37-42.

Griffiths, J.C., 1969, Cybernetics-geomathematics interaction: Geol. Soc. America Sp. Paper 146, p. 87-94.

Griffiths, J.C., 1970, Current trends in geomathematics: Earth Science Rev., v. 6, no. 2, p. 121-140.

Griffiths, J.C., 1972, Some aspects of classification, *in* Protz, R., and Martini, I.P., eds., Classification of soils and sedimentary rocks: Proc. Symp. Dept. of Land Resource Science, Center for Resources Development, Univ. Guelph, p. 123-146.

Griffiths, J.C., 1974, Quantification and the future of geoscience, *in* Merriam, D.F., ed., The impact of quantification on geology: Syracuse Univ. Geol. Contr. 2, p. 83-101.

Griffiths, J.C., 1978, Mineral resource assessment using the unit regional value concept: Jour. Math. Geology, v. 10, no. 5, p. 441-472.

Griffiths, J.C., and Ondrick, C.W., 1969, Modelling the petrology of the detrital sediments, *in* Merriam, D.F., ed., Computer applications in the earth sciences: Plenum Press, New York, p. 73-97.

Hofstedter, D.R., 1979, Goedel, Escher and Bach; an eternal golden braid: Basic Books Inc., New York, 777 p.

Kemeny, J.G., Mirkil, H., Snell, J.L., and Thompson, G.J., 1959, Finite mathematical structures: Prentice-Hall, Inc., New Jersey, 487 p.

Krynine, P.D., 1948, The megascopic study and field classification of sedimentary rocks: Jour. Geology, v. 56, no. 2, p. 130-165.

Krynine, P.D., 1950, Petrology, stratigraphy and origin of the Triassic sedimentary rocks of Connecticut: Connecticut Geol. and Natl. Hist. Survey Bull. 73, 247 p.

Labovitz, M.L., Menzie, W.D., and Griffiths, J.C., 1977, COMOD: A program for standardizing mineral resource commodity data: Computers & Geosciences, v. 3, no. 3, p. 497-537.

Loudon, T.V., 1979, Computer methods in geology: Academic Press, New York, 269 p.

Lucas, J.R., 1961, Minds, machines and Goedel, *in* Sayre, K., and Crossan, F., eds., The modeling of mind: Univ. Notre Dame Press, South Bend, Indiana, p. 112-126.

Menzie, W.D., Labovitz, M.L., and Griffiths, J.C., 1977, Evaluation of mineral resources and the unit regional value concept, *in* Ramani, R.V., ed., Application of computer methods in the mineral industry: Soc. Min. Eng. America, Inst. Min. Metall. and Petrol. Eng. Inc., New York, p. 322-339.

Michie, D., 1961, Trial and error: Science survey, 1961 (pt. 2): Penguin, Harmondsworth, p. 129-145.

Missan, H., Cooper, B.R., el Raba'a, S.M., Griffiths, J.C., and Sweetwood, C., 1978, Workshop on areal value estimation: Jour. Math. Geology, v. 10, no. 5, p. 433-439.

Otto, G.H., 1938, The sedimentation unit and its use in field sampling: Jour. Geology, v. 46, no. 4, p. 569-582.

Robens, RT. Hon., Lord, 1969, People and computers - part 2: Computers and Automation, v. 18, no. 12, p. 53-55.

Rosenbleuth, A., Wiener, N., and Bigelow, J., 1943, Behavior, purpose and teleology: Phil. Sci., v. 10, no. 1, p. 18-24.

Rosenbleuth, A., and Wiener, N., 1950, Purposeful and nonpurposeful behavior: Phil. Sci., v. 17, no. 4, p. 318-326.

Sainsbury, C.L., 1969, Tin resources of the world: U.S. Geol. Survey Bull. 1301, 55 p.

Simon, H.A., 1969, The sciences of the artificial: M.I.T. Press, Boston, Massachusetts, 123 p.

Skinner, B.F., 1971, Beyond freedom and dignity: A.A. Knopf, New York, 225 p.

Vistelius, A.B., 1949, On the question of the mechanism of the formation of strata: Doklady Akad. Nauk, SSSR., v. 65, p. 191-194.

von Bertalanffy, L., 1968, General system theory: G. Braziller, New York, 289 p.

Weinberg, G.M., 1975, An introduction to general systems thinking: John Wiley & Sons, New York, 279 p.

DEVELOPMENTS AT THE MAN-MACHINE INTERFACE

Donald B. McIntyre

Pomona College

ABSTRACT

Because early computers were expensive and had small memories, attention had to be given to using them efficiently, although this made computer languages cumbersome and unnatural for the human beings who used them. Computer programming had little connection with the development of mathematical concepts.

A summary of the evolution of computer hardware shows that the power available for a fixed price has increased dramatically. Changes in computer languages and in operating systems have gone along with changes in hardware. As computers have become larger and cheaper, more of the operating system has been transferred to control storage. Scheduling of the use of computer resources, which used to be an important responsibility of a human operator, is to an ever increasing extent taken over by the machine itself. The complicated paging that is performed to make efficient use of main memory and auxiliary storage is transparent to the users, who therefore are freed to work on the problems that are their proper concern.

Implementation of the concepts of virtual storage and virtual machines allows many users to work simultaneously as if each had access to a dedicated computer with a larger memory than possessed by the real machine. This, combined with terminals that do not interrupt the central processor unnecessarily, has been an important trend in the past ten years. Terminals now are capable of full-screen displays, which can be modified anywhere on the screen. Card decks have become obsolete.

The most important development in language design has been the implementation of the APL language, which continues the historical evolution of mathematical notation. Its power comes from its use of functions and operators (which are distinguished from one another), from its extension of primary or primitive functions (represented by single symbols), from its emphasis on the use of arrays and the avoidance of much of the apparatus of control statements that characterize most of the other languages, and the use of workspaces in which the user defines the environment. Although APL was considered expensive and inefficient on early computers, this view is no longer justified, especially today when emphasis must be on conserving skilled human resources. APL is supported by all the major computer manufacturers and can be obtained on some desk-top computers.

The development of Viewdata systems, notably the British Post Office's Prestel, is bringing computer technology into the homes of those who have not considered purchasing home computers. This trend will decrease significantly the resistance to computers that is engendered by fear of the unknown.

Online bibliographic searching has become a tool of major importance in scientific research. The use of Geo.Ref, GeoArchive, SCI, and SSIE are especially noted. The growth of nonbibliographic online data bases has raised the problem of how these can be indexed and made more generally available.

I conclude that students of geology must know how to use computers if they are to work effectively in a world in which technology is changing so quickly. The developments at the man-machine interface seem never to have been so exciting or so vitally important.

EVOLUTION OF COMPUTER HARDWARE

The most obvious developments in the field of computing have been in hardware. Given the expenditure of a constant sum of money, the computer power of the central processor has risen dramatically, as I can illustrate from my own experience. Because I have been involved especially with IBM computers, and because of their importance in the market, my examples are mainly taken from IBM machines, but parallel development went on at other companies.

In 1960 the Geology Department at Pomona College purchased a Clary computer for $20,000. It had a drum memory capable of storing 32 integers, and its programs consisted of wired boards that controlled the processor when a mechanical device stepped through the sequence of instructions. Logarithms and trigonometric functions were performed by separate cartridges, only one of which could be plugged in at a time. Today our students carry more powerful computers (programmable calculators) in their pockets, and they cost between $100 and $500.

Prior to announcing System/360 in 1964, IBM had three principal types of computers:

(1) Computers that addressed individual characters. The parent of this series was the IBM 702, introduced in 1953, and from this root came such computers as the 705 (1954), the 1401 (1959), and the 7080 (1960).

(2) Computers that operated as true decimal machines. The parents of this series were the IBM 604, an electromechanical calculator introduced in 1948, and the 650, a vacuum-tube machine with a drum memory, introduced in 1953. From these were descended the 7070 (1958), the 7074 (1960), and the small 1620 (1959).

(3) Binary Computers.
These are the computers descended from the IBM 701 (1952) and 704 (1954). They included the 709 (1957), 7090 (1959), 7094 (1962), and 7094-II (1963); also the 7040/44 (1961), with two registers of 6-bit bytes, and the remarkable 7030 or Stretch computer (1960), which had 8-bit bytes and other features that in 1964 were included in the System/360.

In 1965 Pomona College installed one of the first IBM 360s shipped to a customer. It cost about 10 times what the Clary had cost, but it had a card-reader, a line-printer, and a disk drive. Its initial memory was

16K bytes, but this was increased in steps to 250K as usage grew. In 1975 the Geology Department purchased an IBM 5100 desktop computer for about $20,000. The APL language (including a file system) was part of hardware, and the user had a 64K workspace available. Because the 5100 is portable and can give displays on videomonitors, it is useful especially in the classroom. Moreover students can be encouraged to use it freely as the whole system of hardware and software is purchased and there is no need for any type of accounting.

In May 1979 Pomona College replaced the 360 with an IBM 4331, the second one installed at a customer's site. Unlike the 360/40, the 4331 supports terminals, and 30 of IBM's 3278 full-screen terminals are available for free use by students and faculty. Although the 4331 operates as if it were a number of (virtual) machines, the real memory is 1M; that is 1 Megabyte, or 1 million bytes. The College accepted the 4331 as an interim processor to be used until the more powerful 4341 could be delivered. The 4341 has a memory of 4M, but this can be increased in the field to 8M.

In less than 15 years the real computer memory available in this small liberal arts college has increased 500 times, and the increase in disk storage is equally spectacular.

EVOLUTION OF COMPUTER LANGUAGES AND OPERATING SYSTEMS

The first computers could be programmed only in their own machine language. But when assemblers were developed, much of the detail was turned over to the machine, thus sparing the human programmer. For example, by using assembly language the programmer could refer to a memory location by its symbolic address instead of having to specify an absolute location, although registers had absolute addresses. In assembly language, instructions could be given in mnemonic code (such as CLA for Clear and Add to the Accumulator) with enormously improved readability of the code. Macros could be written in which several assembly-language instructions simulated a new and specialized machine instruction. The development of higher level languages continued this trend allowing the programmer to specify in a single statement what the compiler would translate into many machine instructions. More and more of the work was given to the machine to perform, so that the programmer could work at a level more suitable for the human mind and more appropriate to the problem presented by the user. FORTRAN, designed for scientific work, was developed for binary computers similar to the 704 and the 7090 series. COBOL, suitable for commercial work, was developed for computers which address individual characters, such as the 705 and its descendents. The history of development of these and other high-level languages has been documented by the Association of Computing Machinery (1978). No language of present interest was written for a decimal machine, although the popular 1620 used FORTRAN.

The computer systems that supported FORTRAN and COBOL not only invoked the language compilers needed to translate source code into object code that could run on the particular computer, but automatically fetched the subroutines that were referred to in the program. They also provided a linkage editor that built a coherent package from the modules gathered by the system "librarian", which was part of the operating system.

Before 1964 the computing world was divided into two parts, the commercial and the scientific. There were two types of machines which had entirely different internal structures and which supported dissimilar high-level languages, but the announcement of System/360 on 7 April 1964 changed this: a single general-purpose computer, appropriately termed a 360 (for 'full-circle'), could perform both commercial and scientific computing. It could address an individual character stored in a byte, whose size of 8 bits permitted an extension of the basic character set from 64 (6-bit Binary Coded Decimal) to 256 (Extended BCD). And it also could address a group of contiguous bytes to form a Half Word, a Full Word, or a Double Word (2, 4, or 8 bytes respectively). Moreover all models of System/360 had a common instruction set, so that a program written for any one model could run on any other, provided that time and space imposed no constraints. Unlike previous computers, whose addressing structure necessarily restricted memory to some small size, such as 32K in a 7090, by using the method of addressing by base register and displacement, System/360 had the ability to address memories as large as 16M.

Similar to today's pocket calculators, early computers could do only one job at a time. Operating systems (sometimes termed supervisors or monitors) performed

only the simplest of tasks, and the human operator in
charge constantly had to intervene to initiate or terminate jobs, to decide the order in which jobs would be run,
to assign tape drives, and to mount tapes and load card
decks. Jobs were submitted to a batch stream presided
over by a human operator who tried to optimize the use
of the resources. But when a small job was running,
the operator was unable to allocate unused memory to
another program. Because every program loaded at an
absolute address, no flexibility was possible. But the
360 was a true System whose architecture made provision
for an operating system that enabled the computer to
direct its own activity. For example, it had a PSW
(Program Status Word) and 16 General Purpose Registers,
whereas the 7090 had had an Instruction Counter and three
Index Registers, which were not general purpose but used
only to control addressing.

When Pomona's 360 was delivered, neither we nor the
IBM employees with whom we worked understood the significance of the designation, System/360. We did not even
have an operator's console, and the only manner we could
run was under Basic Assembler, which required the card
deck to pass through the reader-punch twice. Because we
had no operating system, we used all the registers as
high-speed storage, and used privileged instructions,
such as LPSW (Load PSW), with complete freedom. We did
not appreciate that the reason for having "privileged"
instructions was to give the System control over the
work of several simultaneous users. Finally when we
realized what we would gain by giving up some of our
freedom to use parts of the hardware, we began to understand the revolution in the man-machine interface implied
by System/360.

Multiprogramming was possible on the 360 because the
system managed storage, and could assign job streams to
different partitions of memory. Because all addresses
were relative, the loader was free to put a program into
a different part of memory if it was resubmitted. Relative addressing, storage protection, and privileged instructions (used only by the system), were all needed
to provide the flexibility that made better use of the
machine.

VIRTUAL STORAGE AND VIRTUAL MACHINES

In 1970 IBM announced its 370/145, which not only had monolithic storage instead of core memory, but had control storage hardware. Addressing was provided for 16 control registers, although only 4 were implemented in 1970. The implication of the announcement was that IBM was preparing to announce Virtual Storage (VS). This meant that the System was about to take a major step forward. Whereas the operator of the 360 or 370 previously had to decide which jobs would run in different fixed partitions of real memory (foreground and background), the system would henceforth divide memory into "pages", which would be swapped between main memory and auxiliary storage as needed, so that many jobs could be serviced simultaneously and the resources of the machine used efficiently. The operation of the computer had become too complex for a human operator, and the System had to be given the physical resources to do the job itself.

When computers had small memories, a large job required segmentation and overlays, so that when one part of the program had been completed it would call into memory the next part, and destroy itself in the process. The program for automatic contouring published by the Kansas Geological Survey (McIntyre, Pollard and Smith, 1968), originally written for a 7090 and subsequently modified for a small 360, had no fewer than 12 overlays or phases. But VS permits the programmer to ignore the restrictions of a finite real memory, and as long as there is adequate disk memory, the virtual system, using Dynamic Address Translation hardware, can bring into real memory the pages as they are demanded by the program. The System, rather than the human operator or programmer, takes care of the addressing requirements. Programming of large jobs therefore is greatly simplified. Not only is there no need for overlays, but every user can work with the full compiler, instead of with only a subset as was often the situation formerly. And debug aids are more powerful and more available to the average user.

RCA announced VS in September 1970, but a year later abandoned the manufacture of computers. In 1972 IBM announced VS and VM (Virtual Machines), so that the latent power of the 370/145 was invoked. VS, or Virtual Storage, allows the programmer to proceed as if addressable storage extended to 16M, although the real size of main memory is less than this. The system takes care of swapping information between main memory and auxiliary storage, usually on disk drives. Thus, the extended hardware achieves more efficiently the overlaying of the program or data, which previously consumed considerable human effort and ingenuity. On the other hand, with VM a single computer simulates the concurrent operation of many independent machines, each with its own operating system. The user's terminal becomes the operator's console of a private computer. For example, the user can display and modify the contents of the PSW and registers of his virtual machine, exactly as the computer operator in the machine room used to do for the real machine. But this activity on a virtual machine does not affect other users. When Pomona's 360 was new, I taught students how to work in machine language by using the switches on the console to control the contents of the registers and PSW. We could step through a sequence of instructions and observe the operation of the machine in detail. This admirable teaching method required total dedication of the computer, and when the work load on the 360 increased, we could no longer afford this luxury. But the user of a virtual machine can perform these operations without interfering with the work of other users. Thus, everyone has the advantage of what seems to be a dedicated machine, unless, of course, the real resources are inadequate for the total work involved. VM permits more efficient use of the computer's power, and adds to the security of individual users. With VM, interactive users can do their work simultaneously with batch processing.

To implement VM the computer must have Control Storage (including Dynamic Address Translation) and a Control Program. Today's control storage is likely to be considerably larger than yesterday's main memory, and more and more of the operating system is moving from software to control storage.

IBM 4300, DESK-TOP COMPUTERS, AND FULL-SCREEN TERMINALS

On 30 January 1979, IBM announced its long awaited 4300 series of middle-sized processors. The 360 (1964) had used core memory and integrated logic circuits; the 370/145 (1970) had introduced the hardware to perform Dynamic Address Translation, and used monolithic memory in place of core; the 370/158 (1972) used Mosfet technology (Metal oxide silicon field effect transistor) and implemented both VS and VM. The effect of the 4300 announcement was rather to decrease component size and power requirements and to bring the cost down dramatically. Component chips went from a density of 2K to 64K at a single step, and main memory for an IBM computer was reduced from $75,000 to $15,000 for 1 Megabyte. It was said that the number of first-day orders for the 4300 series exceeded the number of computers then existing.

IBM's 5100 desk-top computer, which became available in 1975, demonstrated that the manufacturers of large computers also intended to participate in the rapidly increasing market for desk-top minicomputers. Hewlett-Packard claimed in October 1979 that its desk-top models operated about 10 times faster than their 1974 counterparts and offered users memories that were 14 times larger. It was predicted that by 1984 desk-top systems would reach 2M in user main memory, with 5M on floppy disks, and 120M on hard disks. In 1979 Texas Instruments reported the development of bubble memories with access times twice as fast as previously, and memory chips with more than one-million bits of storage, with automatic error correction, in less than one square centimeter.

Already some students come to college with their own computers, and wish to link them to the campus computer. It is obvious that desk-top computers will become available in almost every home, where they will be taken for granted, as the telephone and television are today. Central computers will be more powerful and appropriate aids for human thought. Terminals will be more elaborate, but perhaps cheaper. And as computers become more numerous, there will be a greater premium on those who know how to use them well.

From 1965 on it became increasingly usual for users to communicate with computers interactively through terminals rather than by submitting a deck of cards to a batch system. The IBM 2741 terminal was especially well known, although there were numerous other brands. The Selectric ball (typing element) permitted a wide range of fonts, obtainable also by the later introduction of the Diablo "daisy wheel". The 2741 was similar to an IBM Selectric typewriter, and its characteristic sound was familiar, not only to computer users, but to the many people who heard them in action at airline counters. The output was typed, and enormous quantities of paper were consumed, usually only to end up in the waste basket when a particularly intractable problem was being explored. Consequently many manufacturers introduced

Cathode-Ray Tube (CRT) displays that eliminated the need to print the entire dialogue with the computer.

All these terminals operated in a start/stop mode; that is, every action necessitated an interrupt on the Central Processing Unit (CPU). To make the interaction with the system more efficient, IBM replaced the 2741 with the 3270 type of terminal, which have CRT screens, but which operate differently from the older, start/stop terminals. The user does not have to work one line at a time, but rather can move the cursor to any part of the screen to enter or modify data. Thus, the user may spend several minutes entering data on the screen, without having to wait for the computer to respond to the many interrupts that would have occurred under the old system. Moreover, only the changes that have been made in the screen are transmitted to and from the computer. Thus, by providing more power in the terminals and their controllers, the system allows the CPU more machine cycles to do its proper job.

Terminals that behave in this manner are termed "full screen" terminals, and they are supported by an operating system, such as IBM's Conversational Monitor System (CMS), that provides full-screen editing and other utilities, such as file management. The result is greatly increased productivity. Because several terminals are supported by a single controller, it is economical to have several terminals in a cluster. They then can share a printer, such as an IBM 3287. With such a system available, punched cards are obsolete.

IBM now has color terminals, and users are just beginning to explore the possibilities, just as TV producers did when color TV became popular. Graphic display also is becoming more accessible to users, although the term "computer graphics" has different meanings to different people. Plotting of graphs of all types is no longer a problem, and students in beginning chemistry classes, for example, can use the computer to plot the results of their experiments. A few firms, such as Evans and Sutherland, have carried true interactive graphics to an extraordinary degree of sophistication, and it is impossible to convey in writing the spectacular results that can be achieved: under the command of a light pen, objects can be "picked up" and moved on the screen, scales can be changed, and the objects can be rotated in three dimensions. This type of application usually is supported on a minicomputer, such as a PDP-11, dedicated to the purpose.

THE APL LANGUAGE AS AN INTELLECTUAL TOOL

We have seen that the evolution of computers in the past 20 years has been towards more complex operating systems that take much of the burden of allocating machine resources away from the human operator. The complicated addressing needed to support paging is transparent to the user. Thus, computers are constantly becoming simpler to use, and one can expect that users will find it easy to write their own programs instead of delegating the job to a programmer who is not an expert in the user's own field. In short the computer should be used as a tool of thought.

The earliest computers had small memories, and only those who knew the internal structure intimately could hope to program and use one of them. As a result of this physical constraint, programming became detached from the mainstream of development of mathematical thought. During the 19th Century, mathematicians such as Cayley and Sylvester developed what the latter termed the "Algebra of Multiple Quantity". They emphasized how important it was to escape from the tyranny of scalars by conceiving of and working with arrays. Sir William Rowan Hamilton was the first to use "vector" in its modern sense, and Sylvester introduced the term "matrix" into mathematics.

Unfortunately much of this experience and tradition was necessarily set aside when computers were developing. Languages such as FORTRAN require the user to manipulate scalars. Loops are built into the program, so that what should be thought of as an array is treated as a succession of operations on scalars. John Backus (1978), the leader of the FORTRAN design team, has said that the development of languages of this type held back the progress of computing for 20 years. The programmer had to concentrate on control statements, loops, and Dimension and Declaration statements, and much of a typical program had little apparent relevance to the subject matter that is properly the user's concern.

The APL language originated as an extension and systematization of mathematical notation, and it is the only computer language that existed independently of the computer and would continue to be used were computers to disappear. In its earliest form, it was used by its inventor, K.E. Iverson, as a method of describing complex systems (Iverson, 1973). It came to the attention of a general audience in Iverson's book "A Programming Language" (1962), in which the first example is an exact account of the architecture of IBM's 7090, then one of the principal computers used for scientific work.

Iverson notation was used in 1964 to give a formal description of the newly announced System/360, and in 1965 the notation was implemented on a 7090 as the language that came to be known as APL. The language is distinguished by its extension of the number of functions recognized in mathematics by a single symbol, such as + -. The ball on the Selectric typewriter provided an introduction to these symbols on the machine, and the extension of the character set on the 360, achieved by going from a 6-bit to an 8-bit code, made internal storage of the symbols possible. Today APL is supported not only by IBM but by almost every major computer manufacturer.

Being dynamic, APL does not bother the user with Declarations and Dimensions, and instead of huge programs, good APL typically consists of single-line functions consistent with mathematical tradition (McIntyre, 1978). Because APL distinguishes between functions and operators (which modify functions), the language possesses extraordinary power and brevity. APL characteristically works on arrays, which can be of any shape and rank, and consequently loops and control statements are usually absent. The user of APL learns to think in terms of arrays, just as Sylvester urged mathematicians to do a century ago. For example, an APL function to compute the mean should work on a matrix or an array of higher rank by summing over the last axis: thus

$$MEAN:(+/\omega)\div 0\bot\rho\omega$$

APL is consistent, more so than ordinary mathematical notation, and it provides a generality that ensures uniform treatment of special situations, such as empty arrays. Through APL, boolean functions become an integral part of algebra, where they properly belong, and unlike other computer languages, APL has an intellectual content that is well worth studying for the insight it brings (McIntyre, 1979, 1980).

The manner in which APL is implemented also is different from that of other languages. Not only is it dynamic and interactive, but it provides an environment, termed a Workspace, which can contain the user's functions and variables, as well as control of the index

MAN-MACHINE INTERFACE 35

origin, print precision, etc. A workspace can be saved on disk, and individual objects can be copied from it. Moreover, although APL could claim to be the highest level of all existing languages, its power enables APL to be used for tasks that would be considered close to machine language. For example, one of my students has used APL (with auxiliary processors for shared variables) to read and use a star-catalog tape issued by the Smithsonian Institution, although it is written in the internal BCD code of a 7094 computer.

My own experience in teaching the use of computers to students and colleagues in many different disciplines is that APL increases productivity because, once the problem is clearly defined, the APL solution is usually evident. It therefore is possible to begin writing and testing the solution at the same time that one endeavors to formulate the problem. This is one of the most important developments at the man-machine interface.

VIEWDATA, CEEFAX, ORACLE, AND PRESTEL SYSTEMS

Almost every home in a country such as Britain has a telephone and at least one television set, which with little or no modification could be used as the display screen for a computer. Indeed it is possible now to buy a keyboard that can use an ordinary TV set for its display, and which can act as a terminal to a remote computer through a simple modem and the existing telephone line. However it takes some knowledge to make the connection to a remote computer, and the online charge for use of the computer probably is too great for most potential users.

A step in this direction was taken by the BBC when it introduced its Ceefax service in 1974, and this was followed in 1975 by the Oracle service produced by Britain's independent television company ITA. A television picture with 575 lines, each with 700 picture elements, and a gray scale of 64 levels (which can be represented by 6 bits), is equivalent to 2.4M bits. Large amounts of information therefore are transmitted over a television channel. Because not all the possible channels are used currently for transmission of pictures, the BBC is able to transmit frames of information that can be displayed on a slightly modified TV set. This is the Ceefax system. About 100 frames are transmitted in a continuous cycle, so that the user who wishes a particular frame has to wait for an average of about 12 seconds before that frame comes round. Once the frame is displayed, it remains on the screen as long as the viewer wishes.

In 1976 there were 500 sets in Britain that could display the frames transmitted by the BBC and ITA. By 1980 it was possible to purchase or lease a set in any major city in the country, and it is forecast that in ten years 25 percent of the homes will be using the service.

Although it must be emphasized that Teletext systems such as Ceefax and Oracle are not interactive, they are, however, free (once the equipment is installed), and they serve as a powerful introduction to computing in the home. It is certain that any family accustomed to these systems will not be afraid of a computer terminal.

Of greater significance is Prestel, a product of the British General Post Office, which also is responsible for Britain's telephone service. The user of Prestel has an account, which similar to a telephone account, charges for use made of the system. The modified television set is linked to the user's telephone, and a connection can be made to the Post Office's computer at any time that the television and telephone are not otherwise being used. A remote control unit, rather similar to a pocket calculator, enables the user to activate the system in the same manner that a change might be made in the TV channel being viewed. Some 200,000 frames, from more than 200 suppliers, are accessible online. If the user has a directory, he can request the display of a particular frame by entering its code number. Otherwise he can display the first frame of a menu and locate the desired frames by a tree (root) system with 10 branches at each node. Each frame contains 880 characters arranged in 22 rows and 40 columns, so that the library of 200,000 frames is equivalent to 176 million characters. This compares with an estimated 55 million characters in the London telephone directory. However the amount of useful information is less. In order to achieve readability, there are probably less than 100 words on an average frame. Because all frames are accessible constantly, access time is better than with Ceefax and Oracle, which depend on cyclic transmission.

Prestel marks an electronic revolution in the transmission of information. The user can obtain information on such diverse topics as Air Travel, Ancient Monuments, Automobiles, Betting, Bible Society, Careers, Contraception, Dishwashers, Divorce, Jobs, Jokes, Legal Advice, Maternity Benefits, Medieval Banquets, Night Life, Pregnancy, Pubs, Science, Sports, and Weather.[1] Some frames are provided free of charge, whereas for others there is a charge. For example, the car buying guide costs 3P per frame and the maximum price per frame is 50P (Cawkell, 1977b; Berkovitch, 1979).

Development of Prestel started in the early 1970s and a public test began in 1976. I was told that in 1979 there were about 1500 users, presumably all using the test site in the London area. It is claimed that by the end of 1980, 60 percent of the people in Britain will be able to connect to Prestel with a local call.

Prestel provides information at the user's request, but it is not truly interactive. Of course, Prestel could be adapted to provide computing power for its users, but the load on the system probably would be intolerable if more than simple calculations were allowed. As the system grows in popularity and computers become cheaper, regional centers may be created where Prestel's information service can be combined with true computing power for the users.

Although the British Post Office has been the leader in developing a computer-based information service for its customers, there is widespread interest in the idea. Germany and Holland have purchased the Prestel service, and rival systems have been created in other countries. Because Prestel is a registered trademark, it is necessary to have a generic name for systems similar to it, and the term Viewdata (Videotex in the US) is used in this sense. The first international conference and exhibition of Viewdata was held in London in March 1980. The British Radio and Electrical Manufacturers Association predict that eventually all television sets made in Britain will be able to be used for Viewdata reception.

---

[1]The Prestel Users Guide and Directory: Eastern Countries Newspaper Group Ltd., Norwich, England.

ONLINE BIBLIOGRAPHIC SEARCHING

Online bibliographic searching has grown in recent years to be a major research tool and marks an important step in the development of man-machine interaction. The principal suppliers of the service are Systems Development Corporation, in Santa Monica, California, and Lockheed Information Systems, Palo Alto, California. These firms use their computer power and knowledge to make available data bases created by a variety of suppliers. The number of these data bases is great and constantly growing. To make use of this service one needs to have an account number and a terminal, such as the portable Texas Instruments Silent-700. Charges are made only when the service is actually used, and then the price, which differs with the data base, is about $1.20 per minute of connect time. If a large bibliography is desired, it is more economical to have it printed offline and mailed.

The principal geological data base used in the United States is Geo.Ref, which is produced by the American Geological Institute. Geo.Ref has been available only through SDC's Orbit System, but Lockheed plans to make Geo.Ref available on its Dialog System early in 1981.

GeoArchive is similar in intent to Geo.Ref. It is produced by GeoSystems, London, England, and has been available in the United States only through Lockheed's Dialog System. This is not the place to discuss the relative merits of Geo.Ref and GeoArchive, the details of search strategy, or the Orbit and Dialog commands, but for those interested I recommend that some of the same references be found on both data bases and printed out in full. One then can see which search strategies would recover these articles in each data base, and which strategies would fail. The user should be equipped with the Thesaurus and Guide for the data base in order to work effectively, but it is remarkable how successful one can be in using a new data base without these aids provided that one is familiar with the computer system.

Other bibliographic data bases that I find particularly useful are Science Citation Index, SCI, and the Smithsonian index to funded research, SSIE. SCI is an excellent data base to search if you know a key reference, because you can find out who has since cited it. SSIE is valuable if you wish to know who currently is doing research on a topic.

As a simple example of a Geo.Ref search, I entered "1755" as the key for searching, and retrieved the following references, among others:

Isoseismal map of the 1755 Lisbon earthquake (1979)

A discussion of the 1755 Lisbon earthquake (1977)

The earthquake of November 1, 1755 (1968)

This is hardly a sophisticated search, and it did yield some papers that were not relevant to Lisbon (such as one on Moon rock number 1755), but it is certainly successful in gaining entry into published modern views about the Lisbon earthquake of 1755. With these successful "hits" in hand, one then can consult the references that they in turn cite.

During the Geochautauqua, one of the speakers referred to the work of Crain on the distribution of craters on Mars. Being interested in the topic, but not knowing Crain's work (or even the spelling of his name), I used a portable terminal to search Geo.Ref during a break in the sessions.

I asked for any author "Crane" or "Crain" that was combined with the word "Mars". I got one hit, namely the paper by Ian K. Crain on "Statistical Methods for Geotectonic Analysis" that was an abstract in the International Geological Congress resumes (1972).

I next used SCI to see who had cited any paper published by "Crain IK" in 1972, and found the paper by Buckley and Buckley on "The Packing of Royal Tern Nests", published in Auk 94 (1977) 36-43.

Returning to Geo.Ref, I asked what variants of the author "Crain I" were present, and by combining "Crain, I.K." with "Crain, Ian K." I found 25 papers. One of these was Crain's paper on "The Monte-Carlo Generation of Random Polygons", in Computers & Geosciences 4 (1978) 131-141. I also found that the author had moved from Canberra, Australia, to Ottawa, Canada. It took only a few minutes to obtain this information.

In order to find out more about Prestel, Viewdata, Teletext, Ceefax, and Oracle, I used the INSPEC file on SDC. I found papers such as the following, all with short abstracts which I do not reproduce here:

> Prestel - the UK Post Office's Viewdata service (1978)
> Strengths and weaknesses of Prestel (1979)
> Teletext systems: a review (1979)
> Rivals of Viewdata and Teletext in the International field (1978)
> The coming of age of Viewdata (1978)
> The technical side of Viewdata (1978)

Although I had not previously used the INSPEC data base, and did not have the thesaurus or the manual, I was nevertheless able to find a number of informative recent articles on the subject of my search.

Interesting uses of the Citation Index are given by Garfield (1974), Cawkell (1977a), and Scrutton (1977), who show how SCI can be a tool in the study of interrelationships in scientific work.

NONBIBLIOGRAPHIC DATA BASES

Nonbibliographic data bases are important also, but I have found it difficult to discover the existence of many of those that are available. The only one that I use myself is the file of X-ray Diffraction data for crystalline substances available from the Joint Committee on Powder Diffraction Standards. The magnetic tape is available for purchase, and I have described a simple search program that can be used to help in the identification of minerals by X-ray diffraction methods (Glazner and McIntyre, 1979).

An example of a large nonbibliographic data base that is available online from a commercial time-sharing firm is IMPORTS, available from I.P. Sharp Associates, Toronto. It gives information on every shipment of crude oil or petroleum products into the U.S.

I should note here that Felix Chayes is engaged in a praiseworthy project to construct a worldwide data base on the chemical compositions of igneous rocks. The difficulty in having access to data of this type is that the information is not supported financially by commercial applications.

Indexes are available for online data bases, and I have found the following to be useful:

Directory of Online Databases
Cuadra Associates

Information Sources
Information Industry Association

But there is a great need for a compilation of non-bibliographic data bases relevant to geology. The National Oceanic and Atmospheric Administration (NOAA) is one example of a source that can supply machine-readable geologic data, such as gravity anomalies, heat-flow values, etc.

CONCLUSION

Computers have become a vital part of modern culture, and their use affects many aspects of geologic work. I believe that it is essential for students of geology who hope to survive in this fast changing and exciting world to be taught how to work at this interface between man and machine, so that they will put these powerful tools to proper use and avoid the fears, pitfalls, and abuses that come from ignorance.

REFERENCES

ACM, 1978, History of Programming Languages Conference, Los Angeles, California (June 1-3, 1978): ACM Sigplan Notices, v. 13, no. 8, 310 p.

Backus, J., 1978, Can programming be liberated from the Von Neumann style?: Comm. ACM, v. 21, no. 8, p. 613-641.

Berkovitch, I., 1979, Building a science magazine within Prestel: Phys. Technol., v. 10, 3 p.

Cawkell, A.E., 1977a, Science perceived through the Science Index. Endeavour: New Ser. 1, no. 2, p. 57-58.

Cawkell, A.E., 1977b, Developments in interactive on-line television systems and Teletext information services in the home: On-Line Review, v. 1, p. 31-38.

Glazner, A.F., and McIntyre, D.B., 1979, Computer-aided X-ray diffraction identification of minerals in mixtures: Am. Miner. v. 64, p. 902-905.

Iverson, K.E., 1973, APL in exposition: IBM Tech. Rept. 320-3010 (Available from APL Press), 61 p.

McIntyre, D.B., 1978, Experience with direct definition one-liners in writing APL applications: I.P. Sharp Assoc. Ltd., An APL Users Meeting, Proceedings, p. 281-297.

McIntyre, D.B., 1979, Computer Corner: Computers & Geosciences, v. 5, no. 2, p. 273-275.

McIntyre, D.B., 1980, APL in a Liberal Arts College: I.P. Sharp Assoc. Ltd., An APL Users Conference, Proceedings, p. 544-574.

McIntyre, D.B., Pollard, D.D., and Smith, R., 1968, Computer programs for automatic contouring: Kansas Geol. Survey Computer Contr. 23, 75 p.

Scrutton, R.A., 1977, Fragments of the earth's continental lithosphere: Endeavour, New Ser. 1, no. 2, p. 58-62.

COMPUTERS AS AN AID IN MINERAL-RESOURCE EVALUATION

Frederik P. Agterberg

Geological Survey of Canada

ABSTRACT

Separate flowcharts have been constructed for (1) some computer-based techniques for mineral-resource estimation; (2) different types of input for computer-based mineral-resource estimation and statistical exploration; and (3) quantification and analysis of geoscience map data. Several examples are presented to illustrate the geostatistical modeling of results obtained by image analysis, and the interpretation of probability index maps derived by multivariate statistical analysis of systematically quantified data on the geological framework of a region.

INTRODUCTION

For planning purposes it may be necessary to acquire knowledge about the mineral potential of regions where undiscovered ore deposits may be found in the course of future exploration operations. This paper is concerned with the contribution that computers can make to resource evaluation. Special attention will be paid to the subjects of quantification and multivariate analysis of geoscience map data.

The information for a region consists not only of published geoscience maps and reports but also of an intricate network of concepts regarding the geological history of the region considered and the modes of occurrence of the undiscovered deposits. In the past, these concepts were frequently in conflict with one another. For example, one geologist would advocate the hydrothermal origin of a given type of mineral deposit. In exploration operations he would emphasize the significance of deep-seated fractures. At the same time another geologist, assuming a synsedimentary origin for this deposit type, may have stressed the importance of lithological features. Especially if the evidence is scarce, it remains necessary to entertain mutually contradictory genetic concepts, and scientific investigations then should be guided by Chamberlin's method of multiple working hypotheses. However, concepts on the genesis of ore deposits are being refined continuously through the acquisition of new information and through better insight into geological processes. This leads to a reduction of the variance of the opinions and allows the usage of more precise metallogenic concepts in statistical resource estimation.

Because of the continuing advancement of geophysical and geochemical methods which produce systematic data, it is tempting to base quantitative resource appraisals on these measurements only. However, this would seem to be a risky procedure. Although certain types of mineral deposits may have direct geophysical signatures - as, for example, radiometric measurements for uranium and thorium, aeromagnetic responses for iron deposits, and airborne electromagnetic methods for massive sulphide deposits - the inherent difficulties in interpreting such signatures directly in terms of economic concentrations of minerals are well known. In the example of indirect interpretations of such data there are difficulties also. Although many types of quantitative geophysical and geochemical measurements are continuous, they generally are influenced by the characteristics of the near-surface geology which, in any given area, is likely to be nonuniform because of the presence of discontinuities separating different rock types. The heterogeneous nature of the geological framework will be reflected in the quantitative measurements. Although the boundaries between geological rock units may not be exposed, the available knowledge about them should be used as much as possible and integrated with the geophysical and geochemical data. In general thus, it seems that, as is the situation in effective exploration for mineral deposits, effective and realistic resource appraisal based on predictive models will require a judicious blend of geological, geophysical, and geochemical data on a base of sound metallogenic concepts.

In the past, mineral-resource appraisals usually were made by experienced geologists who, in general, formulated their opinions using the language of subjective probabilities. For example, an electromagnetic anomaly may indicate the existence of a massive sulphide deposit which is rich in copper and zinc but simple probabilities would suggest that this is not generally the situation. It might indicate equally the occurrence of a graphite deposit or perhaps the presence of a pyrite body without copper and zinc sulphides, or even merely a geological structure such as a fault. Hence it would be unwise to use the language of certainty when asked whether an undiscovered orebody is present. Either a subjective or an objective probabilistic answer is required.

During the past ten years, computers have been employed increasingly as a tool in mineral-resource evaluation. They are being used for file-building and information management, manipulation of subjective probabilities, image analysis of geoscience map data, and multivariate statistical analysis. These applications can be considered as components of resource analysis, a relatively new topic where the more speculative geological concepts regarding the genesis of mineral deposits are joined with the inductive logic of mathematical statistics and the data-processing capabilities of digital computers.

COMPUTER-BASED MINERAL-RESOURCE ESTIMATION AND STATISTICAL EXPLORATION

Some computer-based techniques for mineral-resource estimation are shown in Figure 1. Different types of input for resource evaluation and statistical exploration are listed in Figure 2. Most of the techniques of Figure 1 have been discussed in more detail in a review of statistical exploration methods by Agterberg and David (1979). Other recent publications concerned with the subject include a review of resource-appraisal methods by Harris (in press), a compendium of Russian resource-estimation techniques by Rundkvist and others (1979), and the final report on "Prospector", a computer-based consultant for mineral exploration which consists of semantic networks encoding subjective models developed by economic geologists (Duda and others, 1979). The close relationship between quantitative regional mineral-resource estimation and exploration which is aimed specifically at the discovery of new deposits is shown schematically in Figure 2. Findlay and Walsh (1979) have argued that the

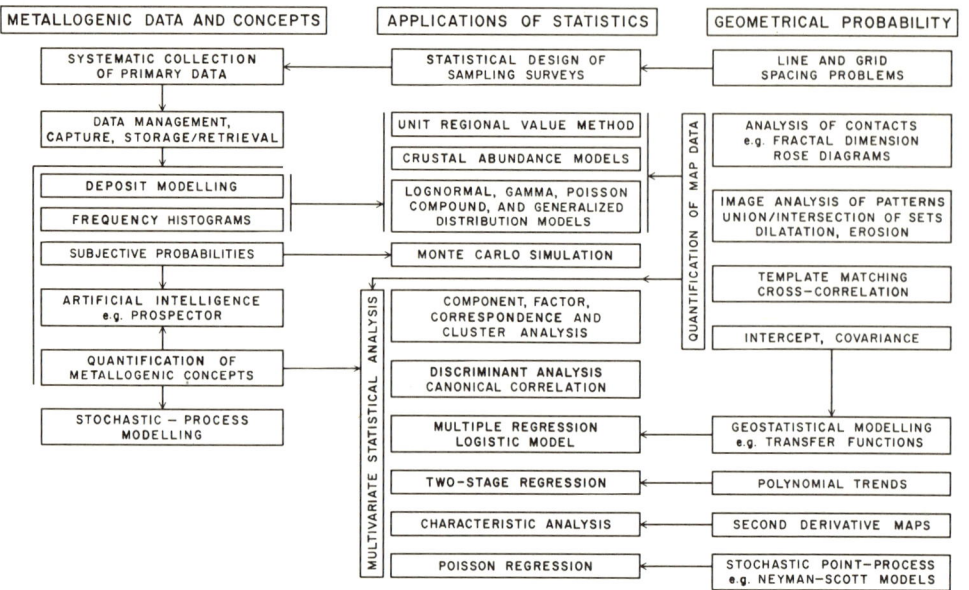

Figure 1. Some computer-based techniques for mineral-resource estimation.

application of computer-based statistical techniques is valid only towards the end stages of a regional-resource analysis after a firm data base has been established. These authors also have pointed out that the relationships between the processes leading to exploration on the one hand and resource estimation on the other are close but the end-products are different because exploration hopefully results in discovery and the other merely places prognostications in quantitative terms.

The headings of Figure 1 are (1) metallogenic data and concepts, (2) applications of statistics, and (3) geometrical probability. It can be attempted to use the geoscience map data of a region for a resource appraisal in which little or no use is made of metallogenic concepts. Then the procedure which can be followed consists of two stages: (a) quantification of map data; and (b) multivariate statistical analysis. We would be concerned primarily with geometrical probabilities and statistical techniques. The advantage of this approach is restricted to its objectivity; however, in practical applications, it turns out that the number of combinations of map patterns that can be tested systematically for their

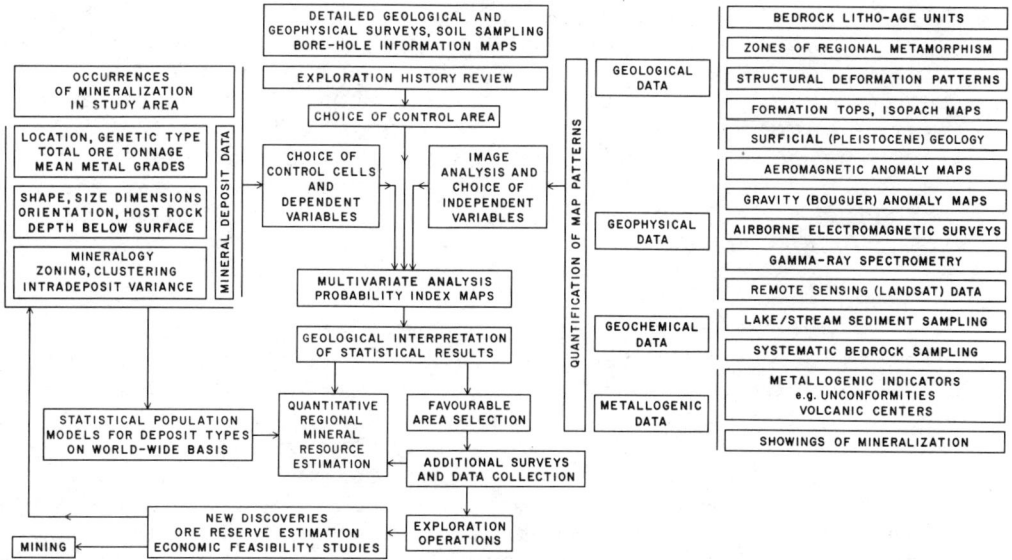

Figure 2. Different types of input for computer-based mineral-resource estimation and statistical estimation.

correlation with a pattern for occurrences of mineral deposits is limited severely. In general, the choice of variables for statistical analysis should be guided by metallogenic concepts in order to obtain improved results. This approach is shown schematically in Figure 1 by the arrow leading from the quantification of metallogenic concepts to multivariate statistical analysis which consists of the manipulation of quantified map data and other inputs derived from concepts of geometrical probability. Studies such as the one by Divi, Thorpe, and Franklin (1979) in which metallogenic concepts are modeled and tested statistically will be helpful in resource evaluation for deciding on the types of variables to be used and the choice of control areas.

It can be attempted to perform resource appraisals on the basis of metallogenic data and concepts without any use of statistical models. In fact, this procedure is close to a conventional geological practice of which the objective is to attempt to identify target areas that duplicate as much as possible with the available data the areas in which deposits are known. Other geologists with knowledge of the same area generally would appreciate that any geometrical configuration proposed in the absence of sufficient control remains hypothetical and does not represent necessarily the exact true which is unknow. It is, however, difficult to transmit this type of uncertainty to nongeologists. In comparison, statistical models have the advantage of their built-in uncertainty which can be estimated. The step from deterministic geological modeling to statistical modeling is illustrated in the following example.

Suppose that in a well-developed area, n mineral deposits with T tons of ore are known to occur in association with a metallotect of which the combined surface area amounts to A km^2. In the target area which is to be appraised, the surface area of the metallotect is equal to B km^2. It then can be useful to argue that there are nB/A tons of ore. This estimate neglects the undiscovered potential of the control area. It also may be necessary to qualify it by geological arguments, for example there may be the possibility that the "metallotect" in the target area is not comparable completely with its counterpart in the control area because of the absence of a critical but unknown factor that controlled the mineralization.

The preceding deterministic model can be made probabilistic by introducing additional assumptions regarding the nature of the random distribution according to which the mineral deposits are distributed with respect to the metallotect. These assumptions can be tested statistically for a control area. For example, suppose that the deposits are distributed randomly across the metallotect according to a simple Poisson process. This model can be visualized easily as follows. Let the surface area of the metallotect be subdivided into many small equal-area cells by superimposing a grid. A random distribution according to the simple Poisson process then indicates that each of the small cells has an equal probability of containing a mineral deposit. Under these conditions the expected number of deposits $\lambda_x = \lambda x$ for an area of x km^2 underlain by the metallotect remains equal to that computed by the deterministic model. In this formula λ is a constant which is independent of the size of the area underlain by the metallotect. We have $\lambda_x = n = \lambda A$ for the control area, and $\lambda_x = nB/A = \lambda B$ for the target area.

Although the expected number of deposits calculated by our statistical model is equal to that of the deterministic model, the method of prediction is entirely different because the number of deposits is assumed to be a random variable K with the Poisson distribution:

$$P(K = k) = e^{-\lambda_x} \lambda_x^k / k! \qquad (1)$$

For example, if $\lambda_x = nB/A = 2.7$ in the target area, then the statistical model predicts that there is a probability $P(K=3) = 0.220$ that there are exactly 3 undiscovered deposits. Suppose that after development of the target area it turns out that it contains only one mineral deposit. This outcome would prove that the deterministic model is wrong but it is compatible with the statistical model because $P(K=1) = 0.181$. Of course, the statistical model could lead to the acceptance of an erroneous hypothesis regarding the analogy between target and control areas. For this reason, it should be attempted not only to test the goodness of fit of the statistical model but also to perform geological interpretation of estimates of expected values resulting from the statistical analysis.

In order to predict the total tonnage of ore in the target area, we can assume that the ore tonnage of a single deposit is a random variable of which the mean value is equal to T/n (=average amount of ore per deposit in the control area). If the random variables for deposit density and ore tonnage are statistically independent, the new statistical model will yield an expected ore tonnage of TB/A for the target area which is equal to the amount predicted by means of the deterministic model. However, the statistical estimate is subject to a considerable uncertainty depending on the form (e.g. lognormal) assumed for the deposit size distribution.

The preceding statistical model for prediction of total amount of ore in the larger area makes use of a so-called generalized random variable (cf. Fig. 1). A so-called compound random variable (also see Fig. 1) arises if it can be assumed that the amount of metallotect per larger equal-area cell also satisfies a random variable whereas the number of deposits per unit of area underlain by the metallotect remains controlled by a simple Poisson process. When the random variable for amount of metallotect has the gamma distribution, the compound random variable for number of mineral deposits per larger equal-area cell satisfies the negative binomial model (cf. Agterberg, 1977).

In some applications of multivariate analysis to data quantified from different types of maps, it is useful to separate local features (e.g. anomalies) from regional trends or gradients. Agterberg and Cabilio (1969) fitted polynomial trend surfaces to a number of lithological variables showing that favorable environments for precious metal telluride ores in the Abitibi volcanic belt of the Canadian Shield are defined more precisely by residuals after elimination of regional trends. For similar reasons, Botbol and others (1979) have applied characteristic analysis which is a special type of principal-components analysis to binary patterns extracted from second derivative maps for many geochemical variables. It would be useful to attempt when the conceptual models will have been developed further to model the genesis of some types of ore deposits as stochastic processes in a manner previously employed for sedimentological processes (Schwarzacher, 1975). This stochastic-process modeling (cf. Fig. 1) also may become helpful for deciding on the types of transformations and combinations to be applied to the variables used in multivariate analysis.

Figure 2 shows that the primary inputs for resource analysis are of three different types. The information on the mineral deposits leads to the definition of patterns of cells with known deposits and dependent variables which are correlated with the independent variables consisting of combinations of patterns for geological, geophysical, and geochemical data. The choice of control area which is based on an exploration history review is of great importance during the multivariate analysis. However, as discussed in Agterberg (1974), the quantification of the amount of exploration per cell can present a difficult problem because our knowledge about the presence or absence of mineral deposits in a given volume of rock is itself a function of the independent variables.

QUANTIFICATION OF MAP DATA

Methods for coding geoscience data for large regions are shown in Figure 3. Traditionally, the spatial variability of almost any type of variable is displayed on a map. Although this practice facilitates our understanding, it may involve a significant amount of interpretation and generalization. In many types of statistical analysis, the map data are coded for equal-area cells belonging to a grid. For example, Agterberg and others (1972) coded the pattern of acidic volcanic rocks in

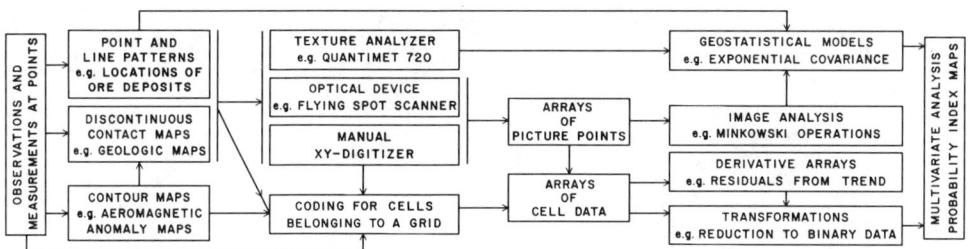

Figure 3. Quantification and analysis of geoscience map data.

east-central Ontario shown in Figure 4 for cells measuring 10 km on a side. Every 10 km cell was subdivided into 400 subcells measuring 500 m on a side and presence of the rock type at the centers of these subcells was point-counted. During the past 10 years significant progress has been made in the field of image analysis and map patterns now can be quantified to give arrays of picture points which, in turn, can be processed using digital computers (cf. Fabbri, and Kasvand, 1978). An example of texture analysis applied to geological map data (from Agterberg, 1978) is shown in Figure 4. The pattern of Figure 4A was used as input for a Quantimet 720 with linear correlator module. Geometrical covariances measured on the Quantimet for the east-west and north-south directions are shown in Figures 4C and 4D. In order to obtain these results the pattern of Figure 4A was shifted with respect to itself for a sequence of distances (d = ka) equal to multiples (k = 1,2,...,28) of a constant sampling interval equal to a = 4.694 km. The area of overlap between the original pattern and the shifted pattern was measured after each shift yielding the covariance which is expressed in number of picture points divided by 500,000 in Figures 4C and 4D. Apart from measurement errors, the covariance for k = 0 provides a measure of the proportion (= 7.2 percent) of the total area of Figure 4A which is underlain by acidic volcanics. The geometrical covariance can be transformed into an autocorrelation coefficient (r_d) shown in Figure 4B in four directions for small shifts only using a logarithmic scale in the vertical direction. The exponential model provides a good fit to these results. This enables us to use Matheron's geostatistical formulae (see e.g. Journel and Huijbregts, 1978) for estimating the variance of the amount of the rock type contained in a cell of

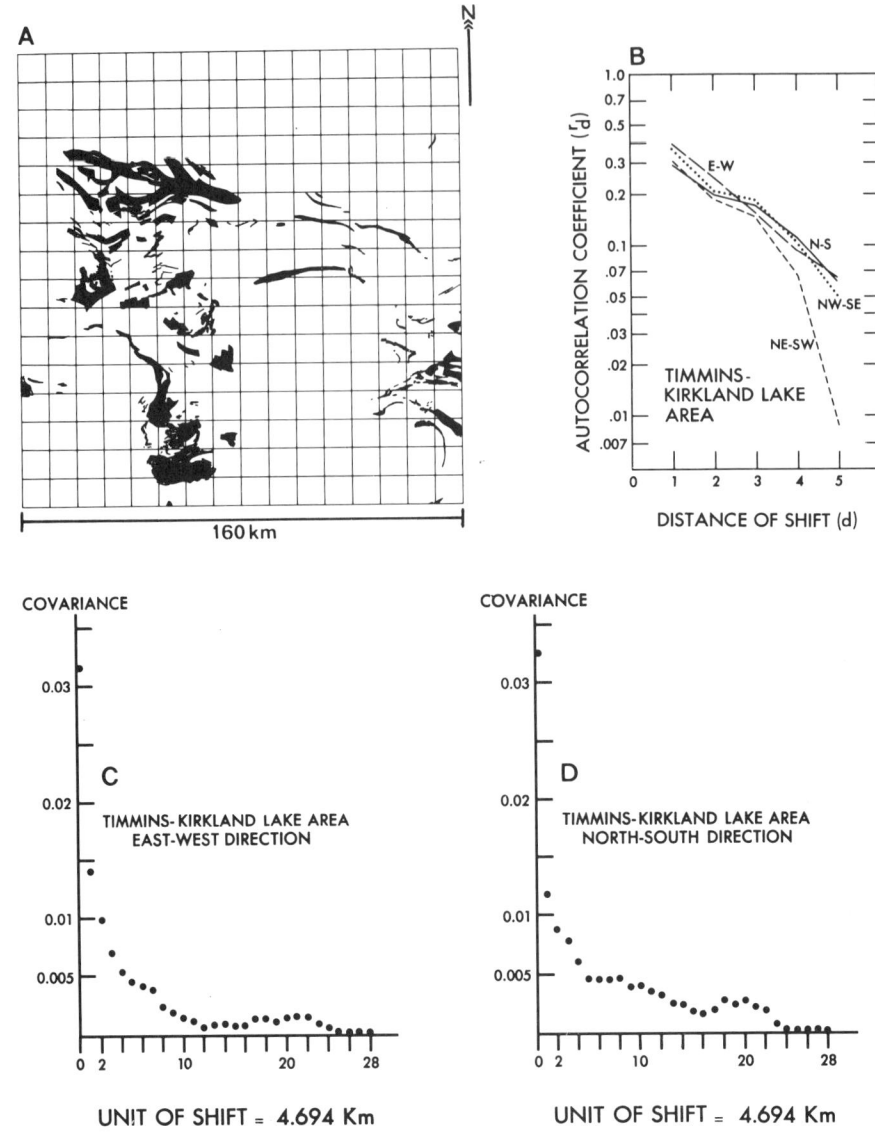

Figure 4. Image analysis of acidic volcanics in Timmins-Kirkland Lake area, Ontario (from Agterberg, 1978). (A) Original pattern with superimposed grid (10 km cells). (B) Autocorrelation coefficients for four directions for pattern of Figure 4A. Distances of shifts as on left side of Figures 4C and 4D. (C) Covariance (=area of overlap after shift) in picture points divided by 500,000 for east-west direction. (D) Ditto for north-south direction.

variable size and shape randomly superimposed on the pattern.

The geostatistical modeling illustrated in Figure 4 can be taken a step further as shown by Agterberg and Fabbri (1978) and Agterberg (in press). For the purpose of multivariate statistical analysis of map data quantified for cells belonging to a grid, we are interested in the frequency distributions of variables such as cell proportion underlain by a specific rock type. If the cells are small, this frequency distribution is U-shaped. For the somewhat larger (e.g. 10 km) cells used in multivariate analysis, it is h-shaped, assuming zero value in cells where the rock type is absent and a positive value less than one in cells where it is present in the region. These sequences of frequency distributions can be modeled if the variance is known. In Figure 5, observed frequency distributions and estimated model distributions are shown for nine examples including the examples of Agterberg and Fabbri (1978) for percentage of acidic volcanies in 10 km and 20 km cells in the Bathurst area of New Brunswick. Suppose that a rock type percentage value x (as plotted in the vertical direction of Fig. 5) is transformed into a a value y by the inverse of the relationship $x = \Phi(y)$ where Φ denotes a fractile of the standard normal distribution. Then

$$y = (rz - b)(1 - r^2)^{-\frac{1}{2}} \qquad (2)$$

where z denotes a value obtained by application of the same type of transformation to $p = \Phi(z)$ with p being the frequency percentage value as plotted in the horizontal direction of Figure 5. The parameter b in Equation (2) satisfies $\bar{x}=1-\Phi(b)$ where $\bar{x}$ denotes the average cell proportion value. The parameter r follows from $\bar{x}$ and the variance $s^2(x)$. A nomogram for the relationship between r, $\bar{x}$, and $s^2(x)$ is given in Agterberg (in press).

Simply stated, the experimental data of Figure 5 imply that a "probit" transformation of cell proportion data helps to normalize them. Because of the similarity between "logits" and "probits" (see e.g. Fisher and Yates, 1963), the transformation of cell proportion data into logits would have a similar effect. This type of transformation is applied to the dependent variable only in the logistic model of multivariate analysis to be discussed in the next section.

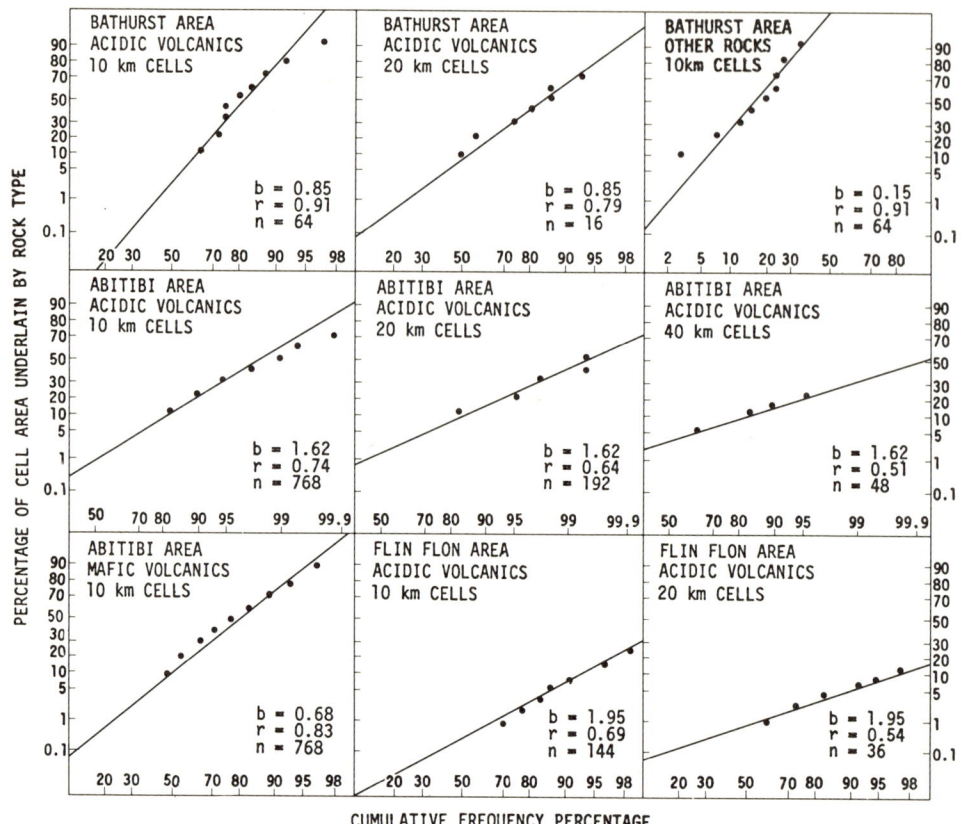

Figure 5. Straight-line fits for various rock types in three areas in Canada. Lines satisfy Eq. (2) with b and r computed from mean $\bar{x}$ and variance $s^2(x)$ of proportion values in n cells. Both axes have normal probability scales.

MULTIVARIATE STATISTICAL ANALYSIS OF CELL DATA

Some of the multivariate techniques that can be used to correlate the occurrences and characteristics of mineral deposits to variables systematically quantified for the geological framework in a region were shown in Figure 1. Figure 6 represents a contoured probability index map for copper deposits constructed by Agterberg and others (1972) using multiple regression analysis. The quantified map data consisted of rock types including the acidic volcanics of Figure 4A and geophysical (Bouguer and aeromagnetic anomaly) data.

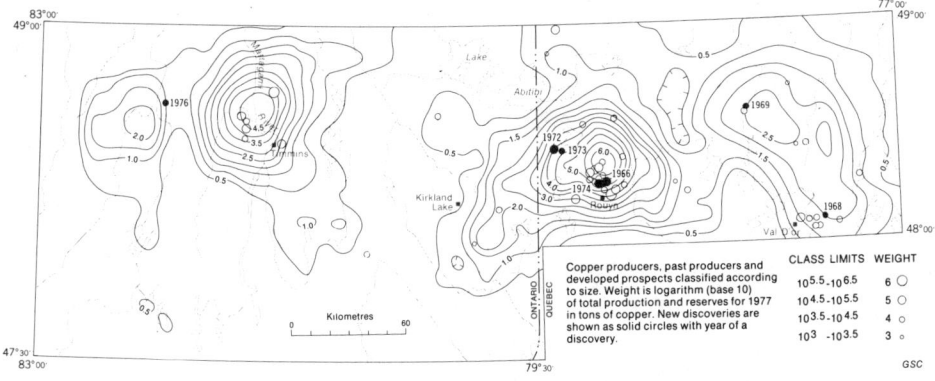

Figure 6. Contoured probability index map for number of events per 40 km unit area in Abitibi area of Canadian Shield (after Agterberg, and others, 1972) with "new" discoveries shown for comparison. Event represents one or more copper deposits per 10 km cell.

Also shown in Figure 6 are "new" discoveries of copper deposits on which no data were available when the map was constructed. A more detailed discussion of this hindsight study has been given in Agterberg and David (1979). In this section, we will investigate in more detail the basic assumptions which have to be made in order to construct a probability index map for occurrences of both known and undiscovered deposits.

Figure 7A shows the locations of a number of stratiform massive sulphide deposits in Archean rocks in the vicinity of Noranda, western Quebec. It also shows the locations of the 10 km cells of the grid (UTM grid) used for multivariate statistical analysis. Sets of probabilities computed by using (stepwise) multiple regression and the logistic model are shown in Figures 7B and 7C. These cells were used as "control cells". In the linear model of multiple regression, the dependent variable was set equal to one for control cells while it was kept equal to zero in all other cells. It is likely that a number of these other cells contain undiscovered deposits and this complicates the statistical estimation procedure. Recently, some properties of the estimators resulting in this situation have been studied by Chung and Agterberg (1980).

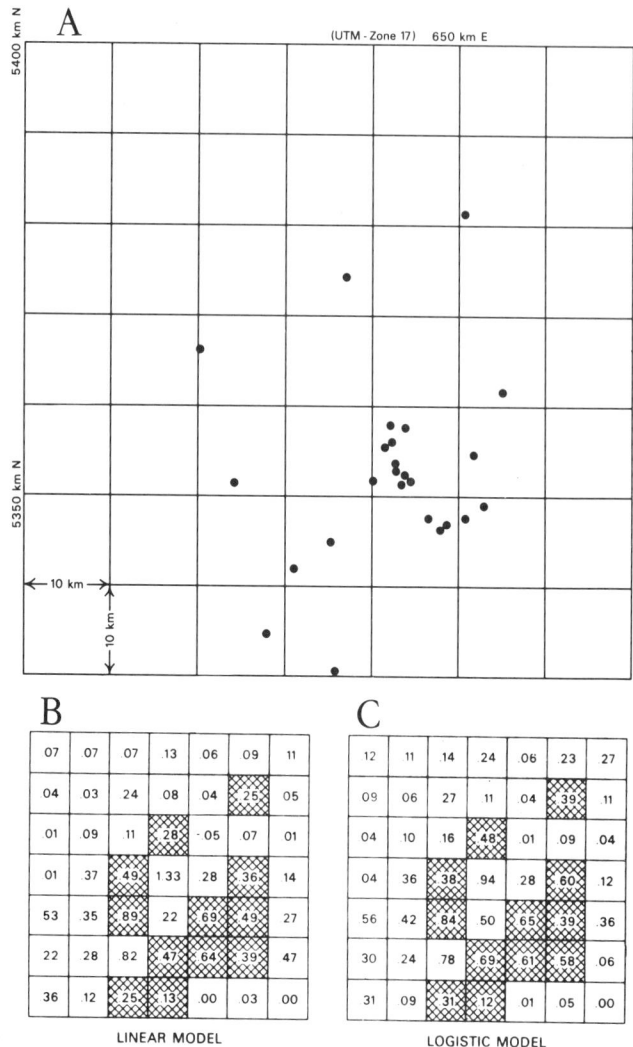

Figure 7. (A) Stratiform massive sulphide deposits in vicinity of Noranda, Quebec with 10 km cells for multivariate analysis. (B) Each number represents probability that a 10 km cell contains one or more deposits computed by multiple regression. (C) Ditto for logistic model. Hatched pattern in Figures 7B and 7C indicates control cells with one or more known deposits (from Agterberg, 1975).

The estimated probabilities arising from the linear model were divided by a constant (f) before plotting them in Figure 7B. The constant f was set equal to the sum of all estimated probabilities in a control area divided by the total number of control cells in this control area. It can be regarded as an approximation of the probability of discovery in the entire area of study. An analogous method was used to correct the probabilities initially resulting from the logistic model. Figures 7B and 7C represent, for each 10 km cell, the probability of occurrence of an "event" which consists of one or more deposits. By using Poisson regression, each of the deposits shown in Figure 7A could be considered separately. However, it may be advantageous to reduce the weights of individual deposits when there is a tendency toward relatively strong local clustering as in the situation of Figure 7. The linear model can be used as an approximation for either the logistic model of the Poisson model. The logistic model results in probabilities (Fig. 7C) which cannot be negative or greater than one whereas unconstrained multiple regression can give negative values or values greater than one as shown in Figure 7B. The estimated probabilities of events can be interpreted as the expected values of random variables. Consequently, they can be added for larger unit areas and this may yield contourable patterns as the one in Figure 6.

The contours of Figure 6 are for the expected value of a random variable representing number of events per square unit area measuring 40 km on a side. This random variable was interpreted as a positive binomial variable in Agterberg and others (1972) and as a Poisson variable in Agterberg (1975). These two interpretations yield sequences of probability values for actual numbers of events per unit which are approximately equal to one another in most practical applications. Neither model can be satisfied exactly in a strict sense. The Poisson model has the advantage that it is fully additive as illustrated by the following example. Suppose that the number of deposits in some area A has a Poisson distribution with parameter λ_a and that in area B it has a Poisson distribution with parameter λ_b. Then the number of deposits in the combined area C(=A+B) satisfies a Poisson distribution with paramter $\lambda_c = \lambda_a + \lambda_b$. In the Poisson model, a 10 km cell can contain more than one deposit. In the situation of Figure 6, however, the effect of local clustering was reduced by defining "events". Suppose that a 40 km unit area in Figure 6 is subdivided into sixteen 10 km cells none of which could be indicative of more than a single event. It follows that the Poisson model cannot be exactly satisifed for the 40 km unit area.

If it is assumed that the logistic model is valid, then the occurrence of an event consisting of one or more deposits per 10 km cell is controlled by a Bernouilli variable. The combination of a number of Bernouilli variables would give a so-called "subnormal" variation of the binomial distribution of which some properties have been summarized by Johnson and Kotz (1969, p. 80). It is interesting to compare the preceding three models for the random variable of which the expected value is shown on the contour map. From a practical point of view the Poisson and binomial models are easier to use than the subnormal binomial model.

For a larger region, the preceding three models can be approximated by normal distribution models which facilitates the comparison. For example, the sum of all $n = 49$ probabilities shown in Figure 7C is $\Sigma p_i = 13.75$. Consequently, according to each of the three models, about 14 events are expected in the area of Figure 7 and this is rather close to the observation of 12 control cells in this area of 49 cells. According to the Poisson model, the variance is equal to 13.75. The positive binomial approximation would have parameter $p = \Sigma p_i/n = 0.2806$ because $n = 49$. Hence its variance would be $np(1-p) = 9.89$ which is less than the variance of the Poisson variable.

The subnormal binomial distribution has variance $np(1-p) - n \cdot \text{var}(p_i) = 7.05$ where $\text{var}(p_i)$ denotes the variance among the p_i $(i = 1, 2, \ldots, n)$ values. If the three discrete distributions are approximated by normal distributions with mean values equal to 13.75, the standard deviations which are equal to the square roots of the preceding variances become 3.7, 3.1, and 2.7, respectively. These values are relatively close to one another, especially in view of other uncertainties associated with the selection of variables and choice of control area in the multivariate statistical approach.

CONCLUDING REMARKS

Practitioners of mineral-resource analysis should appreciate the great uncertainty generally associated with any attempt to predict the occurrence of undiscovered deposits. This applies even to relatively simple applications of geological analogy for the extrapolation from an area with known mineral deposits to a similar area without known deposits of the same type.

Computers can facilitate greatly the work of the geoscientist engaged in resource analysis by providing (1) access to a more extensive data base; (2) automating quantification of map features; and (3) accessing many types of statistical techniques. In multivariate analysis, it is imperative to test each model by changing the input variables in order to investigate the stability or lack of stability of the results. These repetitive tasks can be performed only with the aid of computers. The choice of variables and the interpretation of statistical results should as much as possible be based on metallogenic concepts.

REFERENCES

Agterberg, F.P., 1974, Geomathematics: Elsevier, Amsterdam, 596 p.

Agterberg, F.P., 1975, Statistical models for regional occurrence of mineral deposits, *in* Schriften fur Operations Research und Datenverarbeitung im Bergbau 4: Gluckauf, Essen, p. C-1, 1-15.

Agterberg, F.P., 1977, Frequency distributions and spatial variability of geological variables, *in* Proc. 14th Symp. Applications of Computer Methods in the Mineral Industry: Soc. Min. Eng. AIME, New York, p. 287-298.

Agterberg, F.P., 1978, Quantification and statistical analysis of geological variables for mineral resource evaluation, *in* Sciences de la Terre et Mesures, Colloque Jean Goguel: Bur. Rech. Geol. Min. Memoire No. 91, p. 399-406.

Agterberg, F.P., 1979, Mineral resource estimation and statistical exploration, *in* Facts and Concepts of World Oil Occurrence: Can. Soc. Petr. Geol. Mem. 6, p. 301-318.

Agterberg, F.P., in press, Cell value distribution models in spatial pattern analysis, *in* Future Trends of Geomathematics, J.C. Griffiths Volume: Pion, Ltd., London.

Agterberg, F.P., and Cabilio, P., 1969, Two-stage least squares model for the relationship between mappable geological variables: Jour. Math. Geology, v. 1, no. 2, p. 137-153.

Agterberg, F.P., Chung, C.F., Fabbri, A.G., Kelly, A.M., and Springer, J.S., 1972, Geomathematical evaluation of copper and zinc potential of the Abitibi area, Ontario and Quebec: Geol. Survey Canada Paper 71-41, 55 p.

Agterberg, F.P., and David, M., 1979, Statistical exploration, *in* Computer Methods for the 80's: Soc. Min. Eng. AIME, New York, p. 90-115.

Agterberg, F.P., and Fabbri, A.G., 1978, Spatial correlation of stratigraphic units quantified from geological maps: Computers & Geosciences, v. 4, no. 3, p. 285-294.

Botbol, J.M., Sinding-Larsen, R., McCammon, R.B., and Gott, G.B., 1978, A regionalized multivariate approach to target selection in geochemical exploration: Econ. Geology, v. 73, no. 4, p. 534-546.

Chung, C.F., and Agterberg, F.P., 1980, Regression models for estimating mineral resources from geological map data: Jour. Math. Geology, v. 12, no. 5, p. 473-488.

Divi, S.R., Thorpe, R.I., and Franklin, J.M., 1979, Application of discriminant analysis to evaluate compositional controls of stratiform massive sulfide deposits in Canada: Jour. Math. Geology, v. 11, no. 4, p. 391-406.

Duda, R.O., Hart, P.E., Konolige, K., and Reboh, R., 1979, A computer-based consultant for mineral exploration: Final Rpt. Project 6415, SRI International, Menlo Park, Cal., 185 p.

Fabbri, A.G., and Kasvand, T., 1978, Picture processing of geological images: Geol. Survey Canada Paper 78-1B, p. 169-174.

Findlay, D.C., and Walsh, J.H., 1979, Canadian applications in resource analysis: Resources Policy, March 1979, p. 61-70.

Fisher, R.A., and Yates, F., 1963, Statistical tables for biological, agricultural and medical research (6th ed.): Hafner, New York, 145 p.

Harris, D.C., in press, Mineral endowment - geostatistical theory and methods for appraisal: Paper prepared for Symposium on Methods for Broad Mineral Resource Appraisal, Univ. West Virginia, Main 1978, 85 p. (in press, Univ. West Virginia Press).

Johnson, N.L., and Kotz, S., 1969, Distributions in statistics, discrete distributions: Houghton Mifflin, Boston, 328 p.

Journel, A.G., and Huijbregts, Ch. J., 1978, Mining geostatistics: Academic Press, London, 600 p.

Rundkvist, D.V., ed.-in-chief, 1979, Quantitative forecasting in regional metallogenic investigations: VSEGEI, Leningrad, 88 p. (in Russian).

Schwarzacher, W., 1975, Sedimentation models and quantitative stratigraphy: Elsevier, Amsterdam, 382 p.

QUANTITATIVE BIOSTRATIGRAPHY, 1830-1980

James C. Brower

Syracuse University

ABSTRACT

Charles Lyell (1830-1833) became the first quantitative biostratigrapher when he proposed his method for determining the relative age of Tertiary fossil assemblages by calculating the percent of living species in them. Quantitative biostratigraphy languished for about 125 years but a major renaissance began in the late 1950's which is continuing at an accelerating pace.

Methods of quantitative biostratigraphy can be grouped into several categories. The first category consists of the quantification of the index-fossil concept including measurement of the attributes of an index fossil as well as the relative biostratigraphic values of the species concerned.

The second category represents the treatment of assemblage zones with multivariate analysis. The basic data may represent proportions of different species in a series of samples but presence-absence data also are used. The unweighted presence-absence data may be analyzed but the presences also can be weighted by the amount of biostratigraphic information conveyed by the species involved. Most of the techniques applied here are well-known multivariate methods such as cluster analysis, multidimensional scaling and some forms of archeological seriation.

In the third category, the biostratigrapher is faced with a plethora of methods for determining the most likely sequence of biostratigraphic events based on observations from numerous stratigraphic sections. Most schemes are concerned with point events which consist of the highest and lowest occurrences of the species treated. Several techniques are concerned with entire range zones. Sequencing algorithms produce either average sequences or are intended to give the stratigraphically highest possible estimate of the top of a range zone and the stratigraphically lowest possible estimate of the base of that range zone. Once the most likely sequence of events is ascertained, it can serve to correlate the individual samples and stratigraphic sections. Some of the techniques are elaborate whereas others are simplicity carried to the ultimate.

Although numerous methods of quantitative biostratigraphy have been proposed which do produce excellent results, most biostratigraphers have resisted successfully the impact of quantification. There are several reasons for this. First, most practicing biostratigraphers are basically nonquantitative. Secondly, many of the methods involve logic and algorithms which are not familiar to biostratigraphy. Thirdly, biostratigraphers learn that nonquantitative techniques produce acceptable results although only after long periods of time and much effort. Essentially quantitative biostratigraphy has proven unpalatable to the intended consumers.

Therefore, in the fourth category, we have cast about for super-simple methods of quantitative biostratigraphy, almost without numbers, which closely replicate the logic of biostratigraphers. One such technique is archeological seriation which can work directly on a species by samples data matrix to simultaneously produce a range chart as well as correlation of the samples. Both additive and nonadditive models are applicable. Another is a sequencing method which involves comparison of lists of events in a series of stratigraphic sections and resolving the inconsistencies between the different sections in such a manner as to produce a composite zonation.

Lastly, paleontologists have studied evolutionary sequences in a numerical context using a smorgasbord of techniques for many years. The statistical methods range from the simplest univariate to the most complicated multivariate types with or without the aid of time-series analysis. Although evolution provides the basis for biostratigraphy, it is surprising that many evolutionary sequences are approached through a strictly biological point of view rather than in a biostratigraphic context.

Through the past 150 years, biostratigraphers have proposed many numerical schemes for quantitative correlation. It is unfortunate that most of these methods have not been tested rigorously on numerous actual and simulated data sets. Hopefully, during the next decade quantitative biostratigraphers will evaluate systematically the available algorithms by case studies to ascertain which techniques provide the best results with various types of data. Thus the 1980's should represent an interval of consolidation instead of a decade in which a new horde of algorithms will be invented.

INTRODUCTION

Biostratigraphic correlation provides the foundation for many studies in stratigraphy, both of an applied and theoretical nature. The basis of historical geology is the relative time scale which is founded on biostratigraphy. Without this time scale, earth history could not be reconstructed because any historical science involves the placement of events in a matrix defined by time and space. The time scale is essential to studies in evolutionary biology because ancestors and descendents could not be recognized without such a scale. Biostratigraphic correlation also is useful in the search for fossil fuels. Coal, oil, and gas are concentrated in certain zones and depositional environments which are localized in time and space. Biostratigraphy represents one of the main weapons in the search for these elusive targets.

An obvious question is: why bother to quantify biostratigraphy. Three major reasons can be listed. The first is the massive amounts of biostratigraphic data. Data matrices may include hundreds of species and samples and these cannot be comprehended simultaneously by the human mind. Any systematic treatment of biostratigraphic data should use, evaluate, and screen all the information available. This requires efficient algorithms and the memory bank of a large computer. A second is that quantitative methods have the potential for providing a higher degree of biostratigraphic resolution than do qualitative techniques. Thirdly, some of the quantitative methods provide information about the statistical probabilities associated with various biostratigraphic hypotheses. Such data cannot be obtained from qualitative methodology.

It is with this in mind that IGCP (International Geological Correlation Program) Project 148 on Quantitative Stratigraphic Correlation was established several years ago. As far as biostratigraphy is concerned, three goals have been set: first is to devise quantitative techniques for biostratigraphy. The numerous available methods will be outlined subsequently. The second involves the implementation of computer programs for those methods and to make the programs available to the scientific public at large. Third is to utilize case studies of actual and simulated data to test and compare the various methods. Furthermore, it seems reasonable to evaluate quantitative biostratigraphy and to present a prognosis for future work based on these goals and how well they have been realized to date.

HISTORICAL BACKGROUND

Consideration of the history of stratigraphy indicates that a series of breakthroughs have lead to all of the correlation methods, both qualitative and quantitative, used in biostratigraphy. Excellent reviews of the development of biostratigraphy are given by Hancock (1977) and Mallory (1970). The obvious prerequisite for any biostratigraphic methodology is the recognition that fossils could be used for correlation. This concept originated with William Smith and was published in tables, maps, and writings in 1799, 1815, and 1816-1819. Smith realized that the strata could be arranged in stratigraphic order and that each stratum was characterized by an assemblage of fossils. The faunal zones cataloged by Smith thus are assemblage zones.

Frederick Quenstedt (e.g., 1856-1858) was probably the first individual to systematically record detailed data and attempt to compile the range zones of fossils. However, it was left to his student Albert Oppel (1856-1858) to exploit fully the potential of range zones. Oppel advocated compiling range-zone charts of suites of species. These were scrutinized then to determine intervals characterized by assemblages of taxa that occurred together as well as overlap zones between two or more taxa. This allowed him to refine considerably the zonations proposed by earlier workers. Two major types of quantitative methods can be tied to the concepts first laid out by Smith, Quenstedt, and Oppel. These are the quantification of assemblage zones as well as the analysis of the most likely sequence of events.

Index fossils were known and used to trace and identify zones long before the time of Darwin, but the exact root of the idea is uncertain. According to Eicher (1976) and Donovan (1966), the concept of an index fossil can be traced back to Albert Oppel (1856-1858) who named each of his zones after a particularly abundant and characteristic species which was termed the "index." The origin of the idea could be even earlier. For example, William Smith wrote in 1817 (on p. iv of the Stratigraphical System of Organized Fossils):

> By the tables it will be seen which fossils are peculiar to any stratum and which are repeated in others.

At any rate and from whatever origin, correlation by index fossils became usual practice by the late 19th century and is employed yet by many biostratigraphers.

The last major revolution in biostratigraphy is the doctrine of organic evolution which can be attributed to Charles Darwin in 1859. Evolution obviously provides the mechanism for morphological changes in organisms and therefore is the vital basis for biostratigraphy. It is interesting to note that much biostratigraphical methodology developed long before the doctrine of evolution was accepted by paleontologists (e.g. Gould, 1977). As far as biostratigraphy is concerned, the essential facts lie in the changes in the organisms - not the mechanisms involved. At any rate, once evolution and natural selection began to permeate the paleontological community, paleontologists could start to analyze lineages of fossils in a meaningful manner.

Charles Lyell (1830-1833) probably was the first quantitative biostratigrapher when he defined several of the Tertiary epochs based on percentages of living molluscan species. These percentages consisted of: Eocene, 3 percent; Miocene, 18 percent; and Pliocene, 49 percent. The Lyellian method is statistical for two reasons. First, the percentages were derived from massive amounts of data, namely some 8,000 species and 40,000 specimens. Second, the definitions furnished an identification rule for determining the age of unknown samples. Lyell's zonation of the Tertiary provided the first quantified assemblage zones. Subsequent developments are covered later in this paper.

Several reviews on quantitative techniques in biostratigraphy have been published recently, namely those of Brower and Millendorf (1978) and Hay and Southam (1978). Hazel (1977) summarized the state of the art in the quantification of assemblage zones whereas Miller (1977) outlined the graphical correlation approach of Shaw (1964). Reyre (1974) reviewed many quantitative studies in palynology, most of which were done by European workers. Nevertheless, developments have been proceeding rapidly and these reviews are outdated, at least partially, hence this paper.

Four general approaches have been employed in biostratigraphy. These include: (1) treatment of the index-fossil concept, (2) quantification of assemblage zones, (3) determination of the most likely sequence of biostratigraphic events, and (4) study of lineages.

INDEX-FOSSIL CONCEPT

Index fossils were used before Darwin's time to distinguish and trace biostratigraphic zones, and correlation by such fossils currently is in widespread practice. However, Jeletzky (1965) argued against the possibility of quantified biostratigraphy because he believed it was impossible to express numerically the degree of biochronological usefulness of fossils. He concluded (1965, p. 135), "Any attempt at the quantification of biochronological correlation is, thus, precluded by the fundamentally *qualitative* and *non-statistical* nature of its most valuable data (index fossils)." Subsequent discussion will show that this is not true. The first practical attempt at a quantification of the index fossil concept was that of Hazel (1970). A novel approach was presented by Cockbain (1966) who proposed the entropy function as a numerical measure of the relative biostratigraphic usefulness of a fossil. This method, as McCammon (1970, p. 49) pointed out, is inefficient.

The three attributes of an index fossil, namely facies independence (F), geographical persistence (G), and vertical range (V), are measured easily as discussed by McCammon (1970), Brower, Millendorf, and Dyman, (1978); and Millendorf, Brower, and Dyman (1978).

The work of McCammon (1970) was a major breakthrough in quantification of the index-fossil concept and provided a method for determining a numerical index of the amount of biostratigraphic information conveyed by the presence of a particular fossil taxon or species. These indices are termed relative biostratigraphic value (RBV) and they are scaled through a range of 1.0 for the ideal index fossil to 0.0 for a species with no useful biostratigraphic information. The index designed by McCammon (1970) weighted all three parameters equally. Two types of taxa could have RBV's of almost 1.0. In the first instance, a classic index fossil would be valued for three attributes, that is a high degree of facies independence and geographical persistence in conjunction with a short vertical range. Secondly, a geographically widespread species that is restricted to a single facies also will have a high RBV. Such a taxon will be useful for its ability to trace a biofacies laterally and vertically and the RBV of this form is not influenced strongly by its vertical range. Thus, the McCammon index represents a compromise between time-stratigraphic correlation on one hand, and establishing the physical continuity and geographical persistence of a particular biofacies on the other.

Brower, Millendorf, and Dyman (1978) advocated another index in which the parameters were weighted in a different manner. The geographical persistence and facies independence were weighted equally but the vertical range was given double weight relative to either of the other two parameters. This index was designed specifically to identify taxa that would be most useful for time-stratigraphic correlation. Brower, Millendorf, and Dyman (1978) also point out that other RBV's can be structured which incorporate the desired properties.

Once the index fossils have been quantified, the information can be used in several contexts. The appropriate number of index and near-index fossils can be selected to construct a reasonable zonation of biostratigraphic events or samples. For example, McCammon (1970) demonstrated that only about 10 percent of the species with the highest RBV's were needed to provide a useful zonation of some Tertiary strata from the Gulf Coast. This essentially is a search for parsimony, that is a reasonably effective zonation based on the smallest possible number of taxa. McCammon (1970) also devised simple classification or identification functions to assign samples to the faunal zones. In another approach, Brower, Millendorf, and Dyman (1978) and Millendorf,

Brower, and Dyman (1978) successfully used several of these RBV's to weight the data for species in the multivariate analysis of assemblage zones as discussed later. The RBV's and the biostratigraphic attributes also can be used to test biostratigraphic hypotheses. For example, are the biostratigraphic properties of bottom dwellers different from those of pelagic organisms? Have the biostratigraphic parameters of one or more groups of organisms changed with time? Do species living in different types of habitats have the same properties?

QUANTIFICATION OF ASSEMBLAGE ZONES

Assemblage zones were the first type of zone to be recognized in biostratigraphy. As mentioned before the zones of William Smith and Charles Lyell basically are assemblage zones. After the quantitative definition of the Tertiary epochs by Lyell, no advances were made in the quantification of assemblage zones for about 125 years (see Mallory, 1970, and Hancock, 1977 for reviews). This interval of quiescence was ended when Simpson (1947, 1960) and Sorgenfrei (1958) initiated work on similarity coefficients to measure faunal resemblance. Although Simpson (1947, 1960) studied these coefficients in the context of biogeography, he was aware of the implications for biostratigraphy. Similar indices were used by Sorgenfrei (1958) for correlation of the Middle Miocene of Jutland with the Miocene of other areas. These pioneering works have given rise to numerous studies that have attempted to define time zones or ecological zones which can be employed in correlation.

Assemblage zones are characterized by a particular suite of taxa regardless of their ranges, and yield typically a combination of stratigraphical and ecological information. Assemblage zones show the distribution of faunal discontinuities in time and space. The techniques for depicting such zones are numerous and many are rooted in conventional multivariate analysis and numerical taxonomy. Comprehensive papers on biostratigraphy which outline some of the methods, present examples of applications, and compare results from various techniques include Hazel (1970, 1971, 1977), Brower, Millendorf, and Dyman (1978), and Millendorf, Brower, and Dyman (1978). The reader should note that many of the techniques of quantitative paleoecology and biogeography are similar to those of biostratigraphy; examples of such allied publications include Ali, Lindemann, and Feldhausen

(1976), Blackith and Reyment (1971), Buzas (1970), Ellison (1963), Feldhausen (1970), Fox (1968), Gill, Boehm, and Erez (1976), Henderson and Heron (1977), Imbrie (1964), Kaesler (1966), Lynts (1971, 1972), Lynts and Stehman (1971), McCammon (1966), Oltz (1969, 1971), Park (1974), Raup and Crick, (1979), Rowell and McBride (1972), Rowell, McBride, and Palmer (1973), Scott (1970), Simpson (1960), Sorgenfrei (1958), Stone (1967, 1973), Symons and DeMeuter (1974), Valentine and Peddicord (1967), and the volume edited by Scott and West (1976).

The primary data matrices of biostratigraphy are rectangular with species or taxa along one axis and samples arrayed on the other. Generally the data consist of presences and absences of the taxa in the samples but Reyre (1972, 1974) analyzed relative abundances or proportions of various species. However, presence-absence data contain all the information needed for most biostratigraphic studies. For simplicity, it is easiest to discuss the analysis of assemblage zones in the following order: Q-mode analyses between samples, R-mode analyses between taxa, and methods which treat both taxa and samples in a single sequence of operations.

Q-Mode Analyses

Study of the relationships between samples involves a series of steps, each of which presents decisions that must be made by the biostratigrapher.

In Step 1, if the data are presence-absence, as is the usual situation, one must decide whether to convert the data to the range-through form or work with the original data. In the range-through method of data treatment, a taxon is listed as present in all samples within its local range zone for each stratigraphic section. This helps to eliminate sampling problems and minimizes bias toward whatever other factors, such as environmental parameters, that could dictate the presence of a particular organism. The time duration of the species is of interest, not the vagaries of its distribution. This type of data is used by most biostratigraphers. In some situations the range-through method can lose pertinent information such as where several biofacies intertongue repeatedly. The range-through technique is not applicable for data consisting of relative abundances.

The question for Step 2 is whether to calculate the relative biostratigraphic values (RBV). The course of action involved here depends on the purpose of the analysis. If stratigraphical groupings are desired, the RBV's can be useful. If the RBV's are to be determined, one must measure the attributes of an index fossil and elect the RBV index of interest (see previous discussion). The RBV's can be employed in several manners. The presences can be weighted by the RBV's. If so the presence of a species will contribute an amount of similarity between samples that is proportional to its RBV. This usually produces clusters that are tighter and stratigraphically more homogeneous. Another strategy is to discard the taxa with lower RBV's and to complete the analysis only on the species with the highest RBV's. This results in a parsimonious solution consisting of a reasonable zonation based on the smallest possible number of species (McCammon, 1970).

Step 3 comprises the computation of a similarity or difference matrix between all pairs of samples. For binary data, the presences and absences may suffice or the presences can be weighted by the RBV's. Absences cannot be evaluated completely for biostratigraphic data because a taxon can be absent from a sample for various reasons. Therefore similarity coefficients based on mutual presences or positive matches are preferred by most authors. Various similarity coefficients are employed such as the Jaccard, Dice, Otsuka, or Simpson although middle-of-the road types such as the Dice coefficient usually are recommended. However other types of coefficients have been applied in quantitative biostratigraphy such as the correlation coefficient of the Pearson product-moment type. Useful reviews of these coefficients are available in Cheetham and Hazel (1969), Hohn (1976), Sepkoski (1974), and Sneath and Sokal (1973). Recently, several probabilistic similarity coefficients have been formulated by Henderson and Heron (1977) and Raup and Crick (1979). Although not tested on biostratigraphic data, these may yield excellent results in the quantification of assemblage zones.

Many varieties of similarity and difference coefficients are available for data on relative abundances. Coefficients that are used in paleoecologic studies and could be used in biostratigraphy consist of the quantified association coefficients of Sepkoski (1974), correlation coefficients, such as that of Sorenson and the Pearson product-moment type, and a legion of distance coefficients.

The last step in the analysis is to extract the main themes from the similarity or difference matrix. Methods that have been employed for biostratigraphic or paleoecologic data include agglomerative cluster analysis (Hazel, 1977; Millendorf, Brower, and Dyman, 1978), divisive clustering in the form of association analysis (Gill, Boehm, and Erez, 1976) and dissimilarity analysis (Gill and Tipper, 1978; Tipper, 1979), principal components (Hazel, 1977), factor analysis (Symons and DeMeuter, 1974), correspondence analysis (see Reyre, 1974), principal coordinates (Hazel, 1977), nonlinear mapping and multidimensional scaling (Millendorf, Brower, and Dyman, 1978), and lateral tracing (Millendorf, Brower, and Dyman, 1978).

Plant ecologists have developed some methods which are useful especially for nonlinear data (Whittaker, 1973 outlines some of these techniques). Cisne and Rabe (1978) applied several of these direct and indirect ordination algorithms (polar ordination, principal components, and reciprocal averaging) to numerous samples from Middle Ordovician rocks in New York which could be correlated in time by position relative to a series of bentonites. The data consist of percentages of common taxa, all of which range throughout the time interval studied. The problem thus is basically paleoecological. The different samples could be characterized by their position on a depth gradient. Plots of ordination scores on a time-stratigraphic diagram permit the identification of transgressions and regressions which could be used for time correlation. The ordination scores also serve to calculate rates of change with respect to distance along transects and to estimate the slope of the basin. Other geological examples of such ordination techniques are presented by Ali, Lindemann, and Feldhausen (1976), Feldhausen (1970), and Park (1974).

Shier (1978) outlines a simple technique termed sample ordering which produces a one-dimensional ordination of samples that can be used for paleoecological analysis. This strategy also could be applied to biostratigraphic data.

R. Christopher (1978, pers. comm.) has contoured matrices of similarity coefficients derived from samples in two stratigraphic sections (see also Reyre, 1974).

Biostratigraphers profitably might borrow some of the techniques developed by archeologists. For many years archeologists have faced problems closely allied to those of biostratigraphy. Samples containing various objects must be arranged into a sequence which represents time or an evolutionary series. Examples of such problems are numerous and embrace such fascinating case studies as artifacts in graves, sentence structure, and word frequency in manuscripts written through the ages, etc. Strangely enough, few of these methods have been noticed by biostratigraphers. General discussions of these techniques are in Cowgill (1972), Doran and Hodson (1975), Hodson, Kendall, and Tautu (1971, especially the articles by Gelfand and Kendall), Johnson (1972), and Marquardt (1978). These methods range from the exceedingly simple where one just rearranges the rows and columns of the matrix to concentrate the most similar samples along the diagonal to exceedingly complex methods which combine multidimensional scaling with principal components. Schuey and others (1978) calculate an evolutionary scale from similarity coefficients based on an autocorrelation function.

For the studies done by Millendorf, Brower, and Dyman (1978), the best results were obtained from agglomerative cluster analysis of the UPGM type, multidimensional scaling shows the major features of the faunal zones and discontinuities between these whereas lateral tracing provides detailed correlations within the zones. Techniques restricted to linear relationships such as principal components are not adequate in many instances. Most authors (e.g. Hazel, 1970, 1971, 1977) have worked only with the entire data set. However, Brower, Millendorf, and Dyman (1978) and Millendorf, Brower, and Dyman (1978) adopted a more sophisticated two-fold approach. First, the entire data set is treated in order to gain an overall picture. Hopefully, the desired correlations or zones can be extracted from the clusters or ordinations. Second, the analysis focuses on pairs of adjacent sections to prepare a line of sections or a fence diagram showing more detailed correlations.

R-Mode Analyses

An R-mode analysis can be performed on the taxa with the same techniques except for several details. For example, lateral tracing is not applicable in the R-mode. The data can be presence-absence, either in the original or range-through form. There is no purpose in weighting the data by the relative biostratigraphic values. Relative abundance data also can be processed for the species.

R-mode results can be difficult to interpret in terms of the stratigraphic distribution of the taxa.

Hazel (1970, 1977) presents some results in the form of dendrograms and principal components. Species which are confined to one assemblage zone generally cluster or group together. However long-ranging forms which occur in several or more assemblage zones obscure the structure of the data and introduce distortion into the dendrogram or ordination plot. Another problem with R-mode analyses is that they mainly disclose information about the "center of gravity" or "centroid" of the individual range zones rather than the endpoints or upper and lower limits of the taxa. Thus the endpoints of the range zones may be of more interest to biostratigraphers than the centroids.

Sequential Analyses in R- and Q-Modes

The only example of a dual-space technique in biostratigraphy known to me is that of Hazel (1977). A matrix of correlation coefficients was computed for presence-absence data of the range-through type for 24 species of Cambrian trilobites. Extraction of the principal components of this matrix produced four rather loosely structured assemblage zones along with several species that could not be definitely assigned. The first three principal components account for 59 percent of the variance in the correlation matrix. Although the plot of the first three principal components does show the assemblage zones, these are not in stratigraphical order unless they lie along a rather complex spiral. Principal-component scores array the relationships between the samples. Generally, principal components seem to have been adequate for this rather simple data set consisting of 24 species and 65 samples. However, such eigenvector methods probably would not be adequate for more complex data with ordination axes that are strongly nonlinear. The results obtained from other similar techniques, such as factor and correspondence analysis, probably would be similar to those derived from principal components.

Archeologist Cowgill (1972) outlines a seriation technique which might be useful in biostratigraphy. The data matrix includes the presence-absence of a series of objects in numerous samples or units. A similarity matrix was formed for the units, after which two axes were obtained from this matrix by multidimensional scaling. The two axes were reduced to a single dimension by "graphical regression" although formal regression or principal components would have served the same purpose (Kendall, 1971). This produced coordinates for a one-dimensional seriation of the objects. For biostratigraphical data of the range-through type, the coordinates would give roughly the midpoints of the zones. Scores are determined for each sample based on the seriation coordinates for the objects, the number of samples containing each object, and the objects present and absent in each sample. These scores arrange the samples with respect to the seriation sequence which should give the desired stratigraphy. The experience of Millendorf, Brower, and Dyman (1978) with multidimensional scaling in the Q-mode suggests that this scheme might be successful for biostratigraphical information.

MOST LIKELY SEQUENCE OF EVENTS

As discussed earlier, the idea of using sequences of biostratigraphic events for correlation dates from at least the days of Quenstedt (1856-1858) and Oppel (1856-1858). Numerous methods are available to the biostratigrapher for determining sequences of events that are observed in many stratigraphic sections. These can be divided into two categories based on the type of events to be sequenced. The first set of schemes is concerned with point events, consisting of the highest and lowest occurrences of any given species. This has the advantage that one can develop separate zonations based on either highest or lowest occurrences of the species concerned. The second category involves methods designed to determine sequences of entire range zones.

Sequencing methods either produce average sequences or are designed to give the stratigraphically highest possible estimate of the top of a range zone and the stratigraphically lowest possible estimate of the base of a range zone. I will term the latter as "conservative zonations." Both philosophies are subject to mixed advantages and disadvantages. An average sequence gives the most probable zonation for a particular suite of data; such sequences of events are reasonable in many situations and operational within a particular area. Much statistical theory is applicable directly to average zonations which therefore are on reasonably firm statistical ground. For example, one can estimate confidence intervals and reliabilities for some of the average zonations as discussed later (e.g. Hay, 1972; Southam, Hay, and Worsley, 1975; Hay and Southam, 1978). Alternatively, maximizing the tops and minimizing the basis of range zones yields a conservative zonation which is consistent with the nature of error in biostratigraphic data. The reasoning is as follows.

If reworking and misidentification are ignored as sources of error, the only possible estimates of a range zone will be either the true range zone, an unlikely situation, or an underestimate thereof. For an underestimate, either the base of the range zone will be too high in the sequence or the top of the range zone will be too low in the sequence. It therefore follows that the best estimate of a true sequence of biostratigraphic events is the one that places the tops of the range zones as high as possible in the sequence and assigns the bases as far down as possible in the sequence of events (e.g., Shaw, 1964; Edwards, 1978; Rubel, 1978). Thus conservative zonations are intuitively and theoretically appealing. Unfortunately, the methods for most conservative zonations tend to be rather ad hoc schemes which cannot be tied directly to statistical theory. After the sequence of events has been determined, it can be used to correlate the samples in the various stratigraphic sections (e.g., Blank, 1979; Shaw, 1964).

Average Sequences for Point Events

W.W. Hay in conjunction with John Southam and Thomas Worsley devised a technique based on binomial probability (Hay, 1972; Southam, Hay, and Worsley, 1975; Southam and Hay, 1978). For any two events, say I and J, only two possibilities are allowed. Event I is over J or J is over I. If the two events are located at the same level, the occurrence is ignored. The frequencies are recorded in matrix form for all possible pairs of events. The information required includes the number of stratigraphic sections in which event I is over J and the number of sections containing both events. These data then are structured into a matrix which gives the probabilities for row events over column events. The events can be sequenced in different manners which are used by statisticians in paired comparisons. For example, the row sums of the probability matrix or a matrix containing the frequencies of event I over J could be used. Worsley and Jorgens (1977) and Blank (1979) employed simple row and column operations on the latter matrix. If element a_{ij} exceeds a_{ji} then row j is interchanged with column i. Approximate confidence limits can be assigned to all pairs of events based on incomplete beta functions, etc. These data are used to determine the most reliable zonation consisting of the required number of events.

One practical problem may arise with this technique. Biostratigraphic data are typically incomplete. Also, Hay, Southam, and Worsley ignore ties where events I and J occur together which introduces additional missing data into the system. Consequently, cycles of three or more inconsistent events may occur. An example with three events, A, B, and C, would be where A is over B which is over C where C also is over A. Misidentifications and other sources of error can result in inconsistencies. These inconsistencies produce great difficulties in determination of the most likely sequence of events.

F. Gradstein and F.P. Agterberg (manuscript in preparation) have improved the binomial probability method. First, if events I and J occur at the same level, the probabilities are allocated on the basis of 0.5 for I over J and 0.5 for I under J. This reduces the number of blanks in the data. In addition, cycles of inconsistent events are identified and eliminated during the sequencing of the data. The second modification is concerned with the method of reporting the sequence of events. The usual practice is simply to list the sequence from oldest to youngest. However, Gradstein and Agterberg have applied probability theory to estimate the distances between events along the relative time scale. All events are assumed to have the same variance; the variance of the events is set arbitrarily equal to one because the time scale is relative. Thirdly these authors perform cluster analysis on the distance scale between all of the events. These clusters then outline assemblage zones which can be employed for large-scale correlation.

Edwards and Beaver (1978) outlined an algorithm which allows three possibilities for events I and J. I would be over J or J could occur above I. Both events could occur at the same level, a possibility that is not tolerated by the binomial method. Trinomial probability is clearly a logical extention of the binomial method. A modification of the Bradley-Terry model for paired comparisons serves to determine the position of the individual events in the entire sequence. Although theoretically and intuitively appealing, there are several practical problems to this technique. The equations for the Bradley-Terry model are cumbersome and complex which restricts this technique to relatively small numbers of events. Also any event that is always located in the same position in the sequence must be deleted from the analysis.

Millendorf and Brower (unpublished data) have attempted to create composite sections based on simple averages of footages or proportional distributions of events in stratigraphic sections. These authors also experimented with ranking the events and treating the information with nonparametric statistics. Both methods resulted in reasonable zonations.

I. Dienes has formalized definitions for stratigraphic and biostratigraphic phenomena in the rubric of set theory. In addition, he has proposed several methods for sequencing point events. One of these advocates multi-dimensional integer kriging as a solution whereas the other uses spatial and temporal precedence matrices (Dienes, 1977, 1978).

Conservative Zonations for Point Events Based on Correlation and Regression

In one of the first works on quantitative biostratigraphy, Shaw (1964, see recent review by Miller, 1977) applied correlation and regression in a graphical context. The data consist of the vertical distances above a particular datum plane for the upper and lower occurrences of a group of taxa in a set of stratigraphic sections. In the first step data from two stratigraphic sections (X and Y) are plotted on a bivariate graph with the section on the X axis serving as the reference. Changes in rates of sedimentation, unconformities, faults, etc. produce various types of segments or "doglegs" on the graph. The data are segmented visually although computer algorithms could be used for this operation. Originally, straight lines of the least-squares type were fitted to the linear segments on the graph although the errors of the data points do not fit the assumptions made by least squares (Shaw, 1964). However as currently practiced by many individuals, the endpoints for the line segments are selected by the biostratigrapher (e.g., Miller, 1977).

An individual data point, that is either a highest or lowest occurrence of one particular taxon, might be subject to two types of error. A lowest occurrence could represent an unfilled base which is too high in the reference section. Alternatively, the highest occurrence could be too low in the reference section. These errors are updated by projecting the errant points onto the trend line and eventually onto the reference section. All projections are parallel to the X and Y axes. The purpose of such updating of the data is to place the tops of the range zones as high as possible in the sequence of events and move the lowest occurrences or bases as far down as possible in the sequence of events. In this manner, the method attempts to provide the most likely true sequence of events in the two sections in which the updated reference section becomes the composite for the two sections.

In the next and subsequent stages, new sections are incorporated gradually into the composite for the first two sections until a composite is created for all sections. As new sections are added, the composite or reference section is updated so that the base and top of the range zone of each taxon are minimized and maximized respectively as just mentioned for the first two sections. The composite section gives the order or sequence for all events as well as their elevations above the datum in composite section units. Once formulated, the composite can serve to correlate unknown sections and samples.

Edwards (1978; see also Murphy and Edwards, 1977) notes that in many instances only the relative order of the events is necessary and that the absolute spacing between the events need not be determined. Consequently, Edwards presented a nonparametric version of Shaw's (1964) graphical correlation approach which avoids the problem of fitting regression lines. The algorithm results in a simple sequence of events of the conservative type. Similar to Shaw and his followers, Edwards treats two sequences of events at a time which are plotted on the X and Y axes of a bivariate scatter plot. Unfortunately, this method of presentation is difficult to visualize. As mentioned later, displaying the two sequences of events side by side yields data which are more familiar to biostratigraphers.

Hohn (1978) generalized Shaw's method by using principal components. His approach is ingenious in terms of computer-core requirements because the principal components are extracted from a correlation matrix between the stratigraphic sections. Missing data are interpolated and the scores for the first principal component give the zonation of the events. An alternative and less efficien approach tried by Brower (unpublished data) is to calculate a correlation matrix between the events; the first principal component of this matrix would list the necessary zonation. It is important to observe that principal components result in an average zonation rather than the conservative zonation of the Shaw algorithm.

Sequences of Range Zones

The next group of methods is designed to organize sequences of entire range zones. At least some of the concepts of set theory are applied to the operations.

Guex (1977, 1978a, 1978b, 1979; Davaud and Guex, 1978) proposed a simple technique which has some features of archeological seriation. A taxon by taxon matrix of associations, here termed A, is formed. If taxa I and J occur together in one or more of the stratigraphic sections studied, a 1 is recorded for a_{ij}. If I and J are not known to occur together a_{ij} becomes 0. The association matrix then is analyzed in a manner that is essentially the same as in some of the simpler types of archeological seriation in which the rows and columns of the matrix are reordered. This concentrates the 1's and the associated pairs of taxa along the diagonal of the association matrix. The 0's which denote taxa that do not occur together are placed in the off-diagonal elements. The concentration principle also is organized to produce the largest possible square submatrices which contain blocks of species that are consistently associated. After these row and column operations, range charts of taxa and assemblages of species are read directly from the matrix.

Rubel (1976, 1978) devised a scheme for treating entire range zones in terms of set theory. Four possibilities are visualized for the range zones of I and J. I can be over J or vice versa. The two range zones can overlap or intersect in different manners. For example the top of I could occur above that of J or vice versa; also one range zone might occur within the other one or the two range zones might be located at the same horizon or horizons. All of these possibilities fall into the category of overlapping or intersection. Also one or both range zones could be missing. The range zone data are recorded for all sections separately and then summed for all sections following simple algebraic rules derived from set theory that were developed by Rubel. Thus summation provides the most likely sequence of range zones. The algebraic rules for summation are designed so that the estimated positions of the tops of the zones are placed as high as possible in the sequence of events whereas the bases are minimized and placed as low as possible. The algorithm is not probabilistic and reliabilities or confidence limits are not assigned to the zones. Also Rubel did not present any algorithm designed to extract systematically the sequence of range zones.

Davaud (Davaud and Guex, 1978) combined some aspects of the methods of Guex and Rubel with some new ideas in an ingenious manner. As done by Guex, a species by species association matrix is calculated in the first step. Unlike Guex and similar to Rubel, Davaud recognizes four types of association: taxon I over J, taxon I under J, coexistence of I and J if the two occur together at least once, and the indeterminate situation where I and J do not occur in the same stratigraphic section. Although the association matrix is asymmetrical, only the upper half need to be treated because the information in the lower half of the matrix is redundant. The next stage resolves as many of the indeterminate situations as possible. The remaining indeterminations are discarded. In the third step, the association matrix is reorganized in a similar fashion to that employed in the Guex technique. Lastly the range charts and associations of taxa can be tabulated in stratigraphic order.

Several common denominators underlie these three algorithms. The methods are not probabilistic in the sense that estimates of the accuracy or reliability of the zonation of the events are not presented. Also the techniques can lose some pertinent information. As mentioned before, if taxa I and J occur together, a 1 is assigned for this relationship. However it might be the situation that the base of the range zone of I could be consistently above that of J. This and similar information is not always recovered by these methods.

SIMPLE METHODS FOR QUANTITATIVE BIOSTRATIGRAPHY

Most nonquantitative biostratigraphers have resisted stubbornly numerical methods. Essentially, quantitative biostratigraphy has proved unpalatable to the intended consumers, in spite of the fact that many methods of quantitative biostratigraphy have produced effective results with large, cumbersome, and complex data sets. I believe the lack of palatability is due to two causes. First, many of the quantitative methods are complex whereas most biostratigraphers are not quantitatively oriented. Secondly, most of the quantitative techniques utilize methodologies that basically are foreign to biostratigraphers.

Consequently, through the past year I have concentrated my attention on simple methods of quantitative biostratigraphy. Some of these techniques literally almost remove the numbers from quantitative biostratigraphy. It is hoped that these simple techniques will prove more palatable to biostratigraphers than the more elaborate methods that have been developed to date. Furthermore, these simple methods share at least some features in common with other biostratigraphic methods.

Method 1

Begin by considering the true range zone of a taxon. If reworking and misidentification can be discarded as possible sources of error, then, the only possible estimates of the range zone are either the actual or true range or an underestimate thereof. Underestimates are obviously the most likely possibility. In underestimates, the base or lowest occurrence of the range zone can be placed too high or the top of the range zone can be placed too low. Thus if a relative zonation of the events is considered, the most likely estimate of a series of range zones is that which places the tops of the range zones as high as possible in the zonation and places the bases of the range zones as low as possible in the zonation. Shaw (1964) and Edwards (1978) made use of this reasoning in their parametric (Shaw) and nonparametric (Edwards) correlation and regression-based approaches. This method is a somewhat different version of the Edwards (1978) scheme. Point events are treated, that is highest and lowest occurrences of a series of taxa. One lists the sequence of events, from lowest to highest in the n stratigraphic sections.

The algorithm is exceedingly simple. First a composite sequence of the events is determined. For any two stratigraphic sections, the lists of events are placed side by side and the similar events are connected with tie-lines. If the events are in the same order, all tie-lines will be parallel and the events in the two sections are consistent with one another. If the events are not in the same order in the two sections, some of the tie-lines will cross indicating that the events are inconsistent with one another in the two sections. The resolution of the inconsistencies is conducted by placing the bases of the range zones (lowest occurrence events) as low as possible in the zonation and the tops of the range zones (highest occurrence events) as high as possible in the zonation. This revised or updated information is used to create a composite zonation of events for the two sections. Next the first composite then is matched with the series of events in a third stratigraphic section and a second composite section created as outlined; the second composite contains the revised zonation for the first three sections. Information from all other sections gradually is assembled into the data until a composite series of events is ascertained for all stratigraphic sections.

In the second step, the composite serves as a standard to correlate all of the individual sections, much as done by Shaw (1964) and Edwards (1978). The events are matched between the composite and the individual sections and tie-lines are drawn. Tie-lines that cross emphasize the inconsistencies in the sequences of biostratigraphic events between the composite and the individual sections. These "false" correlations then can be ignored. The "goodness of fit" between the individual events in the composite zonation and the local sections can be measured in several manners; for event I the amount of mismatch might be determined by (number of sections in which event I is inconsistent with the composite/ total number of sections) or by a simple type of "standard-deviation" such as the following:

$$\text{"Standard-deviation" for event } I = \sqrt{\frac{\sum_{J=1}^{n}\left(\begin{array}{l}\text{Number of events crossed} \\ \text{by event } I \text{ in section } J\end{array}\right)^2}{(\text{Number of sections, } n)}}$$

Higher "standard deviations" denote the events with lesser consistencies that will be less useful in correlating the sections. This method basically is the same as that used by many practicing nonquantitative biostratigraphers who usually do not update their data in the form of a composite or explicitly delete false correlations. Also, nonquantitative biostratigraphers do not formalize estimates of the error inherent in the zonations which they produce. This method also is essentially the same as that of Edwards (1978). However, Edwards plots the two series of events as the axes of a bivariate scatterplot. As mentioned before, data presented in this fashion are difficult for biostratigraphers to visualize. Plotting the series of events side by side displays the data in a form more familiar to biostratigraphers. Also, Edwards and I resolve some of the inconsistent events in a slightly different manner.

Method 2

This category of techniques is borrowed from archeological seriation. As discussed earlier, archeologists may encounter the problem of arranging various objects, such as graves and Iron Age brooches, into one-dimensional sequences which generally represent either time or "evolutionary" series. Mathematical solutions to this problem date back to Flinders Petrie (1899). Useful reviews of many of these techniques are given in Cowgill (1972), Doran and Hodson (1975), Gelfand (1971), Hodson, Kendall, and Tautu (1971), Kendall (1963, 1971), Johnson (1972), Marquardt (1978), and Wilkinson (1974). One can perform seriation on a $p \times p$ matrix of distances or similarities measured between objects or samples as mentioned earlier under the quantification of assemblage zones. Original data matrices (say $n \times m$) including both objects and samples can be seriated and it is this type of seriation that is of interest here.

Scott (1974) first suggested that archeological seriation could be applied usefully to biostratigraphic data. Doveton (1978, pers. comm.) seriated some paleoecological data on Pennsylvanian conodonts (see also Doveton, Gill, and Tipper, 1976). I thank John Doveton for a copy of his unpublished manuscript and for calling my attention to the possibility of seriation on original data matrices. Tipper (1977) also applied seriation to original data matrices of ecological and paleoecological data. Although the data were discussed in the context of networks, Smith and Fewtrell (1979) seem to have adopted a seriation approach for microfossil data.

The biostratigraphic data to be considered represent presences and absences of various taxa or species in a series of samples observed in different stratigraphic sections. Presences can be scored as 1.0 and absences as 0.0. The data also could be viewed in black and white. The intuitive appeal of black and white data rather than various shades of dingy gray should be obvious. The data are ordered as follows. The presences and absences are arranged with the taxa in the columns and samples in the rows. The samples are put in a preliminary sequence so that the data from the different sections are placed in sequential blocks; for each section, the samples are listed from top to bottom. The range-through method of data recording is used for each stratigraphic section. Thus each taxon is scored as present in all samples within its range zone for each section.

The object of the exercise is simple. Consider, first, the full seriation model which utilizes both first and last occurrences. Here the rows and columns of the matrix are interchanged in such a manner as to concentrate the presences along the diagonal of the matrix; the absences are located as far away from the diagonal as possible. This is referred to as the "concentration principle" by archeologists. It also minimizes the range zones of the taxa among the samples as suggested by Scott (1974). The rearranged matrix has the taxa in the columns and the samples in the rows of the matrix. The taxa with the most similar range zones will be in adjacent columns of the rearranged matrix. Also the most similar samples will be grouped in the adjacent rows. Thus, the technique groups both taxa and samples simultaneously. Two types of of solutions can be obtained. The unconstrained solution ignores all information about the stratigraphic position of the samples and taxa within the stratigraphic sections. Data on the stratigraphic position for the samples within the individual sections are introduced into the analysis for a constrained solution. This constraint forces the samples within each section to remain in stratigraphic order. Such is not true for the unconstrained solution in which two samples within a single section can be seriated out of stratigraphic order.

Some biostratigraphers only work with one type of event, either highest or lowest occurrences. This simplifies the seriation to a largely additive model. The more usual situation, that is concerned with highest occurrences, will be outlined briefly. The data are arranged as before. For each section, a taxon is recorded as present in all samples below its highest occurrence. One simply shuffles or interchanges the rows and columns of the matrix until all the presences are located below the diagonal of the matrix and all absences are concentrated above the diagonal. As for the full seriation model, the rearranged matrix both correlates the samples and gives the sequence of events.

These seriation methods have two major advantages. First they share the virtue of simplicity. Second, they roughly duplicate the procedures followed by many non-quantitative biostratigraphers. It is hoped that non-quantitative biostratigraphers will adopt some of these simple methods if not the more complex algorithms.

EVOLUTIONARY SEQUENCES

The study of lineages in paleontology can be traced back at least as far as 1899 when Rowe systematically examined the evolution of *Micraster* from the Cretaceous chalks of England. Some early examples of statistical analyses of evolutionary sequences are those of Carruthers (1910; see also Swinnerton, 1921) on Carboniferous corals, Trueman (1922) on Jurassic oysters, and the monumental work of Brinkmann (1929) on Jurassic ammonites. The latter study especially is interesting because of the large samples involved and also for the plots of morphology versus stratigraphic position which were used to make correlations and infer the existence of unconformities. The statistical techniques are derived from a broad menu including such recipes as simple univariate and bivariate statistics (e.g., Gingerich, 1976; Kellogg, 1975; and Ozawa, 1975), principal components or factor analysis (see Kaesler, 1970; Malmgren, 1974, 1976; and Rowell, 1970 for simple and clear examples), principal coordinates (Blackith and Reyment, 1971), discriminant analysis (Reyment, 1978a, 1978b, 1978c; Reyment and Banfield, 1976; Campbell and Reyment, 1978; Blackith and Reyment, 1971; Symons and Ringele, 1976), various numerical taxonomic techniques (Cheetham, 1968; Kaesler, 1970; and Rowell, 1970), and cladistic analysis (Kesling and Sigler, 1969). An evolutionary sequence shows the distribution of size and shape in a framework defined by time and space. Therefore other methods also are important such as time-series analysis (Southam and Hay, 1978; Reyment, 1978a, 1971), Fourier analysis of shapes (e.g., Waters, 1977; Christopher and Waters, 1974; Kaesler and Waters, 1972), optical-data processing (see Srivastava, 1977 for applications to geological maps; the same scheme also could be used for fossils), quantified transformation grids (e.g. Sneath, 1967), and many other methods used for pattern analysis.

One of the main problems concerning the analysis of lineages and its utility in biostratigraphy is the nature of evolutionary changes. Two end-member models, "punctuated equilibria" and "phyletic gradualism," have been proposed. The concepts and many case studies are reviewed by Eldredge and Gould (1977) and Gould and Eldredge (1977). These authors strongly advocate the punctuated equilibria model but Gingerich (1976, 1979) argues persuasively for phyletic gradualism. According to punctuated equilibria evolutionary changes are abrupt. If so, then lineages can be segmented definitely and used to show chronostratigraphy. Conversely if species tend to change gradually as predicted by the phyletic gradualism model, then lineages become difficult to breakdown into their component species or other morphological units because any divisions of a true continuum are ultimately arbitrary. Either of these patterns can be complicated by the fact that the appearance and disappearance of a species or morphological trait need not be synchronous throughout the area inhabited by a lineage. At least some examples of both phyletic gradualism and punctuated equilibria are known. For example, even Gould and Eldredge (1977) concede that the proloculus of a Permian foraminifer studied by Ozawa (1975) constitutes an example of phyletic gradualism. Gingerich (1976, 1979) presents other examples of gradualistic change but these are debated by Gould and Eldredge (1977). Situations which probably fit the punctuational model are Pleistocene land snails and hominids (Gould, 1969, 1976) and many fossil brachiopods (e.g., Johnson, 1975).

Evolution obviously provides the basis for biostratigraphy. Ironically, many evolutionary sequences are approached in a paleobiologic context rather than from a biostratigraphical point of view (for example, see most of the studies in Gould and Eldredge, 1977). However, some studies of lineages have been oriented strongly toward biostratigraphy. One example of such is the examination of the Cretaceous of the Western Interior of the USA by Kauffmann (e.g., 1970, 1977) and various colleagues including Cobban (e.g., 1951, 1958) and Sohl (1977). Detailed study of rapidly evolving lineages can result in well-defined species that are short lived in time. Kauffmann (1977) reports average durations of Cretaceous species of ammonites and pelecypods that range from 0.19 to 3.1 million years. Based on a combination of the lineages and range charts, Kauffmann (1970, 1977) documents regional composite assemblage zones in these Cretaceous rocks that range from 0.08 to 0.5 million years in duration as determined by radioactive dating. The data treatment may be qualitative or relatively simple statistics are involved. More powerful statistical algorithms might yield a higher degree of resolution of both the lineages and the range charts.

Reyment has been developing methodology for the analysis of morphometric chronoclines. A brief review was published in 1978 and full details are in a book on "Morphometric Methods in Quantitative Biostratigraphy" (Reyment, 1980). The data studied represent such things as morphometric variables on a lineage, ecologically caused changes in size, shape, or frequency of various taxa, etc. Typically multivariate statistics are combined with time-series techniques such as cross-correlation and slotting (e.g., Gordon and Reyment, 1978). Southam and Hay (1978) also are experimenting with time series for paleontological data.

SUMMARY

Numerical methods in biostratigraphy can be grouped into four basic categories. Firstly, although index fossils were used long before the time of Darwin, the measurement of the attributes of an index fossil did not begin until the 1960's. The quantification of index fossils allows the identification of the taxa which convey the largest amounts of biostratigraphic information. These data can be employed in several manners.

The second one represents the analysis of assemblage zones, typically by multivariate statistics. Quantified assemblage zones give data about both paleoecology and stratigraphy. The locations of faunal discontinuities in time and space are emphasized. Correlations, both between and within the assemblage zones, can be determined.

The third group embraces the numerous methods which are available for determining the most likely sequence of biostratigraphic events recorded in different stratigraphic sections. The zonations can be of either the average or conservative types. Assemblage zones, overlap zones, etc., are derived easily from dissecting the series of all events. The sequencing methods produce lists of events which are efficient for correlation but yield little information about environments. Once ascertained, the sequence of events can be used to correlate the various stratigraphic sections.

The analysis of evolutionary sequences is placed in Category 4. A complete numerical approach requires morphemetric statistics which are used in conjunction with time-series algorithms. Careful examination of lineages may pay off with well-documented taxa that are short ranged in time. Another advantage is that the endpoints of the range zones can be known definitely.

Although they belong in the other categories, several simple techniques are treated separately in the hope that they may entice nonquantitative biostratigraphers to try numerical techniques. *Method 1* determines a conservative sequence of events. The techniques of *Method 2* are borrowed from archeological seriation and these schemes simultaneously produce range charts and correlations of the samples.

The annotations given earlier indicate that the various algorithms are data oriented. They require different types of input data, make different assumptions and usually produce results which are more or less different. The biostratigraphic information which can be derived from the various techniques covers a wide spectrum, ranging from the biostratigraphic properties of taxa or biostratigraphic events, to the distribution of taxa and morphology in an evolutionary and time-stratigraphic context.

The previous discussion denotes that quantitative biostratigraphers have fulfilled the first two goals of IGCP Project 148 reasonably well. To date, numerous algorithms have been developed and computer programs for many of these generally are available. Unfortunately, most of these techniques have only been tried on a few case studies. Only a few data sets have been examined with more than one technique. One of the exceptions is the Cambrian fossils of Shaw (1964, graphical correlation) which have been treated by Edwards (1978, pers. comm., nonparametric "regression"), Edwards and Beaver (1978, trinomial probability), Guex (1977, a seriation-type method), and Hohn (1978, principal components). Hazel (1977) and Millendorf, Brower, and Dyman (1978) applied a battery of techniques to several case studies on assemblage zones.

A PREDICTION FOR THE 1980's

Thus far, the status of quantitative methods in biostratigraphy has been reviewed up to 1980. The next question is what will the future bring?

The 1960's and 1970's have brought a plethora of algorithms. Many of these work effectively, particularly on certain types of data, but the problems of biostratigraphy remain far from being resolved. In one sense, this period constitutes an adaptive radiation of techniques. It is possible that a major algorithm, which is yet to be written and may be capable of resolving all of the trials and tribulations of biostratigraphical data, is just about to be discovered. However, writing as a self-acknowledged skeptic, I doubt it. (I may eventually regret these words.) Rather than continue to create another horde of algorithms in the 1980's, I would prefer to see this decade as an interval of consolidation. The adaptive radiation of techniques should cease. Now is the time to begin to weed out the methods that are determined unfit by rigorous case studies. Accordingly, I hope the 1980's will concentrate on the third goal of IGCP Project 148, namely to evaluate the methods of quantitative biostratigraphy through case studies of both real and simulated data.

REFERENCES

Ali, S.A., Lindemann, R.H., and Feldhausen, P.H., 1976, A multivariate sedimentary environment analysis of Great South Bay and South Oyster Bay, New York: Jour. Math. Geology, v. 8, no. 3, p. 283-304.

Blackith, R.E., and Reyment, R.A., 1971, Multivariate morphometrics: Academic Press, London, 412 p.

Blank, R.G., 1979, Applications of probabilistic biostratigraphy to chronostratigraphy: Jour. Geology, v. 87, no. 5, p. 647-670.

Brinkmann, R., 1929, Statistisch-biostratigraphische untersuchungen an mitteljurassischen Ammoniten uber Artbegriff und Stammesentwicklung: Abhandlungen der Gesellschaft der Wissenschaften zu Gottingen, Mathematisch-physikalische Klasse, Neue Folge Bd. 13, 249 p.

Brower, J.C., and Millendorf, S.A., 1978, Biostratigraphic correlation within IGCP project 148: Computers & Geosciences, v. 4, no. 3, p. 217-220.

Brower, J.C., Millendorf, S.A., and Dyman, T.S., 1978, Quantification of assemblage zones based on multivariate analysis of weighted and unweighted data: Computers & Geosciences, v. 4, no. 3, p. 221-227.

Buzas, M.A., 1970, On the quantification of biofacies: Proc. North Am. Paleont. Conv., Chicago, 1969, pt. B, p. 101-116.

Campbell, N.A., and Reyment, R.A., 1978, Discriminant analysis of a Cretaceous foraminifer using shrunken estimators: Jour. Math. Geology, v. 10, no. 4, p. 347-359.

Carruthers, R.G., 1910, On the evolution of *Zaphrentis delanouei* in Lower Carboniferous times: Geol. Soc. London Quart. Jour., v. 66, p. 523-538.

Cheetham, A.H., 1968, Morphology and systematics of the bryozoan genus *Metrarabdotos*: Smithsonian Misc. Coll., v. 153, no. 1, Publ. 4733, 121 p.

Cheetham, A.H., and Deboo, P.B., 1963, A numerical index for biostratigraphic zonation in the mid-Tertiary of the eastern Gulf: Gulf Coast Assoc. Geol. Soc. Trans., v. 13, p. 139-147.

Cheetham, A.H., and Hazel, J.E., 1969, Binary (presence-absence) similarity coefficients: Jour. Paleontology, v. 43, no. 5, p. 1130-1136.

Christopher, R.A., and Waters, J.A., 1974, Fourier series as a quantitative descriptor of microspore shape: Jour. Paleontology, v. 48, no. 4, p. 697-709.

Cisne, J.L., and Rabe, B.L., 1978, Coenocorrelation: gradient analysis of fossil communities and its applications in stratigraphy: Lethaia, v. 11, no. 4, p. 341-364.

Cobban, W.A., 1951, Scaphitoid cephalopods of the Colorado Group: U.S. Geol. Survey Prof. Paper 239, 39 p.

Cobban, W.A., 1958, Late Cretaceous fossil zones of the Powder River Basin, Wyoming and Montana, *in* Wyoming Geol. Assoc. Guidebook, 13th Ann. Field Conf., 1958, Powder River Basin, p. 114-119.

Cockbain, A.E., 1966, An attempt to measure the relative biostratigraphic usefulness of fossils: Jour. Paleontology, v. 40, no. 1, p. 206-207.

Cowgill, G.L., 1972, Models, methods and techniques for seriation, *in* Models in archaeology, Clarke, D.L., ed.,: Methuen and Co. Ltd., London, p. 381-424.

Davaud, E., and Guex, J., 1978, Traitement analytique (manuel) et algorithmique de problemes de correlations biochronologiques: Eclogae Geol. Helvetiae, v. 71, no. 3, p. 581-610.

Dienes, I., 1977, Formalized stratigraphy and its use on the Dorog Basin area, *in* Matematicke Metody v Geologii: Hornicka Pribram ve vede-technice (Pribram, Czechoslovakia), p. 626-645.

Dienes, I., 1978, Methods of plotting temporal range charts and their application in age estimation: Computers & Geosciences, v. 4, no. 3, p. 269-272.

Donovan, D.T., 1966, Stratigraphy: an introduction to principles: Rand McNally and Co., Chicago, 199 p.

Doran, J.E., and Hodson, F.R., 1975, Mathematics and computers in archaeology: Harvard Univ. Press, Cambridge, 381 p.

Doveton, J.H., Gill, D., and Tipper, J.C., 1976, Conodont distributions in the Upper Pennsylvanian of eastern Kansas; binary pattern analyses and their paleoecological implications (abst.): Geol. Soc. America, Abstracts with Programs, v. 8, no. 7, p. 842.

Edwards, L.E., 1978, Range charts and no-space graphs: Computers & Geosciences, v. 4, no. 3, p. 247-255.

Edwards, L.E., and Beaver, R.J., 1978, The use of a paired comparison model in ordering stratigraphic events: Jour. Math. Geology, v. 10, no. 3, p. 261-272.

Eicher, D.L., 1976, Geologic time (2nd ed.): Prentice-Hall, Inc., New Jersey, 150 p.

Eldredge, N., and Gould, S.J., 1977, Evolutionary models and biostratigraphic strategies, *in* Concepts and methods of biostratigraphy, Kauffman, E.G., and Hazel, J.E., eds.: Dowden, Hutchinson & Ross, Stroudsburg, Pennsylvania, p. 25-40.

Ellison, R.L., 1963, Faunas of the Mahantango Formation in south-central Pennsylvania: Pennsylvania Topog. and Geol. Survey, General Geology Rept. G 39, p. 201-212.

Feldhausen, P.H., 1970, Ordination of sediments from the Cape Hatteras continental margin: Jour. Math. Geology, v. 2, no. 2, p. 113-129.

Fox, W.T., 1968, Quantitative paleoecologic analysis of fossil communities in the Richmond Group: Jour. Geology, v. 76, no. 6, p. 613-640.

Gelfand, A.E., 1971, Rapid seriation methods with archaeological applications, *in* Mathematics in the archaeological and historical sciences, Hodson, F.R., Kendall, D.G., and Tautu, P., eds.: Edinburgh Univ. Press, Edinburgh, p. 186-201.

Gill, D., Boehm, S., and Erez, Y., 1976, ASSOCA: FORTRAN IV program for Williams and Lambert association analysis with printed dendrograms: Computers & Geosciences, v. 2, no. 2, p. 219-248.

Gill, D., and Tipper, J.C., 1978, The adequacy of non-metric data in geology: tests using a divisive-omnithetic clustering technique: Jour. Geology, v. 86, no. 2, p. 241-259.

Gingerich, P.D., 1976, Paleontology and phylogeny: patterns of evolution at the species level in early Tertiary mammals: Am. Jour. Sci., v. 276, no. 1, p. 1-28

Gingerich, P.D., 1979, The stratophenetic approach to phylogeny: reconstruction in vertebrate paleontology, *in* Phylogenetic analysis and paleontology, Cracraft, J., and Eldredge, N., eds.: Columbia Univ. Press, New York, p. 41-77.

Gordon, A.D., and Reyment, R.A., 1979, Slotting of borehole sequences: Jour. Math. Geology, v. 11, no. 3, p. 309-327.

Gould, S.J., 1969, An evolutionary microcosm: Pleistocene and Recent history of the land snail *P. (Poecilozonites)* in Bermuda: Bull. Mus. Comp. Zoology, v. 138, no. 7, p. 407-531.

Gould, S.J., 1976, Ladders, bushes, and human evolution: Nat. Hist., v. 85, no. 4, p. 24-31.

Gould, S.J., 1977, Ontogeny and phylogeny: The Belknap Press of Harvard Univ. Press, Cambridge, Massachusetts, 498 p.

Gould, S.J., and Eldredge, N., 1977, Punctuated equilibria: the tempo and mode of evolution reconsidered: Paleobiology, v. 3, no. 2, p. 115-151.

Guex, J., 1977, Une nouvelle methode d'analyse biochronologique: note preliminaire: Bull. Soc. Vaud. Sci. Nat., v. 73, no. 351, p. 309-322.

Guex, J., 1978a, Le Trias inferieur des Salt Ranges (Pakistan): problemes biochronologiques: Eclogae Geol. Helvetiae, v. 71, no. 1, p. 105-141.

Guex, J., 1978b, Influence du confinement geographique des especes fossiles sur l'elaboration d'echelles biochronologiques et sur les correlations: Bull. Soc. Vaud. Sci. Nat., v. 74, no. 354, p. 115-124.

Guex, J., 1979, Terminologie et methodes de la biostratigraphie moderne: commentaires critiques et propositions: Bull. Soc. Vaud. Sci. Nat., v. 74, no. 355, p. 169-216.

Hancock, J.M., 1977, The historic development of concepts of biostratigraphic correlation, *in* Concepts and methods of biostratigraphy, Kauffman, E.G., and Hazel, J.E., eds.: Dowden, Hutchinson & Ross, Inc., Stroudsburg, Pennsylvania, p. 3-22.

Hay, W.W., 1972, Probabilistic stratigraphy: Eclogae Geol. Helvetiae, v. 65, no. 2, p. 255-266.

Hay, W.W., and Southam, J.R., 1978, Quantifying biostratigraphic correlation: Ann. Rev. Earth Planet. Sci., v. 6, p. 353-375.

Hazel, J.E., 1970, Binary coefficients and clustering in biostratigraphy: Geol. Soc. America Bull., v. 81, no. 11, p. 3237-3252.

Hazel, J.E., 1971, Ostracode biostratigraphy of the Yorktown Formation (Upper Miocene and Lower Pliocene) of Virginia and North Carolina: U.S. Geol. Survey Prof. Paper 204, 13 p.

Hazel, J.E., 1977, Use of certain multivariate and other techniques in assemblage zonal biostratigraphy: examples utilizing Cambrian, Cretaceous, and Tertiary benthic invertebrates, *in* Concepts and methods of biostratigraphy, Kauffman, E.G., and Hazel, J.E., eds.: Dowden, Hutchinson & Ross, Inc., Stroudsburg, Pennsylvania, p. 187-212.

Henderson, R.A., and Heron, M.L., 1977, A probabilistic method of paleobiogeographic analysis: Lethaia, v. 10, no. 1, p. 1-15.

Hodson, F.R., Kendall, D.G., and Tautu, P., eds., 1971, Mathematics in the archaeological and historical sciences: Edinburgh Univ. Press, Edinburgh, 565 p.

Hohn, M.E., 1976, Binary coeffients: a theoretical and empirical study: Jour. Math. Geology, v. 8, no. 2, p. 137-150.

Hohn, M.E., 1978, Stratigraphic correlation by principal components: effects of missing data: Jour. Geology, v. 86, no. 4, p. 524-532.

Imbrie, J., 1964, Factor analytic model in paleoecology, *in* Approaches to paleoecology, Imbrie, J., and Newell, N., eds.: John Wiley & Sons, Inc., New York, p. 407-422.

Jeletzky, J.A., 1965, Is it possible to quantify biochronological correlation?: Jour. Paleontology, v. 39, no. 1, p. 135-140.

Johnson, J.G., 1975, Allopatric speciation in fossil brachiopods: Jour. Paleontology, v. 49, no. 4, p. 646-661.

Johnson, L., 1972, Introduction to imaginary models for archaeological scaling and clustering, *in* Models in archaeology, Clarke, D.L., ed.: Methuen and Co. Ltd., London, p. 309-379.

Kaesler, R.L., 1966, Quantitative re-evaluation of ecology and distribution of Recent Foraminifera and Ostracoda of Todos Santos Bay, Baja, California, Mexico: Kansas Univ. Paleont. Contr., Paper 10, 50 p.

Kaesler, R.L., 1970, Numerical taxonomy in paleontology: classification, ordination and reconstruction of phylogenies: Proc. North Am. Paleont. Conv., Chicago, 1969, pt. B, p. 84-100.

Kaesler, R.L., and Waters, J.A., 1972, Fourier analysis of the ostracode margin: Geol. Soc. America Bull., v. 83, no. 4, p. 1169-1178.

Kauffman, E.G., 1970, Population systematics, radiometrics, and zonation - a new biostratigraphy: Proc. North Am. Paleont. Conv., Chicago, 1969, pt. F, p. 612-666.

Kauffman, E.G., 1977, Evolutionary rates and biostratigraphy, *in* Concepts and methods of biostratigraphy, Kauffman, E.G., and Hazel, J.E., eds.: Dowden, Hutchinson & Ross, Inc., Stroudsburg, Pennsylvania, p. 109-141.

Kellogg, D.E., 1975, The role of phyletic change in the evolution of *Pseudocubus vema* (Radiolaria): Paleobiology, v. 1, p. 359-370.

Kendall, D.G., 1963, A statistical approach to Flinders Petrie's sequence dating: Intern. Statistical Inst. Bull., no. 40, p. 657-680.

Kendall, D.G., 1971, Seriation from abundance matrices, *in* Mathematics in the archaeological and historical sciences, Hodson, F.R., Kendall, D.G., and Tautu, P., eds.: Edinburgh Univ. Press, Edinburgh, p. 215-252.

Kesling, R.V., and Sigler, J.P., 1969, *Cunctocrinus*, a new Middle Devonian calceocrinid crinoid from the Silica Shale of Ohio: Univ. Michigan, Mus. Paleontology Contr., v. 22, no. 24, p. 339-360.

Lyell, C., 1830-33, Principles of geology: J. Murray, London, 3 vols.

Lynts, G.W., 1971, Analysis of the planktonic Foraminifera fauna of core 6275, Tongue of the ocean, Bahamas, Micropaleontology, v. 17, no. 2, p. 152-166.

Lynts, G.W., 1972, Factor-vector analysis models in ecology and paleoecology: 21st Intern. Geol. Congress (Montreal) Sect. 7, Paleontology, p. 227-237.

Lynts, G.W., and Stehman, C.F., 1971, Factor-vector models of middle Eocene planktonic foraminiferal fauna of core 6282, Northeast Providence Channel, Bahamas: Rev. Espanola Micropaleontol., v. 3, p. 205-213.

Mallory, V.S., 1970, Biostratigraphy - a major basis for paleontologic correlation: Proc. North Am. Paleont. Conv., Chicago, 1969, pt. F, p. 553-566.

Malmgren, B.A., 1974, Morphometric studies of planktonic foraminifers from the type Danian of southern Scandinavia: Stockholm Contr. Geol., v. 29, p. 1-126.

Malmgren, B.A., 1976, Size and shape variation in the planktonic foraminifer *Heterohelix striata* (Late Cretaceous, southern Scandinavia): Jour. Math. Geology, v. 8, no. 2, p. 165-182.

Marquardt, W.H., 1978, Advances in archaeological seriation, *in* Advances in archaeological method and theory, v. 1, Schiffer, M.B., ed.: Academic Press, London, p. 257-314.

McCammon, R.B., 1966, Principal component analysis and its application in large-scale correlation studies: Jour. Geology, v. 74, no. 5, pt. 2., p. 721-733.

McCammon, R.B., 1970, On estimating the relative biostratigraphic values of fossils: Bull. Geol. Inst. Univ. Uppsala (n.s.), v. 2, p. 49-57.

Millendorf, S.A., Brower, J.C., and Dyman, T.S., 1978, A comparison of methods for the quantification of assemblage zones: Computers & Geosciences, v. 4, no. 3, p. 229-242.

Miller, F.X., 1977, The graphic correlation method in biostratigraphy, *in* Concepts and methods of biostratigraphy, Kauffman, E.G., and Hazel, J.E., eds.: Dowden, Hutchinson & Ross, Inc., Stroudsburg, Pennsylvania, p. 165-186.

Murphy, M.A., and Edwards, L.E., 1977, The Silurian-Devonian Boundary in central Nevada: Univ. Calif., Riverside Campus Museum Contr., no. 4, p. 183-189.

Oltz, D.F., Jr., 1969, Numerical analysis of palynological data from Cretaceous and early Tertiary sediments in east central Montana: Palaeontographica, pt. B, v. 128, p. 90-166.

Oltz, D.F., Jr., 1971, Cluster analyses of Late Cretaceous-Early Tertiary pollen and spore data: Micropaleontology, v. 17, no. 2, p. 221-232.

Oppel, A., 1856-1858, Die Juraformation Englands, Frankreichs und des sudwestlichen Deutschlands, nach ihren einzelnen gliedern eingntheilt und verglichen: von Ebner and Seubert, Stuttgart (originally published in three parts in Abdruck der Wurttemb. naturw. Jahreshefte 12-14; 1856, 1-438; 1857, 439 bis-594 + map; 1858, 695-857 + table).

Ozawa, T., 1975, Evolution of *Lepidolina multiseptata* (Permian foraminifer) in East Asia: Mem. Fac. Sci. Kyushu Univ., Ser. D, Geol., v. 23, p. 117-164.

Park, R.A., 1974, A multivariate analytical strategy for classifying paleoenvironments: Jour. Math. Geology, v. 6, no. 4, p. 333-352.

Petrie, W.M.F., 1899, Sequences in prehistoric remains: Jour. Anthropol. Inst., v. 29, p. 295-301.

Quenstedt, F.A., 1856-1858, Der Jura: H. Lauppschen, Rubingen, 842 p.

Raup, D.M., and Crick, R.E., 1979, Measurement of faunal similarity in paleontology: Jour. Paleontology, v. 53, no. 5, p. 1213-1227.

Reyment, R.A., 1971, Spectral breakdown of morphometric chronoclines: Jour. Math. Geology, v. 2, no. 4, p. 365-376.

Reyment, R.A., 1978a, Biostratigraphical logging methods: Computers & Geosciences, v. 4, no. 3, p. 261-268.

Reyment, R.A., 1978b, Quantitative biostratigraphical analysis exemplified by Moroccan Cretaceous ostracods: Micropaleontology, v. 24, no. 1, p. 24-43.

Reyment, R.A., 1978c, Graphical display of growth-free variation in the Cretaceous benthonic foraminifer *Afrobolivina afra*: Palaeo., Palaeo., Palaeo., v. 25, no. 4, p. 267-276.

Reyment, R.A., 1980, Morphometric methods in biostratigraphy: Academic Press, London, 176 p.

Reyment, R.A., and Banfield, C.F., 1976, Growth-free canonical variates applied to fossil foraminifera: Bull. Geol. Inst. Univ. Uppsala, (n.s.), v. 7, p. 11-21.

Reyre, Y., 1972, Application de l'informatique a la <<gestion>> et a l'interpretation stratigraphique des donnees paleontologiques quantitatives: BRGM Bull. (2nd ser.), Sec. 4, no. 2-1972, p. 49-65.

Reyre, Y., 1974, Les methodes quantitatives en polynologie, *in* Elements de palynologie, applications geologiques, Chateanueuf, J.-J., and Reyre, Y., eds.: BRGM, Orleans, p. 271-312.

Rowe, A.W., 1899, An analysis of the genus *Micraster*, as determined by rigid zonal collecting from the zone of *Rhynchonella Cuvieri* to that of *Micraster coranguinum*: Geol. Soc. London Quart. Jour., v. 55, p. 494-547.

Rowell, A.J., 1970, The contribution of numerical taxonomy to the genus concept: Proc. North Am. Paleont. Conv., Chicago, 1969, pt. C, p. 264-293.

Rowell, A.J., and McBride, D.J., 1972, Faunal variation in the *Elvinia* Zone of the Upper Cambrian of North America - a numerical approach: 21st Intern. Geol. Contress (Montreal), Sect. 7, Paleontology, p. 246-253.

Rowell, A.J., McBride, D.J., and Palmer, A.R., 1973, Quantitative study of Trempealeauian (Latest Cambrian) trilobite distribution in North America: Geol. Soc. America Bull., v. 84, no. 10, p. 3429-3442.

Rubel, M., 1976, On biological construction of time in geology: Eesti NSV Tead. Akad. Toimetised. Keemia. Geloogia, v. 25, nr. 2, p. 136-144 (in Russian).

Rubel, M., 1978, Principles of construction and use of biostratigraphical scales for correlation: Computers & Geosciences, v. 4, no. 3, p. 243-246.

Schuey, R.T., Brown, F.H., Eck, G.G., and Clark, F.C., 1978, A statistical approach to temporal biostratigraphy, *in* Geological background to fossil man: Recent research in the Gregory Rift Valley, East Africa, Bishop, W.W., ed.: publ. for the Geol. Soc. London by Scottish Academic Press and Univ. of Toronto Press, p. 103-124.

Scott, G.H., 1974, Essay review: stratigraphy and seriation: Stratigraphy Newsletter, v. 3, p. 93-100.

Scott, R.W., 1970, Paleoecology and paleontology of the Lower Cretaceous Kiowa Formation, Kansas: Univ. Kansas Paleont. Contr., Art. 52 (Cretaceous 1), 94 p.

Scott, R.W., and West, R.R., eds., 1976, Structure and classification of paleocommunities: Dowden, Hutchinson & Ross, Inc., Stroudsburg, Pennsylvania, 291 p.

Sepkoski, J.J., Jr., 1974, Quantified coefficients of association and measurement of similarity: Jour. Math. Geology, v. 6, no. 2, p. 135-152.

Shaw, A.B., 1964, Time in stratigraphy: McGraw-Hill Book Co., New York, 365 p.

Shier, D.E., 1978, Sample ordering - a new statistical technique for paleoecological analysis: Trans. Gulf Coast Assoc. Geol. Soc., v. 28, p. 461-471.

Simpson, G.G., 1947, Holarctic mammalian faunas and continental relationships during the Cenozoic: Geol. Soc. America Bull., v. 58, no. 7, p. 613-687.

Simpson, G.G., 1960, Notes on the measurement of faunal resemblance: Am. Jour. Sci., v. 258a, p. 300-311.

Smith, D.G., and Fewtrell, M.D., 1979, A use of network diagrams in depicting stratigraphic time-correlation: Geol. Soc. London Quart. Jour., v. 136, pt. 1, p. 21-28.

Smith, W., 1815, A memoir to the map and delineation of the strata of England and Wales, with part of Scotland: John Cary, London, 51 p.

Smith, W., 1816-1819, Strata identified by organized fossils, containing prints on coloured paper of the most characteristic specimens in each stratum: The author, London, 32 p, (parts 1 and 2 in 1816; part 3 in 1817; part 4 in 1819).

Smith, W., 1817, Stratigraphical system of organized fossils, with reference to the specimens of the original geological collection in the British Museum: explaining their state of preservation and their use in identifying the British strata: E. Williams, London, 118 p.

Sneath, P.H.A., 1967, Trend-surface analysis of transformation grids: Jour. Zoology London, v. 151, p. 65-122.

Sneath, P.H.A., and Sokal, R.R., 1973, Numerical taxonomy: W.H. Freeman & Co., San Francisco, 573 p.

Sohl, N.F., 1977, Utility of gastropods in biostratigraphy, *in* Concepts and methods of biostratigraphy, Kauffman, E.G., and Hazel, J.E., eds.: Dowden, Hutchinson & Ross, Inc., Stroudsburg, Pennsylvania, p. 519-539.

Sorgenfrei, T., 1958, Molluscan assemblages from the marine Middle Miocene of South Jutland and their environments: Geol. Survey Denmark (2nd ser.), no. 79, 2 vols., 503 p.

Southam, J.R., and Hay, W.W., 1978, Correlation of stratigraphic sections by continuous variables: Computers & Geosciences, v. 4, no. 3, p, 257-260.

Southam, J.R., Hay, W.W., and Worsley, T.R., 1975, Quantitative formulation of reliability of stratigraphic correlation: Science, v. 188, no. 4186, p. 357-359.

Srivastava, G.S., 1977, Optical processing of structural contour maps: Jour. Math. Geology, v. 9, no. 1, p. 3-38.

Stone, J.F., 1967, Quantitative palynology of a Cretaceous Eagle Ford exposure: Compass, v. 45, no. 1, p. 17-25.

Stone, J.F., 1973, Palynology of the Almond Formation
(Upper Cretaceous), Rock Springs Uplift, Wyoming:
Am. Paleontology Bull., v. 64, no. 278, 135 p.

Swinnerton, H.H., 1921, The use of graphs in palaeontology: Geol. Mag., v. 58, p. 357-364, 397-408.

Symons, F., and De Meuter, F., 1974, Foraminiferal associations of the mid-Tertiary Edegem sands at Terhagen, Belgium: Jour. Math. Geology, v. 6, no. 1, p. 1-15.

Symons, F., and Ringele, A., 1976, Study of time-related variability within the genus *Astarte* (Bivalvia): Jour. Math. Geology, v. 8, no. 2, p. 113-136.

Tipper, J.C., 1977, Some distributional models for fossil marine animals (abst.): Geol. Soc. America, Abstracts with Programs, v. 9, p. 1202.

Tipper, J.C., 1979, An algol program for dissimilarity analysis: a divisive-omnithetic clustering technique: Computers & Geosciences, v. 5, no. 1, p. 1-13.

Trueman, A.E., 1922, The use of *Gryphaea* in the correlation of the Lower Lias: Geol. Mag., v. 59, no. 6, p. 256-268.

Valentine, J.W., and Peddicord, R.G., 1967, Evaluation of fossil assemblages by cluster analysis: Jour. Paleontology, v. 41, no. 2, p. 502-507.

Waters, J.A., 1977, The quantification of shape by use of Fourier analysis: the Mississippian blastoid genus *Pentremites*: Paleobiology, v. 3, no. 3, p. 288-299.

Whittaker, R.H., ed., 1973, Ordination and classification of communities, *in* Handbook of vegetation science, pt. 5: Junk, The Hague, 737 p.

Wilkinson, E.M., 1974, Techniques of data analysis-seriation theory: Archaeo-Physika, v. 5, p. 1-142.

Worsley, T.R., and Jorgens, M.L., 1977, Automated biostratigraphy, *in* Oceanic micropaleontology, Ramsay, A.T.S., ed.: Academic Press, London, p. 1201-1229.

COMPUTERS IN GEOLOGICAL PHOTOINTERPRETATION

K.L. Burns

Syracuse University

ABSTRACT

Computer processing of digital remote-sensing data can produce imagery of high spectral and geometric fidelity without the degradation associated with photographic reproduction. This is a significant advance in quality control in the data-acquisition system.

However progress in the interpretation system lags considerably. In one specific application, interpretation for geological lineaments, there occur low correlations between annotations which have hindered the acceptance of geological photointerpretations as reliable data.

Recently, perception models have been developed which radically alter our understanding of the properties of annotations. In particular the models imply that presence-absence data associated with the existence of lineaments is not a ranked binary variable and correlation measures are meaningless as indicators of data quality.

Computer processing now seems to be essential in geological photointerpretation. The procedures developed to date comprise estimation of operator resolution, digitization of annotations to arrays of cells, fitting perception models, and using the model parameters to assign probability estimates to quality maps written as shade prints or on filmwriters.

INTRODUCTION

At the present time the most widely used method of extracting spatial information from imagery for geological purposes is by human photointerpretation. The method is cheap, simple, and adaptable in terms of targets and operating conditions, and is the main source of information from imagery for a wide range of purposes. The human interpreter, therefore, is an essential link in an information-processing system and his characteristics are important in design of the total system.

In geology, a target of some importance is the discrete, linear feature termed a *lineament*. The interpreter marks their location by a line drawn on the image, so that the result of the interpretation process is a network of intersecting lines termed an *annotation*.

A persistent difficulty with information extracted in this manner is the lack of reproducibility between annotations made on the same image by different interpreters, or by the same interpreter on different occasions. Burns, Huntington, and Green (1977, fig. 1) and Burns and Brown (1978, fig. 2) show repeated interpretations of the same image, and how that although there is some similarity between them, there is nowhere precise agreement. This situation also is illustrated by the comparative plates of Kelley and Clinton (1960) and measurements reported by Podwysocki, Moik, and Sharp (1975) and Siegal and Short (1977). In addition, field examinations at locations indicated by the annotations may yield confirmation in ground-observables of the existence of structures, but sometimes do not. Practical experience is that lineaments interpreted from imagery may or may not be meaningful in terms of the ground geology. As a result of varying experience with annotations, some geologists regard lineaments as in the same class as the Martian Canals - imaginary lines perceived in random spatial noise (Crain, 1972), whereas for others, lineaments are meaningful and their study is incorporated in engineering and mining investigations.

This paper reviews some recent work on the human perception of geological lineaments. Annotations have been shown to have properties rather different to those of signals ordinarily encountered in geophysics, and it is this anomalous behavior that causes much of the difficulty in evaluating their significance. Examination of those properties suggests that annotations could be made more reliable by changes in photointerpretation practice.

PROPERTIES OF ANNOTATIONS

"Resolution"

An annotation may be digitized into a large number of small regions termed *cells*. The appropriate cell size for any annotation is governed by the ability of the observer to discriminate between adjacent lineaments, which is a function of the physical resolution of the system and the human resolution of the observer.

The smallest resolvable object, a criterion used in measurement of physical resolution, is not appropriate for annotations. For example, some high-contrast linear features, such as roads, with widths on the ground of 15 m, have been detected on LANDSAT imagery with pixel dimensions of approximately 30 by 60 m. Their detection rests upon their effect on reflected intensity across the whole of a pixel; the observer cannot locate them with greater precision than "somewhere within" a pixel, so that mere detectability is insufficient as a measure for features with decision boundaries, such as lineaments.

The spatial resolution of an annotation is defined here as the distance between the closest pair of linear features which are visually separable. Burns, Shepherd, and Berman (1976) printed LANDSAT MSS imagery at a scale of 1:500,000 at which scale the pixel size (approximately 0.06 x 0.12 mm) matches human resolution at normal viewing distances (approximately 8 to 10 lines/mm), to eliminate those problems that might arise from a mismatch, and demonstrated two different methods of measuring annotation resolution.

In the first method, interpreters were asked to classify pairs of adjacent lineaments into two states, nonseparable (state 1) or separable (state 2). The number of observations of each state can be written $n_1(a)$ and $n_2(a)$ which are functions of a, the measured distance between features in a pair. A discriminant function, $d(\infty)$ then is erected, where

$$d(\infty) = \frac{\int_0^\alpha n_1(a)\,da + n_2 \int_\alpha^\infty n_2(a)\,da}{\int_0^\infty n_1(a)\,da + \int_0^\infty n_2(a)\,da}$$

which is the empirical probability that two adjacent lineaments can be classified correctly as distinct or not when distance ∞ apart. Measurements yielded a maximum value of $d(\infty)$ at 0.85, 0.90, for LANDSAT imagery corresponding to ∞ = 205, 275 m, respectively, at ground scale (Burns, Shepherd, and Berman, 1976, fig. 5). Thus maximum discrimination was obtained at separations of about 250 m and the accuracy of classification was about 90 percent.

In the second method, the agreement between two different annotations of the same scene was estimated by a parameter R, the "reproducibility" which is a multi-state measure of association discussed further later. Where w is the width of lineaments, arguments can be made that R should increase rapidly as w increases from 0, then dR/dw should decline to be approximately constant, and the resolution of the annotation is the value of w at the beginning of the latter stage. Measurements showed that dR/dw was approximately constant from w=300 to w=600 m, implying a resolution of about 300 m, in reasonable agreement with the result obtained by different methods. This second method is not as precise as the first as it the result is influenced by the geometrical pattern of the features, a factor which can be excluded in the first method. However it does not require observer characterization so is more objective.

At the scale of imagery used for these experiments, the results imply an *interpretation cell* of dimensions approximately 0.5 x 0.5 mm as the area of the image viewed by an observer in deciding whether a pattern is two separate lineaments or only one. This "decision area" corresponds to a "decision level" of about 0.9. It is regarded as significant that the decision area contained approximately 5 x 5 pixels and that this corresponds to the 9 x 6 or 7 x 5 dot matrices used for alphanumeric character representation in dot-matrix printers. The observer makes classifications on the basis of a pattern formed by arrays of resolvable elements.

GEOLOGICAL PHOTOINTERPRETATION

Perception Vectors:

When an interpreter draws a feature on the image he is classifying cells in the field of view into one of two states - a state denoted a_1 for cells which are occupied by a feature and a state denoted a_0 for cells which do not contain a feature. Because lineaments intersect, cells under crossing points are multiply classified (Burns, Shepherd, and Berman, 1976, fig. 1) but the number is small and may be neglected, thus a single annotation is a binary classification of the cells into two states, a_0 and a_1.

If an image is annotated several (say k) times, to give a series of different annotations, each cell is classified several times. The classification state of each cell then may be described by a *perception vector* $v = (v_1, v_2, v_3, \ldots)$ where each component v_i takes the states a_0 or a_1. Thus, for example, the perception vector $v = (a_0, a_1, a_0, a_0, a_1)$ indicates the cell was assigned state a_0 on the first, third, and fourth annotations and state a_1 on the second and fifth.

It is convenient to define a scalar, V, termed the *perception level* as the number of states v_i such that $v_i = a_1$, that is,

$$V = |(v_i | v_i = a_1, v_i \neq a_0)|$$

PERCEPTION MODELS

Burns and Brown (1978) introduced a perception model to explain the process of image interpretation. The image is conceived as consisting of several disjoint sets of cells termed *messages* of different types. The interpreter has a certain probability of perceiving a given message correctly and assigning the correct classification or of not perceiving the message correctly and assigning an incorrect classification.

There could be a number of messages and a number of classification states, and the perception probabilities may change from trial to trial, and complex models can provide close fits to the data (e.g. Burns and Brown, 1978, model B). However it is shown in that paper that an approximation which is adequate for practical purposes is a simple two by two classification (Burns and

Brown, 1978, model A). It was determined that perceived lineaments are not drawn from a mononomial distribution and the simplest model which provides anything approaching an adequate fit is binomial, with the perceived lineaments being drawn from two disjoint distributions with substantially different perception properties. In the simple model, the cells of the image comprise two disjoint sets, termed *positive message* and *negative message* respectively. The positive message is that set of cells which lie on lineaments, the negative message is that set which do not.

This result is contrary to traditional concepts in geological photointerpretation and seems to run counter to line-finding methods which search for a positive message only (e.g. Vanderbrug, 1975) but is in accord with recent work in psychology (Estes, 1975; Sperling and Melchner, 1978).

There are two independent probabilities, a probability P_0 that a cell of the negative message will be assigned the correct perception state a_0 and a probability P_1 that a cell of the positive message will be assigned its correct perception state, a_1. These correspond to the marginal probabilities or empirically estimated attention operating characteristics of Sperling and Melchner (1978, fig. 1). The probabilities of incorrect identification are $(1-P_0)$ and $(1-P_1)$ respectively.

If v is the perception level as defined previously and α is the proportion of positive message in the image, then the simple model may be written as follows:

$$n(k,v) = N\alpha \binom{k}{v} P_0^{k-v}(1-P_0)^v + N(1-\alpha)\binom{k}{v}(1-P_1)^{k-v}P_1^v \quad (1)$$

Where $n(k,v)$ is the number of cells with perception level v after k repeated annotations and N is the total number of cells in the image.

For any particular situation there are two solutions, the correspondence between a solution and its dual being

$$\left. \begin{array}{l} \alpha \iff 1 - \alpha \\ P_0 \iff 1 - P_1 \\ P_1 \iff 1 - P_0 \end{array} \right] \quad (2)$$

The solution is decided between these alternatives on intuitive grounds.

The perception model may be inverted to provide the probability that a message of either type exists at a cell. Burns and Brown (1978) define the *quality* of an annotation as the probability that any cell is classified correctly. For a single interpretation (that is, k = 1) in a typical situation, it is determined that the quality of the annotation is 0.66 for the positive portion and 0.78 for the negative. This indicates that approximately 34 percent of the lineaments drawn on the experimental annotations were spurious, being negative messages interpreted as positive.

This quantifies the field geologist's problem. About one-third of the lineaments he seeks to verify in the field could be "artifacts of the interpreter's imagination" in the same category as the Martian Canals. Because the annotations as usually supplied provide no guide as to which of the lineaments may be in that category, we have the strange situation in mineral exploration where men and machines are moved about in the field to determine which lineaments are projections of an interpreter's imagination and which have substance.

The obvious absurdity of this procedure has led to the introduction by Burns, Huntington, and Green (1977) of methods of processing annotations to produce *quality maps* which show the probability of a message existing at any given location. This is one method of message extraction termed the *perception-level method*.

If a series of k annotations are superposed in such a manner as to generate the set intersection of the lineaments, the size of the intersection set decreases approximately geometrically with k. This is a property of other types of geological targets than the discrete lineaments considered here. This situation corresponds to the application of the condition k = v in expression (1), as follows:

$$n(1,1) = N\alpha(1-P_0) + N(1-\alpha)P_1$$

$$n(2,2) = N\alpha(1-P_0)^2 + N(1-\alpha)P_1^2$$

$$n(3,3) = N\alpha(1-P_0)^3 + N(1-\alpha)P_1^3$$

and so on. As a result, the common area of lineaments dwindles away to nil, in practical situations, when k > 6. This is illustrated in Table 1 and by Burns and Brown (1978, fig. 3).

If a series of k annotations is superposed in such a manner as to take the set union of the lineaments, the size of this union set increases with k. This corresponds to the situation v = 0 in expression (1), as follows

$$N - n(1,0) = N[1-\alpha p_0 - (1-\alpha)(1-P_1)]$$

$$N - n(2,0) = N[1-\alpha P_0^2 - (1-\alpha)(1-P_1)^2]$$

$$N - n(3,0) = N[1-\alpha P_0^3 - (1-\alpha)(1-P_1)^3]$$

and so on. As a result, the combined area of lineaments increases to occupy almost the whole image. This is illustrated by Table 1.

This behavior of the annotations confronts the industrial geologists with a severe problem. Expressed in practical terms, it indicates the more sources of advice he employs, the smaller the area of mutual agreement. It also indicates that each new adviser will canvass new possibilities until eventually they are all covered. He will never obtain a consensus on a limited number of choices. This behavior is in contrast to geophysical signals of other types where repeated acquisition cycles tend to reinforce the signal at the expense of the noise.

However a real signal exists in annotations as shown by the correlation peak at registration obtained by Burns, Shepherd, and Berman (1976, figs. 7 and 8) so the annotations cannot be discarded as random patterns as proposed by Gilluly (1976, p. 1512) but contain real information. At the present time it seems essential to accept the perception process as generating information with curious properties and process them by computer.

REPRODUCIBILITY OF ANNOTATIONS

We consider the simplest situation, which is two different annotations by a single observer. The perception vector for each cell then has the form $V = (v_1, v_2)$

Table 1. Expected proportion of image occupied by lineaments when k annotations are combined by set intersection (column 2) and set union (column 3). In first situation combined area dwindles away to nil as k increases, in second it increases to (eventually) fill whole image. Perception parameters used are $\alpha = 0.6987$, $p_0 = 0.9129$, $p_1 = 0.3883$.

k	$n(k,1)/N$	$[N-n(k,0)]/N$
1	.178	.178
2	.051	.305
3	.018	.399
4	.007	.473
5	.003	.531
6	.001	.580
7	.000	.621
8	.000	.657

where $v_i = a_0$ or a_1. This gives rise to four situations, which are (a_0,a_0), (a_0,a_1), (a_1,a_0), and (a_1a_1). We denote the number of cells in each set by n_{00}, n_{01}, n_{10}, n_{11} respectively. Table 2 shows the values these are expected to take under the perception model.

The general form of the produce-moment correlation coefficient is

$$R_{aa} = \sum_{i=1}^{\ell} (x_i - \bar{x})^2$$

$$R_{bb} = \sum_{i=1}^{\ell} (Y_i - \bar{Y})^2 \qquad (3)$$

$$R_{ab} = \sum_{i=1}^{\ell} (x_i - \bar{x})(Y_i - \bar{Y})$$

where x_i, Y_i are random variables taking any of ℓ states such as $x_1, x_2, \ldots x_\ell$, and similarly for Y.

In this situation, X_i, Y_i take only two values, which for the moment we differentiate by denoting the states of X by a_0, a_1 and the states by Y by b_0, b_1.

The definitions of $\bar{X}$ and $\bar{Y}$ are as follows

$$\bar{X} = [(n_{00}+n_{01})a_0 + (n_{10}+n_{11})a_1]/n$$
$$\bar{Y} = [(n_{00}+n_{10})b_0 + (n_{01}+n_{11})b_1]/n \qquad (4)$$

It then follows, after some manipulation, that

$$R_{aa} = (n_{00}+n_{01})(n_{10}+n_{11})(a_0-a_1)^2/n$$

$$R_{bb} = (n_{00}+n_{10})(n_{01}+n_{11})(b_0-b_1)^2/n$$

$$R_{ab} = (n_{00}n_{11} - n_{01}n_{10})(a_0-a_1)(b_0-b_1)/n$$

This yields the binary reproducibility of Burns, Shepherd, and Berman (1977) which is

$$R = \frac{n_{00}n_{11} - n_{01}n_{10}}{[(n_{00}+n_{01})(n_{10}+n_{11})(n_{00}+n_{10})(n_{01}+n_{11})]^{\frac{1}{2}}} \qquad (5)$$

If we substitute for the perception parameters from Table 2, this becomes

$$R = \frac{\alpha(1-\alpha)[(1-p_1)(1-p_0) - p_1 p_0]^2}{[\alpha p_0 + (1-\alpha)(1-p_1)][\alpha(1-p_0)+(1-\alpha)p_1]} \qquad (6)$$

In a practical situation, $\alpha = 0.6987$, $p_0 = 0.9129$, and $p_1 = 0.3883$, from which we estimate R as 0.1306. This is a typical figure. Usually R lies in the range

Table 2. Expected number of pixels n_{ij} with given perception states $a_0 a_j$, in terms of parameters, N, α, p_0, p_1 of simple perception model.

Perception vector	Symbol	Expected Number	
		Contribution from negative message	Contribution from positive message
a_0	n_0 =	$n\alpha p_0$	+ $N(1-\alpha)(1-p_1)$
a_1	n_1 =	$n\alpha(1-p_0)$	+ $N(1-\alpha)p_1$
$a_0 a_0$	n_{00} =	$N\alpha p_0^2$	+ $N(1-\alpha)(1-p_1)^2$
$a_0 a_1$	n_{01} =	$N\alpha p_0(1-p_0)$	+ $N(1-\alpha)(1-p_1)p_1$
$a_1 a_0$	n_{10} =	$N\alpha(1-p_0)p_0$	+ $N(1-\alpha)p_1(1-p_1)$
$a_1 a_1$	n_{11} =	$N\alpha(1-p_0)^2$	+ $N(1-\alpha)p_1^2$

0.10 to 0.15 with values up to 0.30 representing exceptional agreement.

The reproducibility is an extension of the product-moment correlation coefficient to ordered, multistate variables. As such, it should be comparable to correlation and coincidence measures of Shepherd and Gaskell (1977), Huntington and Raiche (1978), Siegal and Short (1977), and others. It was shown by Burns, Shepherd, and Berman (1976) that R differs with the degree of freedom so that azimuths, lengths, and areas yield different measures. If account is taken of this difference, the results of various authors are comparable and R, or equivalent measures of association, is low if the annotations seem, intuitively, to be in fairly good agreement.

Expression 6 explains the low value for R. It reduces to zero in circumstances when $p_0 + p_1 = 1$, that is, when the probability of correctly perceiving

Table 3. Reproducibility as function of probabilities p_0, p_1 of correct recognition of negative and positive messages, respectively for $\alpha = 0.6987$.

						p_1					
p_0	0	0.1	0.2	0.3	0.4	0.5	0.6	0.7	0.8	0.9	1.0
0	1	.863	.736	.620	.512	.411	.318	.230	.149	.072	0
0.1	.731	.600	.481	.375	.280	.196	.123	.063	.019	0	.032
0.2	.547	.426	.321	.231	.155	.092	.044	.012	0	.015	.070
0.3	.413	.304	.213	.138	.080	.037	.010	0	.011	.046	.114
0.4	.311	.213	.135	.076	.034	.009	0	.009	.038	.089	.167
0.5	.232	.143	.078	.034	.008	0	.008	.034	.078	.143	.232
0.6	.167	.089	.038	.009	0	.009	.034	.076	.135	.213	.311
0.7	.114	.046	.011	0	.010	.037	.080	.138	.213	.304	.413
0.8	.070	.015	0	.012	.044	.092	.155	.231	.321	.426	.547
0.9	.032	0	.019	.063	.123	.196	.280	.375	.481	.600	.731
1.0	0	.072	.149	.230	.318	.411	.812	.620	.736	.863	1

the negative and positive message are complementary. In the example cited, $p_0 + p_1 = 1.3$ and the low reproducibility is due to the proximity of the sum to 1. This is illustrated further in Table 3. The diagonal zeros corresponding to the situation $p_0 + p_1 = 1$ are unaffected by changes in ∞. Table 3 shows that R does not increase monotonically as p_0, p_1 both increase.

DISCUSSION

The obstacles to annotations being accepted as reliable sources of data may be quantified in terms of the low values of the reproducibility R. There seem to be two remedies: replace R with some other method of measuring a comparison that monotonically increases with each of p_0 and p_1 and does not have a minimum at $p_0 + p_1 = 1$, or change the interpretation procedures to take better account of the properties of the information and ensure that the latter condition is avoided.

The first possibility is to redefine the reproducibility. The present definition may be regarded geometrically as an inner product in a space with axes defined by the ordered pairs (a_0, a_1) and (b_0, b_1). However, the perception model shows there is no basis to this ordering. That is, there is no property of the cells incorporated into the model which enables a relation between states of the type $a_0 < a_1$. In previous work, by getting $a_0 = 0$ and $a_1 = 1$ we have incorporated such a relationship. This now is seen to be unjustified. Although the terms in a_0, a_1, b_0, b_1 cancel out in the manipulations leading from expressions (3) to (5) they enter into the treatment. The problem is illustrated by the mean values given by expression (4). The mean is, in fact, meaningless, if a_0 and a_1 cannot be assigned numerical values on some common scale. It is contended that the perception model of Burns and Brown (1978) differentiates the two states and implies that they cannot be ordered on a common scale, so that the mean is an undefinable quantity.

Accordingly, it is considered that the reproducibility should be replaced by quantities which are concerned with each class of information separately. The quality measures introduced by Burns and Brown (1978) serve this purpose and have the advantage of being probabilities. The probability that a given pixel was interpreted correctly as being part of the positive message and assigned the state a_1 is, for k annotations, $Q_1(v,k)$ where

$$Q_1(v,k) = \frac{\binom{k}{v}(1-\alpha)p_1^v(1-p_1)^{k-v}}{\binom{k}{v}(1-\alpha)p_1^v(1-p_1)^{k-v} + \binom{k}{v}\alpha p_0^{k-v}(a-p_0)^v}$$

The corresponding probability for the negative message is

$$Q_0(v,k) = 1 - Q_1(k-v,k)$$

where v is the perception level as defined previously.

The second possibility is to revise the interpretation procedures to avoid the condition $p_0 + p_1 = 1$. These methods may be described as searching the imagery for lineaments with a strong criterion, that is, the most obvious lineaments are found first and annotated. The annotation effectively removes them from further consideration and the criterion then is changed to select more subtle features from the residual area. This process is continued in steps, until subtle features are being extracted. The negative message then is presented as simply the residue after all possible lineaments have been found.

This procedure will be familiar to photointerpreters and I draw attention to one noticeable psychological phenomenon. As soon as all the strongly defined lineaments are covered by annotation which obscures them from vision, there is a discrete shift in perception criteria and a whole new group of lineaments become visible.

The interpretation procedures reflect this psychological phenomenon and with the aim of extracting as much information as possible, advantage is taken of the heightened perception to pursue the signal to low thresholds. Generally, results show that it is pursued too far, and some 34 percent of the positive annotation is lineaments generated from random noise in the negative message area.

This problem may be avoidable by one of two procedures. The first is to establish two strong criteria, one for the recognition of a lineament, and another for the recognition of areas free of lineaments. At a first pass, the two regions are blocked out leaving a residual zone of uncertainty in between. This requires annotating by two methods instead of one, such as a pen to mark lineaments and a brush to shade in areas which are lineament free. The recognition threshold is lowered and a second pass is made over the residual region. This procedure would be repeated several times, with the recognition criteria being made increasingly subtle only as the residual area decreases. This technique may ensure that the two probabilities, p_0 and p_1, are maintained at a comparable level and that p_1 is not raised at the expense of p_0 which is the present situation. The elimination of clearly negative areas by a different annotation procedure should considerably reduce the opportunity for generation of spurious lineaments in the negative region towards the end of the annotation process, which is what seems to happen in present techniques.

Another solution may be search the imagery for the boundary between lineaments and the remaining area. This is a method suitable for geographic themes but does not seem to be practicable for lineaments which tend to be numerous and of small width. Also, there have been no perception models erected for boundary recognition and it is not known what new complications may occur.

It is considered therefore, that the first procedure is to be preferred.

GEOMETRIC CORRELATIONS

For studies of geological structures derived from both ground observation and interpretation of imagery, correlation measures have been used widely to quantify associations between the geometric parameters, such as length and trend, of structural features. Examples from remote sensing include Renner (1968), Schulz and Ingerson (1973), Shepherd and Gaskell (1977), and Huntington and Raiche (1978).

If the annotation is divided into cells and L is the length of lineament in a cell, from Table 2 it may be expected that the length drawn from the positive message is

$$\frac{L(1-\alpha)p_1}{\alpha(1-p_0)+(1-\infty)p_1}$$

whereas the length drawn from the negative message is

$$\frac{L\alpha(1-p_0)}{\alpha(1-p_0)+(1-\alpha)p_1}$$

A problem in geometric correlations has been how to treat cells containing no lineaments. If these are included in measurements of correlation, the correlation becomes extremely high due to the large number of cells of zero length, particularly if the cell size is small. The data are mixed in that it contains two multistate variables (the states of the positive and negative messages) and the parameters, length, and trend, of the lineaments perceived.

If we take the length L as an ordinate, then for a single annotation,

$$\text{for } 0 < L \leq \infty, \quad V = (a_1)$$
$$\text{for } 0 = L, \quad V = (a_0)$$

Hence the region from $0 < L \leq \infty$ corresponds to one-half space in the perception state and the region $L = 0$ to the other. Because geometric measurements are defined only in the former one-half space, they are restricted validly to cells in which L is finite, so that zero-length cells may be ignored.

For geometric correlations between two annotations, if the suffix X refers to the first observer, and Y to the second, the observation space has fields defined as follows:

(i) for $0 < L_x \leq \infty$, $0 < L_y \leq \infty$, $V_x = a_1$, $V_y = b_1$

(ii) for $0 = L_x$, $0 < L_y \leq \infty$, $V_x = a_0$, $V_y = b_1$

(iii) for $0 < L_x \leq \infty$, $0 = L_y$, $V_x = a_1$, $V_y = b_0$

(iv) for $0 = L_x$, $0 = L_y$, $V_x = a_0$, $V_y = b_0$.

It is immaterial to the perception states whether zero cells are discarded or whether cell size is increased until there are no zero cells.

GENERAL IMPLICATIONS

It is general practice in computer processing of mixed-mode data in the natural sciences to quantify presence-absence data by assigning it to a scale in which absent is 0 and present is 1. However, the physical properties of annotations described here and the analysis of Burns and Brown (1978) show that this is not an appropriate assignment for perception data.

Cluster analyses based on similarity measures (e.g. Wishart, 1969; Sepkoski, 1974; Lance and Williams, 1967) therefore should be treated with considerable caution if the procedures require assigning numeric values to multistate perception data (e.g. Wishart, 1969, table 3a).

This caution has wider implications than remote sensing as much geological field data depends upon perception processes (Chadwick, 1975) and the natural sciences generally are characterized by a dependence upon multistate variables where the states are determined by visual inspection of phenomena.

CONCLUSIONS

A model of human perception of geological lineaments in aerial and satellite imagery shows that the output of a photointerpreter (an "annotation") has characteristics which indicate that he perceives two disjoint messages, a *positive message* corresponding to presence of a lineament and a *negative message* corresponding to the absence. This model explains properties of the annotations such as the approximately geometric change in lineament area as multiple annotations are combined, decreasing to zero if the method of combination is set intersection or increasing to fill the whole image if the method is set union, and the anomalously low correlations between intuitively similar annotations. The low opinion of photointerpretations held in some parts of geology is considered to be a consequence of this latter property and not of a lack of meaningful information, although it is shown that some 34 percent of lineaments in a single annotation are "artifacts of the imagination" construed out of spatial noise in regions where probably no lineaments exist.

The two states of cells represented in an annotation, generally "presence" and "absence" of a detectable lineament, cannot be ordered on these properties alone and cannot be assigned values 0 to 1 on a common arithmetic scale. As a result, correlation coefficients are not meaningful measures of comparison, and in addition, do not increase monotonically with increases in the three perception parameters. Annotation quality is better measured in terms of two quality parameters for negative and positive messages respectively, rather than by a single coefficient. Quality parameters which estimate the probability of correct classification of each cell provide a measure of data quality suited to the perception properties of the image.

Interpretation procedures which treat the two types of message separately should improve the reproducibility of the output.

ACKNOWLEDGMENTS

Some of this work was done while the author was Principal Research Scientist at the Division of Mineral Physics, Commonwealth Scientific and Industrial Organization (CSIRO), Sydney, Australia. The author was grateful for comments from Drs. N. Fisher and A. Green of the

CSIRO and W.S. Kowalith of the Department of Applied Earth Sciences, Stanford University, but this acknowledgment does not imply that they necessarily agree with the views expressed here. I am grateful to Dr. N.M. Short of NASA for his encouragement to publish this review.

REFERENCES

Burns, K.L., Shepherd, J., and Berman, M., 1976, Reproducibility of geological lineaments and other discrete features interpreted from imagery: measurement by a coefficient of association: Remote Sensing of the Environment, v. 5, no. 4, p. 267-301.

Burns, K.L., Huntington, J.F., and Green, A.A., 1977, Computer assisted photointerpretation of geological lineaments: perception method, *in* Lynch, A.J., ed., APCOM 77: 15th Intern. symp. on application of computers and operations research in the mineral industries (Brisbane): Aust. Inst. Min. Metall. (Melbourne), p. 275-285.

Burns, K.L., and Brown, G.H., 1978, The human perception of geological lineaments and other discrete features in remote sensing imagery: signal strengths, noise levels and quality: Remote Sensing of the Environment, v. 7, no. 2, p. 163-176.

Chadwick, P.K., 1975, A psychological analysis of observations in geology: Nature, v. 256, no. 5518, p. 570-573.

Crain, I.K., 1972, A statistical approach to the analysis of tectonic elements: unpubl. doctoral dissertation, Australian National Univ., 72 p.

Estes, W.K., 1975, Human behaviour in mathematical perspective: Am. Scientist, v. 63, no. 6, p. 649-655.

Gilluly, J., 1976, Lineaments--ineffective guide to ore deposits: Econ. Geology, v. 71, no. 8, p. 1507-1514.

Huntington, J.F., and Raiche, A.P., 1978, A quantitative method for comparing geological lineament patterns: Remote Sensing of the Environment, v. 7, no. 2, p. 145-161.

Kelley, V.C., and Clinton, N.J., 1960, Fracture systems and tectonic elements of the Colorado Plateau: Univ. New Mexico Publ. in Geology, Univ. New Mexico Press, Publ. No. 6, 104 p.

Lance, G.N., and Williams, W.T., 1967, Mixed-data classificatory programs. I. Agglomerative systems: Aust. Computer Journal, v. 1, no. 1, p. 15-20.

Podwysocki, M.H., Moik, J.G., and Sharp, W.C., 1975, Quantification of geologic lineaments by manual and machine processing techniques: Proc. NASA Earth Res. Symp. v. 1B, NASA TM X-58168, p. 885-904.

Renner, J.G.A., 1968, The structural significance of lineaments in the eastern Monsech area, province of Lerida, Spain: Publ. Intern. Inst. Aerial Survey and Earth Sciences (Ser. B) No. 45, p. 12-25.

Schulz, P.H., and Ingerson, F.E., 1973, Martian lineaments from Mariner 6 and 7 images: Jour. Geophys. Res., v. 78, no. 35, p. 8415-8427.

Sepkoski, J.J., 1974, Quantified coefficients of association and measurement of similarity: Jour. Math. Geology, v. 6, no. 2, p. 135-152.

Shepherd, J., and Gaskell, J.L., 1977, Trend analysis of fractures and fissure vein mineralization in the Drake Volcanics of NSW, Australia: Trans. Inst. Min. Metall. (Sect. B), v. 86, p. 9-16.

Siegal, B.S., and Short, N.M., 1977, Significance of operator variation and the angle of illumination in lineament analysis on synoptic images: Modern Geology, v. 6, no. 1, p. 75-85.

Sperling, G., and Melchner, M.J., 1978, The attention operating characteristic: examples from visual search: Science, v. 202, no. 4365, p. 315-318.

Vanderbrug, G.J., 1975, Line detection in satellite imagery: Symp. Machine Processing of Remotely Sensed Data, Purdue Univ., p. 2A-16 to 2A-21.

Wishart, D., 1969, FORTRAN II programs for 8 methods of cluster analysis (CLUSTAN I): Kansas Geol. Survey Computer Contr. 38, 112 p.

LOOKING HARDER AND FINDING LESS--USE OF THE COMPUTER IN PETROLEUM EXPLORATION

John C. Davis

Kansas Geological Survey

ABSTRACT

The decade of the 70's was marked by steady, if unspectacular, progress in the use of computers by the petroleum industry. Digital seismic processing in the field of geophysics was the single most significant development, and promises to alter many basic procedures in exploration. Programs and techniques for computer mapping have improved significantly during the decade; contour mapping now is probably the most important application of computers to geologic aspects of petroleum exploration. Data banks have grown tremendously, and rapid interactive retrieval systems permit display of combined seismic and geologic information as maps and cross sections. Mini-computer systems are promising to bring dramatic changes to log-interpretation procedures and to the making of subsurface lithofacies maps, increasingly valuable exploration tools for defining stratigraphic traps. Computational advances in the area of petroleum resource assessment allow modeling of economic potential for large areas. These techniques include both economic and geologic considerations.

Data banking, log analysis, and contour mapping are extremely basic procedures, routinely required for almost all exploration activities. Aside from the more exciting geophysical applications, the use of computers in petroleum exploration seems to have advanced during the past decade through the determined refinement of such basic techniques.

INTRODUCTION

The past decade has been one of turmoil in the petroleum industry, marked by shortages, record prices and profits, public and governmental criticism, and Herculean efforts to develop new exploration tools and to find more oil. Without doubt, the most significant development in the application of computers to exploration has been in the geophysical field of digital seismic processing, allowing enhanced resolution of stratigraphy and three-dimensional seismic interpretation and mapping. Use of large-scale color displays of seismic output has proved especially appropriate because the achievements have themselves been so spectacular. Important, although less dramatic, advances also have been made in the use of computers for treatment of geologic (rather than geophysical) data.

Geologic data banks now are at a highly developed state, and include almost all well and seismic information available for most of the petroleum provinces of North America. Sophisticated interactive retrieval programs allow information to be extracted quickly, and displayed as cross sections and maps. Some systems permit the geologist to adjust and modify the contents, and to store them back in the bank. The ability to commingle seismic information with other forms of geologic data is an especially significant advance.

Contour mapping probably is the second most important application of computers to geologic aspects of petroleum exploration, and contour mapping programs have been improved significantly in the past decade. Most programs operate by constructing a regular meshwork of interpolated points across a map area, but new developments include fast routines to create triangular meshes in which each node corresponds to a control point. This insures that the mapped surface passes exactly through each grid point, and minimizes problems of aliasing and smoothing that occur when mapping nonuniformly distributed data. Faults and other discontinuities in the mapped surface can be accommodated easily, a valuable feature which also is included in new versions of contour programs that operate by gridding.

Log interpretation procedures have changed dramatically in recent years, in large part due to the proliferation of interactive minicomputers. Combining a digitizer, minicomputer, disk drive, CRT display screen, and small plotter, these relatively inexpensive systems allow a log analyst to experiment with trial solutions in real time. By using matrix algebra rather than table look-up procedures, fast interpretations can be achieved, even from underdetermined or incomplete log suites. Several wells may be processed simultaneously, and lithofacies or other properties mapped in three dimensions by including trend-surface equations in the matrix of log equations. Because stratigraphic traps are increasingly important as exploration targets, subsurface lithofacies maps based on log interpretations also are increasingly valuable as exploration tools.

Methods of petroleum resource assessment are undergoing active development, at scales ranging from the regional to the prospect. Approaches include use of subjective appraisals, Monte-Carlo simulation, and probabilistic estimation based on multivariate observations. Some statistical models are dependent heavily upon geologic input while others are based on geometric considerations and search theory. To date, no single technique has been established clearly as superior (or even as satisfactory), and there seems to be no relationship between computational complexity and success.

In summary, the past decade has been a time of increasingly effective utilization of computers by explorationists. Geologists now recognize that computers can do certain things extremely well, such as maintaining files, making contour maps, or processing well logs. There is also a realization that no "miracle method," regardless of mathematical elegance, will overcome the shortcomings of most geological data and make the finding of oil an easy task.

DEVELOPMENTS

To determine the decade's most significant advances in computing in oil exploration, the opinions of five persons long associated with the use of computers in exploration were sought. Each expert was requested to provide three nominations. A consulting geologist, the manager of a large independent, a member of the exploration division of a domestic major, a geologist in the research center of another major, and the manager of the long-range planning division of an international oil company offered 10 different suggestions, and one on which all agreed.

There is a single outstanding development in petroleum exploration in the past decade that easily ranks as the most significant advance of all. It came about because of computers, and was an impossibility until computers had advanced to their modern state. This development of course is digital seismic processing. No other technological or methodological advance has so revolutionized modern petroleum exploration. By discrete sampling and digital encoding, modern seismic equipment in principle has unlimited dynamic range. In practice, there are electronic limitations, but nevertheless, digital recording is a tremendous advance over analog methods.

Digital encoding also opened the way for digital signal processing, and it is in this area that exciting developments are occurring. The application of the Fast Fourier Transform (FFT) made complex signal analysis and decomposition a practical reality. The design of digital filters has become a high art in the geophysical profession. The display of specific components of the seismic wavelet has provided new insights, and holds a promise for the eventual direct detection of hydrocarbons. For example, Figure 1 (originally in color in Becquey, Lavergne, and Willm, 1979), shows the acoustic impedance along a $3\frac{1}{2}$ km seismic section over an offshore gas field. The low-impedance gas-bearing sand can be traced easily along the seismic profile, where it stands out in strong contrast to the higher impedance sediments in which it is enclosed. The downward trend in increasing acoustic impedance caused by compaction of the sedimentary column also can be seen on the section.

Another new development is three-dimensional seismic surveying, whereby it is possible to create "pictures" or "maps" of the subsurface at any desired depth, from a processed grid of seismic lines. Interpretations are no longer confined to a seismic profile, but may be displayed as horizontal surfaces, as in Figure 2, originally published in a color advertisement by Geodigit. The illustration shows an area 5 x 6 km; the dips of units at a depth of 3600 m are shown in color. Steepest dips are dark gray, gentlest dips are light gray, and in white areas there is no local gradient. Draping over a channel feature is clearly apparent.

In their original forms, both of these examples also illustrate a hardware advance that has occurred primarily because of the requirements of geophysicists. These are large-size, relatively inexpensive, color-display units

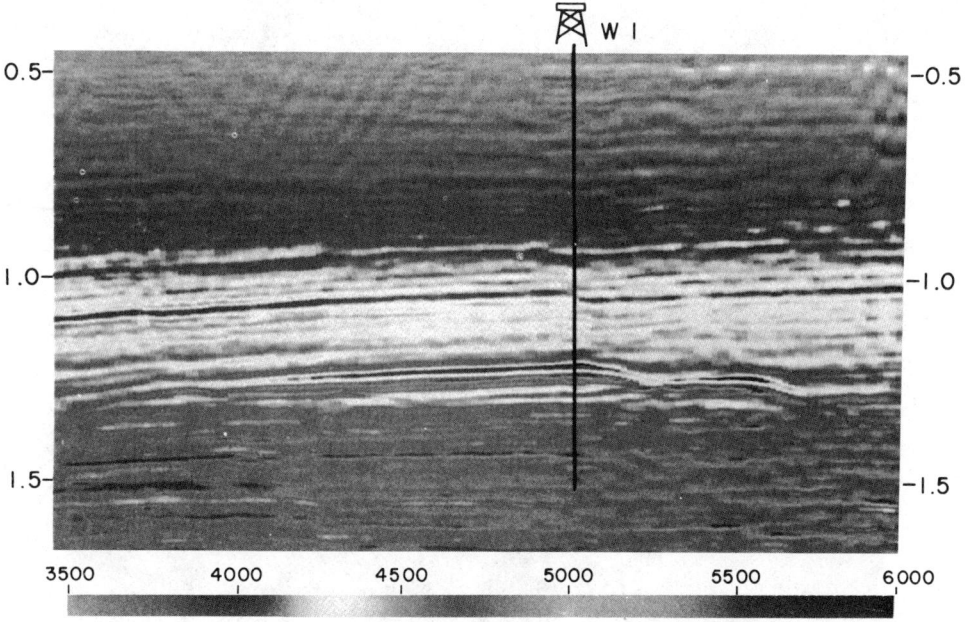

Figure 1. Acoustic impedance along marine seismic traverse across gas field (after Becquey, Lavergne, and Willm, 1979).

in the form of plotters, film and electrostatic printers, and video consoles. These now are becoming available in formats appropriate for other uses, and the routine use of color display devices can be anticipated for all types of geologic applications in the next decade. This will be possible, in large part, because of the market provided by the geophysical industry.

These exciting developments belong to the realm of the geophysicist, and are covered in greater depth by others in this volume (Dobrin, 1980). In contrast, the first among the other nominations for most significant advance in computer applications in exploration did not spring from a dramatic breakthrough in equipment or methodology, does not involve esoteric mathematics, and seldom results in spectacular color displays. However, the evolutionary development of geologic data banks during the 1970's has resulted in significant advances in petroleum exploration, particularly on the domestic scene.

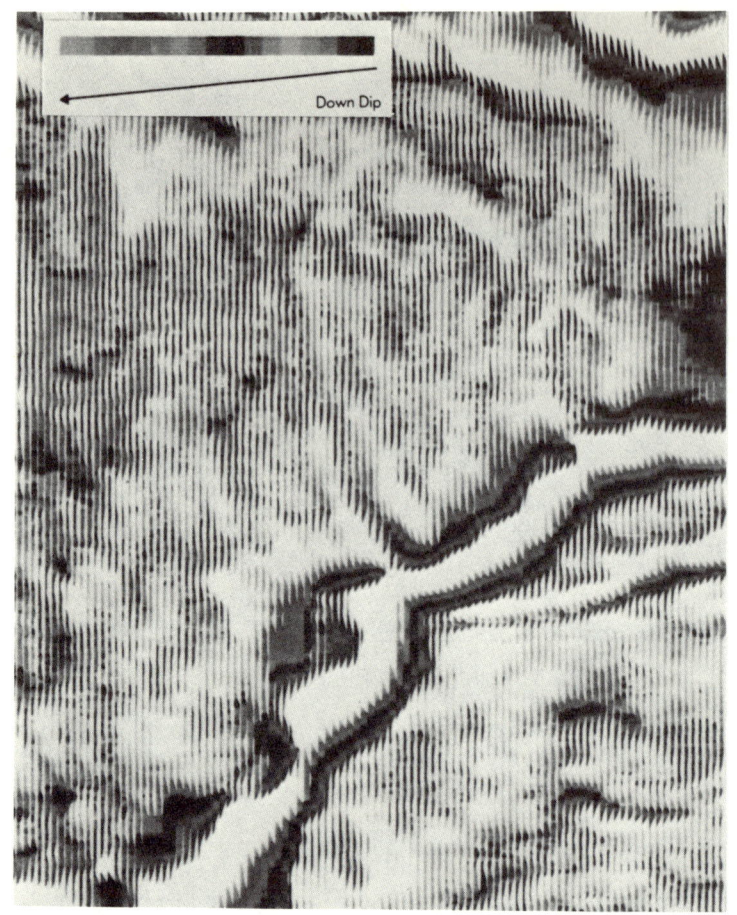

Figure 2. "Map" of subsurface at depth of 3600 m in area 5 x 6 km, obtained by Tomoseis three-dimensional seismic technique (from advertisement by GEODIGIT).

Every major oil company has devoted millions of dollars to develop data storage and retrieval programs, and to encode the geologic, engineering, and geophysical information which they have stored in these banks. Service companies have assembled well-data files, through the cooperation of oil companies, geological surveys, and societies, that cover most of the oil-producing regions of the United States and Canada. Typically, an industrial well-data file will contain about a million records, each of which will include numerous items of information such as the tops of formations, production and DST tests, bottomhole temperatures and pressures, notes on lithologies, geographic coordinates, and legal descriptions. Extensive files now exist containing digital well logs, either of new wells logged in digital mode, or of old wells whose paper traces have been digitized. Such information is becoming increasingly valuable, as procedures used for seismic interpretation are applied to the analysis of acoustic logs.

The need to store and retrieve vast quantities of information quickly and easily has led to the purchase of large computing systems by the oil industry. It is likely that the oil companies operate the largest computers in use outside the military and government spheres, and much of the impetus for the acquisition of these machines has stemmed from the need to handle geological data banks. In fact, some esoteric types of mass storage devices, such as the laser read-only memory, have been developed in direct response to industry needs, and with oil-company support.

A recent innovation in data banking involves the ability to merge disparate types of information, such as well data and seismic profiles, into a single working file. Through an interactive terminal, an exploration geologist then can manipulate the data, correlating and creating subsurface interpretations, which then are incorporated back into the data bank. In 1977, the AAPG sponsored one of its most successful ever research seminars, titled "Exploration Data Synthesis," on the topic of merging different types of exploration data into a single whole within a computer.

Some companies use a large host computer linked to smaller machines or remote job entry (RJE) stations in field offices; often the links extend, via satellite, across international boundaries or even across oceans. This provides the most isolated of field offices with access to the computational power and data-bank resources at the corporate headquarters, as well as an almost instantaneous communications link. Such capabilities are not confined to the large multinational oil companies. Service bureaus and commercial computer networks offer timesharing services at rates which are affordable by even the smallest companies. A small user cannot only purchase computing power, but also can purchase access to software and to commercial data banks. Some of these banks are maintained by commercial organizations; others such as the PDS files at the University of Oklahoma are in the public domain (Tracy, 1978).

The development of petroleum data files has been paralleled through the past decade by an evolution in minicomputer hardware. Ten years ago, a minicomputer was an expensive poor cousin of a mainframe computer. Usually, these machines were limited to 32K core, had a primitive operating system whose high-level language was a feeble subset of that available on larger computers, and were extemely slow by modern standards. Thanks to the technology of LSI - Large Scale Integration - minicomputer hardware now is significantly cheaper in both absolute as well as real terms. Cores have expanded to gargantuan proportions, and operating systems have increased in sophistication to a level rivalling that of the mainframe machines. Indeed, the distinction between a minicomputer and a full-sized machine is mostly arbitrary, and may depend more on pedigree than performance.

These changes have resulted in the explosive proliferation of minicomputers throughout the exploration industry. Usually, these machines are installed originally for a specialized purpose, such as well-log interpretation, but soon become a versatile component of an interactive network. Increasingly, mainframe computers are reserved for mass data storage and retrieval, transmitting working files to peripheral minicomputers for further processing and display under the interactive control of a geologist or geophysicist.

Minicomputers have played a critical role in another advance in exploration geology. Several interactive log analysis systems have been developed by the major oil companies (Smith and Souder, 1975) and by other research organizations. Typically, these systems run on a stand-alone mini, equipped with tape and disk drives, a digitizing table, a small plotter, and a display screen. A log analyst can retrieve digitized well information from the disk, display the various tool responses as cross plots, and estimate the coefficients of the logging equations either by eye or by fitting regressions.

Most log-analysis programs mimic human interpreters in the manner they operate. The programs contain digitized charts, taken from service company chart books, which are stored in core. Interpreted values are found by table look-up procedures, entering data from the digitized logs. However, at least one program (Doveton and Cable, 1979)

uses a more direct approach, setting up and solving a
series of matrix equations which relate log responses to
the interpreted petrophysical quantities. The process is
faster than might be supposed, because the unknown coefficients of the logging equation need only be found once
for an interval of consistent composition. Then, the
entire interval can be interpreted simply by multiplying
the matrix of coefficients by the array of digital log
values.

The program can solve overdetermined suites of logs,
where there are more logging tool responses than unknown
constituents to be estimated, by using least squares. If
the number of digitized logs exactly equals the number of
constitutents being estimated, the matrix equation is
solved directly. Unfortunately, the most common situation occurs when the number of lithologies to be determined exceeds the number of available logs. Approximate
solutions can be determined even in such circumstances
by using a maximum likelihood procedure.

Petrophysical analysis of well records is increasingly important in exploration, especially on the domestic
scene. Much of the remaining undiscovered oil is in
stratigraphic, rather than structural, traps. It is essential that the industry be able to map lithofacies and
changes in porosity in order to locate areas which are
prospective. From interpreted well logs, it is possible
to create many lithologic descriptors which can be mapped
(Bornemann, 1979). These include average lithology, percent sandstone or limestone, number of beds per interval,
degree of mixing or entropy, and the like. These properties can be created easily by a logging program and then
stored in a computer file for later display by a mapping
program (Fig. 3).

A recent development in the mapping of lithofacies
defined from well logs stems from use of the matrix algebra approach to log interpretation. The lithologic components are estimated using a series of simultaneous
equations that relate them to the various log readings.
Trend surfaces are estimated from a set of simultaneous
equations that can relate these same lithologic components to geographic coordinates. Therefore, the two sets
of equations can be combined, and solved at one time.
The result is a trend-surface map of a lithologic component, derived from a set of well logs which are interpreted and mapped in a single operation.

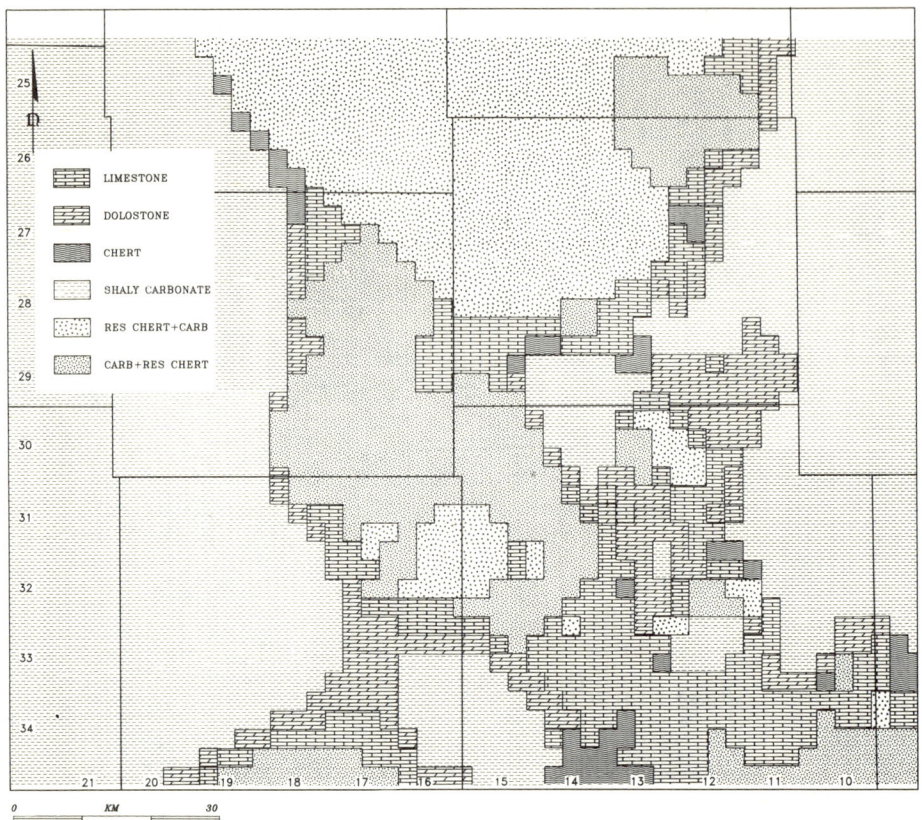

Figure 3. Lithofacies in Viola Limestone (Ordovician) of south-central Kansas. Map drawn by computer from well-log interpretations made using interactive log-analysis system (Bornemann, 1979).

Well-log interpretation programs also are invaluable for purposes other than lithofacies mapping. These include dipmeter analysis, for reservoir studies and paleocurrent mapping; construction of synthetic seismographs from acoustic logs for integrating well information into geophysical sections; and the calculation of porosity/permeability estimates for reservoir studies.

The subject of lithofacies mapping introduces another of the topics nominated by the panel of exploration experts: the construction of maps by computer. A decade ago, computer contour mapping was a controversial operation, regarded with suspicion by many explorationists. Today, computer mapping is a routine part of the preparation of an exploration play, and many of the shortcomings of earlier contouring programs have been overcome. Indeed, second only to data retrieval, mapping is probably the greatest use of computers made by exploration geologists.

Two contouring problems which were especially troublesome in the past now have been at least partially overcome, using a number of different approaches. Contouring programs always have had troubles with geophysical data, because these generally consist of dense arrays of control points along widely spaced lines. Most contouring programs operate by interpolating from the data to a regular grid, through which the contour lines are drawn (Davis, 1975). The interpolation process averages control points in a small neighborhood around each grid node. With typical seismic data, nodes which are near the traverses are close to many data points, resulting in excess averaging and smoothing of the mapped surface. Grid notes in the areas between seismic lines are controlled only poorly, because there is no local information. The resulting maps may have the appearance of contoured egg crates, because of rapid changes in the slope of the map between the seismic lines.

Modern contouring programs approach this problem in different ways. Some generate "pseudopoints" inside the holes between the seismic lines. These points are created by some type of global estimation procedure that includes all of the points along the bounding traverses. Other programs operate by altering the number of grid nodes according to the density of control points in the immediate vicinity. Along the traverses a dense network is created, allowing the surface to be modeled with great fidelity. Between the traverses, only a few grid nodes are used, so the surface, and correctly, has a generalized form.

The other major problem encountered when using contouring programs is representing discontinuous, or faulted, surfaces. Again, several approaches have been developed which can be used to resolve the difficulites which arise, provided the traces of the faults along the mapped surface can be digitized and provided to the contouring algorithm. Detecting unknown faults yet is beyond the capabilities of current procedures.

The most usual procedure involves treating a surface bounded by faults as a separate map, defined only by the control points on that fault plate. In other words, faults are regarded in the same manner as are the margins of the map. This approach, although simple, has certain drawbacks because there may be too few control points on a plate for adequate control. Additional "pseudopoints" may be created along the fault traces by interpolation from points where the throw of the fault is known, usually along seismic traverses (ACI, 1971).

A most sophisticated approach (Berlanga, 1979) for contouring seismic data from faulted surfaces attempts to resolve the system of faulted plates into its original unfaulted form. First, all of the fault intersections along the seismic traverses are connected into a set of fault traces which are defined by spline equations. Next, a linear-programming procedure attempts to iteratively "move" each fault plate until the discontinuities along the seismic traces disappear. Usually, this cannot be accomplished completely, because faulting and folding may not be independent. If this is the situation, the program allocates the residual discontinuities among all of the faults. Then, the restored surfaces are evaluated along the fault traces, using information from both sides of the faults. Finally, the faults are restored, but now the faults are defined on both their upthrown and downthrown sides by an array of control points. This array is included, along with the original set of control points, in contouring. The advantage of this somewhat complicated process is that trends which are evident on one fault plate continue in a logical fashion onto adjacent plates. In the mapping of complexly faulted surfaces, such as those over salt domes, the method creates much more realistic maps than can be generated by other techniques (Fig. 4).

Recent developments in mathematics (Brassel and Reif, 1979) have led to an algorithm which allows irregularly distributed points in a plane to be connected into a unique set of triangles, termed Delauney triangles, that form an optimal partition of the space. The new algorithm (McCullagh and Ross, 1980) is extremely fast, and has permitted revival of the procedure of contouring by triangulation. It has been known for a long time that contouring on an irregular triangular mesh offered significant advantages over conventional gridding procedures (SCA, Inc., 1975). The surface, by definition, would be self-weighted because the grid size conforms exactly to

Figure 4. Contour map of seismic reflection times over part of faulted Sitio Grande structure, Tabasco Basin, Mexico. Contour interval is 0.05 sec, mapped area is 15 x 15 km (Berlanga, 1979).

to the point distribution. Because all control points coincide with grid nodes at the vertices of the triangular mesh, the contoured surface must pass exactly through each control point. The surface itself is defined by triangular spline plates, which blend into adjacent plates (Fig. 5). This produces a smooth surface with continuous first and second derivatives (McCullagh, 1979). Because each part of the surface is defined by a single triangular plate, maps may be joined without discontinuities or changes in trends simply by overlapping a common row of triangles.

Because the surface is defined locally, it is easy to accommodate faults or other discontinuities. Points along the fault trace are flagged, indicating that the conditional requirement for continuity of the surface is

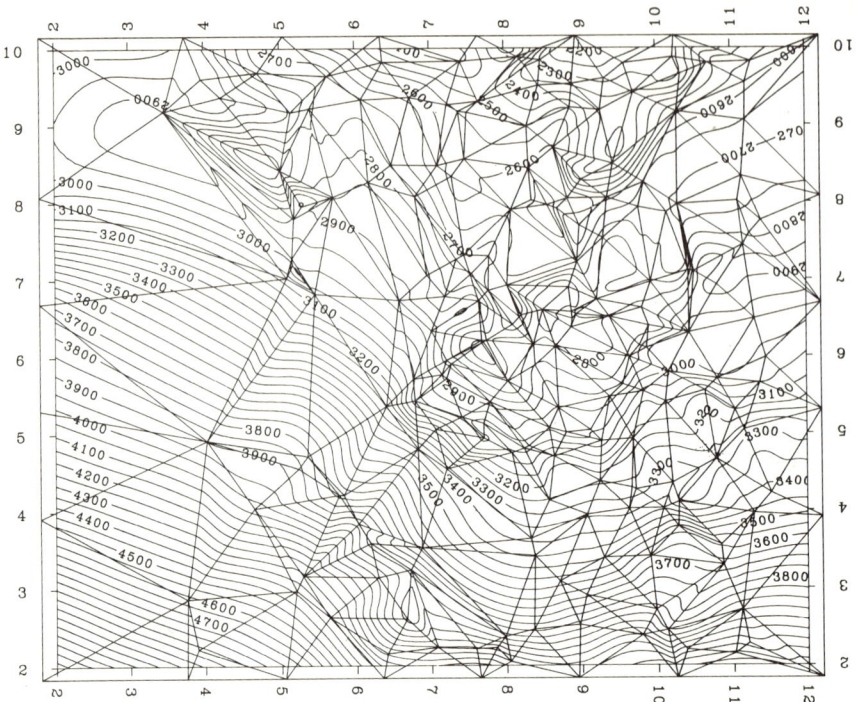

Figure 5. Contour map of subsurface structure on top of Viola Limestone (Ordovician) of south-central Kansas, mapped using contouring program which fits Delaunay triangles to data.

removed. The program then will contour each side of the fault independently, constructing the surfaces from elevations and trends in plates two layers deep on each side of the fault. At the ends of faults, where the discontinuities die out, the surface becomes continuous again.

   This new approach to triangular contouring has two significant, practical advantages. First, it is extremely fast. It is possible to construct a triangular mesh, calculate the surface, and then estimate the values of this surface at a regular grid of points faster than many conventional contouring programs can estimate the grid points directly. Secondly, the program will run on extremely small computers, as small as 32K minicomputers. This is because the surface at any point is defined on the basis of a single triangular plate and the adjacent plates. A large mesh need not be held in core, but can be "rolled through" the computer and processed in strips. The strips will fit together without discontinuities or abrupt changes in slope.

This new program should have a significant impact on petroleum exploration, because it can be run on the numerous minicomputer systems already installed for log interpretation. Any field office which is performing automated log analysis will be able to carry the interpretative process to its next logical step, and to map the results of their interpretations (Bornemann, 1979). This capability will be valuable especially as the search for stratigraphic traps intensifies.

The computational advances discussed so far have been concerned with improving exploration at the prospect scale. That is, the locating and defining of a specific drilling site. Within the past decade, computers also have enjoyed wide use in petroleum exploration at an entirely different scale. This application is in petroleum-resource appraisal; the modeling of large areas, usually equalling or exceeding sedimentary basins in size, for the purpose of estimating their economic potential for the long run. These computational tools are not designed for the individual oil finder, but rather for corporate management, and increasingly, government planners. Because the ultimate concern of management is profitability, and the ultimate concern of government seems to be taxes, these techniques have strong economic, as well as geologic, components.

Economic theory, if not its practice, is better established in quantitative form than is geologic theory. Therefore, there are a wide variety of models in use today, most of which are similar in their economic components. Typically, the economic elements include routines to calculate discounted net cash flows, given assumed costs, expenses, and incomes. Usually, these are broken down into a myriad of individual items, which may be specified as probability distributions to account for the uncertainties in their true values. The costs involved in exploration and development may be forecast with relative assurance. Similarly, income can be forecast reasonably well, given that production is discovered. The great uncertainty is in the amount of oil or gas that will be found, if indeed any is discovered at all.

One of the most widely used procedures is the tract evaluation model used by the U.S. Geological Survey (Akers, 1976) to fix the fair market value of offshore leases.

Similar programs are used by many major oil companies for prospect evaluation. The geologic components of these programs consist of a Monte-Carlo multiplication of various reservoir engineering variables to yield the barrels of recoverable oil in place in a perceived structure. These variables include elements of volume such as height of closure, area of closure, thickness of the reservoir unit, porosity, gas/oil ratio, and the like. The USGS model uses 17 such geologic variables which must be specified in the form of triangular or other probability distributions. The computed volume is weighted by a risk, or "dryhole" factor, to account for the possibility of failure. The expected production distribution then is passed to the economic model, which computes the expected worth of the prospect or lease.

This method is identical to techniques used to evaluate reservoirs, and to test alternative reservoir development schemes (Newendorp, 1977). However, petroleum engineers do not evaluate reservoirs until after they are discovered, at which time something is known about their characteristics. In applying this model to prospects in advance of drilling, or even leasing, the geologist must provide distributions about which he has no information. It should be no surprise that, despite the complexity of the model, it yields imprecise answers in the Louisiana Gulf Coast OCS region (Uman, James, and Tomlinson, 1979). A critique (Davis and Harbaugh, 1980) of this study pointed out that the confidence bands around tract evaluations covered two orders of magnitude. Nevertheless, this model is used widely, and may have a significant effect on federal leasing policy, and on government/industry interaction.

Other techniques for regional assessment include modeling based on assumed distributions of sizes of reservoirs, scattered according to some spatial model within the area being evaluated (Drew, 1974). Models of discovery rate and order of discovery, such as those formulated by Kaufman, Balcer, and Kruyt (1975) are used to generate possible scenerios of exploration in the region. An elaborate model of this type is being used to assess the National Petroleum Reserve in Alaska, where the geographic locations of any reservoirs discovered will have a profound effect on their economic viability (L. White, 1979, personal comm.). The modeling goes so far as to calculate the costs of alternative pipeline systems that would be necessary to produce from this region. A similar but less complex, model has been proposed for the Atlantic OCS, to allow the Department of Energy to experiment with alternative leasing programs without actually implementing these alternatives.

Another alternative assessment procedure has been suggested (Harbaugh, Doveton, and Davis, 1978), and has been implemented by several oil companies in Latin America. This scheme recognizes several troublesome but inescapable facts of regional assessment. First, a guess is just a guess, and terming it a "subjective Bayesian estimate" will not improve its accuracy or likelihood of correctness. Secondly, no prospect is going to be drilled, at least in offshore and frontier regions, unless *something* shows up on the seismic profiles over the prospect location. Therefore, the assessment of a region reduces to the sum of the assessments of the perceived seismic prospects. These seismic prospects have a limited number of attributes, most of which are geometric. They include perceived size, height of closure, area of drainage, and the like.

Thirdly, exploration is a game of reasoning by analogy, either with better explored parts of the same region, or with other regions believed to be similar. Therefore, the suggested regional assessment model (Davis and Harbaugh, 1980) relies upon simple statistical relations between the perceived characteristics of seismic prospects in analogue areas and the results of the subsequent drilling of these seismic prospects. The statistical relationships then are used to predict the outcome of drilling that may be done in the exploration area, using the attributes of seismic prospects in that area (Fig. 6). As exploration proceeds, Bayesian conditional relationships may be used to modify the predictions in light of the experiences gained.

This simple predictive model is not good, but it seems to perform at least as well as alternative evaluation procedures, and its simplicity suggests that it may be inherently better, or at least easier to use. Its greatest drawback is that it requires a modicum of study in areas that have been explored already, in order to establish the initial statistical relationships. For some reason, geologists seem reluctant to rake through the cold ashes of their past adventures. Perhaps it is because they fear that the lessons which they may learn will prove instructive, but in too painful a manner.

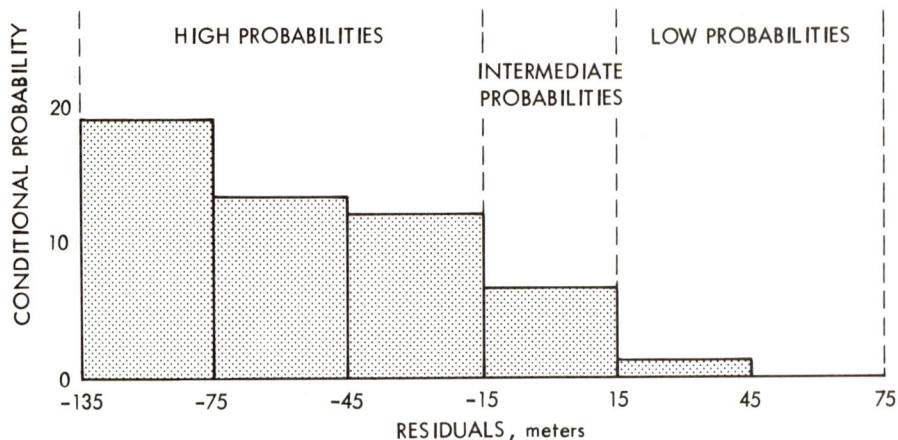

Figure 6. Conditional probability distribution showing likelihood of success in drilling prospects which are defined by seismic residuals in Calafate Basin of Chile. Residuals are computed from seismic interpretations of top of Tobifera Series (Jurassic) which underlies producing interval (internal report, Empresa Nacional del Petroleo, by R.M. Erazo and E.H. Vieytes, 1979).

SUMMARY

There are a multitude of other computer applications which have been made to petroleum exploration, but most have not proved themselves through widespread use. It is significant that the panel of experts nominated procedures which, for the most part, are extremely basic: data banking, log analysis, and contour mapping. This suggests that the use of computers in exploration has advanced in the past decade, not because of esoteric mathematics, but through the determined refinement of basic techniques. Explorationists now realize that there is no computational panacea; rather, the computer is a powerful and useful tool for an exceedingly difficult tasks.

REFERENCES

ACI, 1971, Contouring with faulting: Applications Consultants, Inc., Houston, Texas, 36 p.

Akers, H., Jr., 1977, Monte Carlo range-of-values program description: Internal Memorandum [May 1977], Conservation Division, U.S. Geol. Survey, 73 p.

Becquey, M., Lavergne, M., and Willm, C., 1979, Acoustic impedance logs computed from seismic traces: Geophysics, v. 44, no. 9, p. 1485-1501.

Berlanga, J.M., 1979, Oil exploration outcome probabilities in the Tabasco Basin, Mexico, as estimated by use of seismic information: unpubl. doctoral dissertation, Stanford Univ., 120 p.

Bornemann, E., 1979, Well-log analysis as a tool for lithofacies determination in the Viola Limestone (Ordovician) of south-central Kansas: unpubl. doctoral dissertation, Syracuse Univ., 151 p.

Brassel, K.E., and Reif, D., 1979, A procedure to generate Thiessen polygons: Geographical Analysis, v. 11, no. 3, p. 289-303.

Davis, J.C., 1975, Contouring algorithms, *in* Aangeenbrug, R.T., ed., Auto-Carto II, Proc. Intern. Symp. on Computer-Assisted Cartography: U.S. Dept. Commerce, Bureau of the Census, Washington, D.C., p. 352-359.

Davis, J.C., and Harbaugh, J.W., 1980, Comment on oil and gas in offshore tracts: Estimates before and after drilling: Science, v. 209, no. 4460, p. 1047-1048.

Dobrin, M.B., 1980, Computer processing of seismic reflections in petroleum exploration, *in* Merriam, D.F., ed., Computer applications in the earth sciences, an update of the 70's: Plenum Press, New York, this volume.

Doveton, J.H., and Cable, H.W., 1979, Fast matrix methods for the lithologic interpretation of geophysical logs, *in* Gill, D. and Merriam, D.F., eds., Geomathematical and petrophysical studies in sedimentology: Pergamon Press, Oxford, p. 101-116.

Drew, L.J., 1974, Estimation of petroleum exploration success and the effects of resource exhaustion via a simulation model: U.S. Geol. Survey Bull. 1328, 25 p.

Harbaugh, J.W., Doveton, J.H., and Davis, J.C., 1977, Probability methods in oil exploration: John Wiley & Sons, New York, 269 p.

Kaufman, G.M., Balcer, Y., and Kruyt, D., 1975, A probabilistic model of oil and gas discovery, *in* Haun, J.D., ed., Methods of estimating the volume of undiscovered oil and gas resources: Am. Assoc. Petroleum Geologists Studies in Geology No. 1, p. 113-142.

McCullagh, M.J., 1979, Creation and application of variable density grids to oil exploration data (abst.): Am. Assoc. Petroleum Geologists Bull., v. 63, no. 3, p. 494.

McCullagh, M.J., and Ross, C.G., 1980, The Delaunay triangulation of a random data set: in preparation.

Newendorp, P.D., 1975, Decision analysis for petroleum exploration: Petroleum Publ. Co., Tulsa, Oklahoma, 668 p.

SCA, Inc., 1975, Mapping-contouring system: Scientific Computer Applications, Inc., Tulsa, Oklahoma, 39 p.

Smith, M.B., and Souder, W.W., 1975, Minicomputers for maxi analysis: Soc. Prof. Well Log Analysts, 16th Ann. Logging Symp., R1-13.

Tracy, P.A., 1978, Petroleum Data System--A network of energy information: Proc. 12th Ann. Mtg., Geoscience Information Society, Seattle, Washington, v. 8, p. 25-30.

Uman, M.F., James, W.R., and Tomlinson, H.R., 1979, Oil and gas in offshore tracts: Estimates before and after drilling: Science, v. 205, no. 4405, p. 489-491.

USE OF COMPUTERS IN SEISMIC REFLECTION PROSPECTING

Milton B. Dobrin

University of Houston

ABSTRACT

During the past decade the digital computer has brought about revolutionary improvements in our capability to study geology by the seismic reflection method. New recording and processing technology has brought us much closer than anyone would have predicted ten years ago to our ultimate goal of extracting the same geological information that would be retrievable if there were a borehole to the maximum depth of interest at every shot point.

These improvements in the seismic art have made it possible to map structural features with more accuracy and to present them more correctly; they also have enhanced our ability to identify lithology and deduce strattigraphic relationships from reflection data.

Perhaps the most spectacular development of the past decade in seismic exploration has been the capability under proper conditions of detecting gas deposits on seismic record sections, making use of the fact that reflections from the top of gas-filled sands have a higher amplitude than those from water- or oil-filled sands. Digital recording and processing makes it possible to preserve relative reflection amplitudes on seismic records. This could not be done with analog techniques because of their limited dynamic range.

Geologic structures can be mapped with more precision by automatic techniques for transferring reflections from their apparent positions based on reflection times alone to their true positions in space. Wave-equation and frequency-domain methods are employed for such migration. Three-dimensional recording and processing procedures require computers with high storage capacity but they greatly increase the accuracy with which complex structures, particularly in areas of tectonic disturbances, can be mapped.

New capabilities made possible by digital processing now allow the geophysicist to derive lithological and stratigraphic information of a type that was not obtainable previously from seismic reflection records. Filtering programs are available for extracting from complex source signals simple symmetrical wavelets that enhance resolution of reflections and allow better discrimination of stratigraphic relations. True amplitude registration facilitates determination of reflectivities at lithologic boundaries that permits construction of synthetic velocity logs. Presentation of such logs for all shot points in record-section form allows mapping of velocity, which with proper well ties can be converted to a lithologic cross-section comparable to that which could be obtained from closely spaced boreholes. Such presentations give valuable information on stratigraphy in areas where conditions are favorable. Computer determination of parameters such as instantaneous frequency and instantaneous phase, carried out by complex analysis of waveforms, may provide information on lithology and hydrocarbon content not obtainable from conventional processing.

Important progress has been made in computer modeling of seismic data. This involves ray-path modeling for precise structure interpretation and stratigraphic modeling to relate waveforms and amplitudes to subsurface layering. With the introduction of three-dimensional recording, modeling techniques have been determined particularly useful in the interpretation of the data thus obtained.

INTRODUCTION

At the time of the Lawrence, Kansas conference in 1969, where I last spoke on this subject, the digital computer had been in use for commercial recording and processing of seismic reflection data for less than six years but it already had made a substantial impact on the art of seismic prospecting for oil and gas. Digital filtering had led to significant improvements in data quality of suppression of undesired signals. Programs for determining seismic velocities from reflection records had made it possible to obtain such information with greater accuracy than ever before. In spite of its many capabilities for processing seismic data, however, the computer did not extract any information from seismic signals that was fundamentally different in nature from that which had been obtainable with the analog systems in use at the time digital technology was introduced.

The primary objective of seismic surveys at that time was, as it had been since the earliest days of the art, to map the *geometry* of subsurface boundaries with the object of locating structural entrapments of hydrocarbons. To accomplish this, it was only necessary to observe times of correlative reflection events and to convert these times into depths that could be mapped. Wave forms and amplitudes were only of secondary interest, as they were seldom relevant to the mapping of geologic structures.

Today, thanks to improved digital techniques, it is possible to map structures with considerably greater precision than in 1969. But the most spectacular developments during the past decade have extended the capabilities of seismic reflection beyond the mapping of subsurface geometry into stratigraphic analysis, identification of lithology, and the detection of certain types of hydrocarbons. Many potentially productive stratigraphic features such as pinchouts, truncations, sand bodies and facies changes can now be resolved, when conditions are appropriate, because of better waveform definition and improved suppression of noise, both brought about by computer capabilities introduced over the past decade.

The better structural, stratigraphic, and lithological information which modern digital technology allows us to extract from seismic records makes it seem that we have come much closer to the ultimate goal of seismic prospecting than anyone would have predicted ten years ago. This goal is to obtain geological information equivalent to that retrievable from a series of conceptual boreholes extending to the maximum depth of interest at each shot point.

IMPROVEMENTS IN STRUCTURAL MAPPING

Migration

Let us first consider the improvements that the computer has brought into structural mapping. To obtain a true picture of the subsurface from reflection data, we must correct for distortions in our records that are inherent in the raypath geometry. Conventional record sections show all reflections as originating vertically below the position of the shot-receiver pair where they are generated and recorded, even though the paths of reflections from dipping boundaries, being perpendicular to them, cannot be vertical and the reflection points should be displaced laterally from their plotted positions. Figure 1 illustrates the difference between the *apparent* position of a reflector on a seismic time section and its true position in space. The error thus introduced is usually not significant when dips are gentle, but in areas that are disturbed highly structurally, particularly here there are sharp synclines, the distortions can be so great that the observed geometry on the section will be meaningless structurally, or even misleading. The process of transferring reflections from their apparent position on a record section to a section which shows the events in their true positions is termed *migration*.

The deviation between actual structure and apparent structure may be associated with dipping reflectors in the vertical plane of the shooting line or else with reflecting surfaces not in this plane at all. Reflections from such out-of-plane sources cannot be distinguished from those with vertical ray paths and thus can give rise to erroneous structural interpretations.

During the past decade the computer has made two developments possible for overcoming these limitations. One is automatic migration, the plotting of record sections on which reflection events are represented on each trace at times corresponding to their proper positions in space. The other innovation, three-dimensional recording and migration, usually involves special recording techniques in the field as well as new types of presentation.

Let us first consider the techniques for automatic migration of reflection data shot along a linear profile. In Fresnel, or diffraction, migration is it assumed that each reflecting point along a boundary is the source of a diffraction for which the time of arrival at a surface

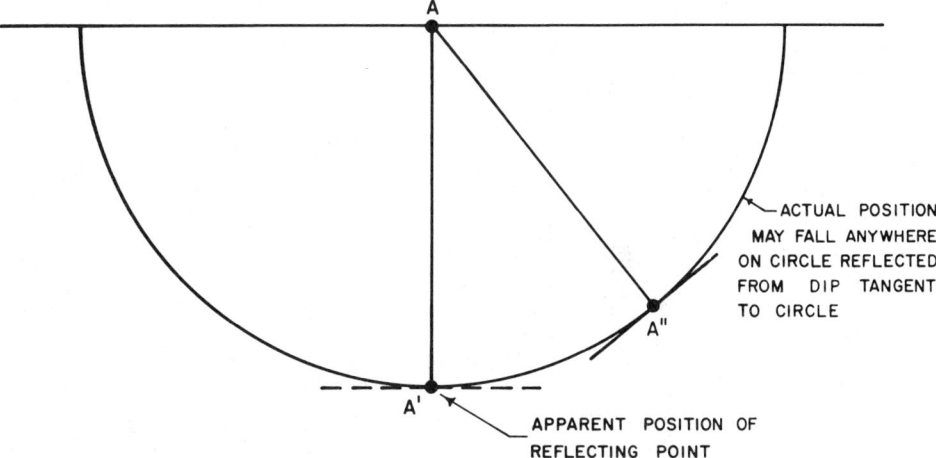

Figure 1. Unmigrated and migrated positions of single reflecting point with source and receiver at coincident position on surface (Lindseth, 1974).

position is related to the horizontal separation of that position and the source point by a hyperbolic curve of the type illustrated in Figure 2. All recorded events which lie along this curve are transferred by computer summation to the vertex of the hyperbola, which is the true position of the source of diffractions on the time section.

If a diffracted event is stored actually at sampled positions which lie along the hyperbola, the summation, by adding values in phase, will yield a high amplitude which is put into storage at the position of the vertex on the record section constituting the output. If no event lies along the line the summation yields an amplitude which will be at background level. If a dipping reflection is stored, each sampled point along it could be considered the source of a new hyperbola. By transfer through summation of each of these hyperbolas to the position of its vertex, the output channels, representing the results of the summation, will show the true position of the reflection along the section.

In the example in Figure 2, amplitudes sampled along the hyperbola on all channels from 1 to 99 are added to give a sum which is plotted at time 1.200 sec on channel 50. The same type of compositing is carried out for all

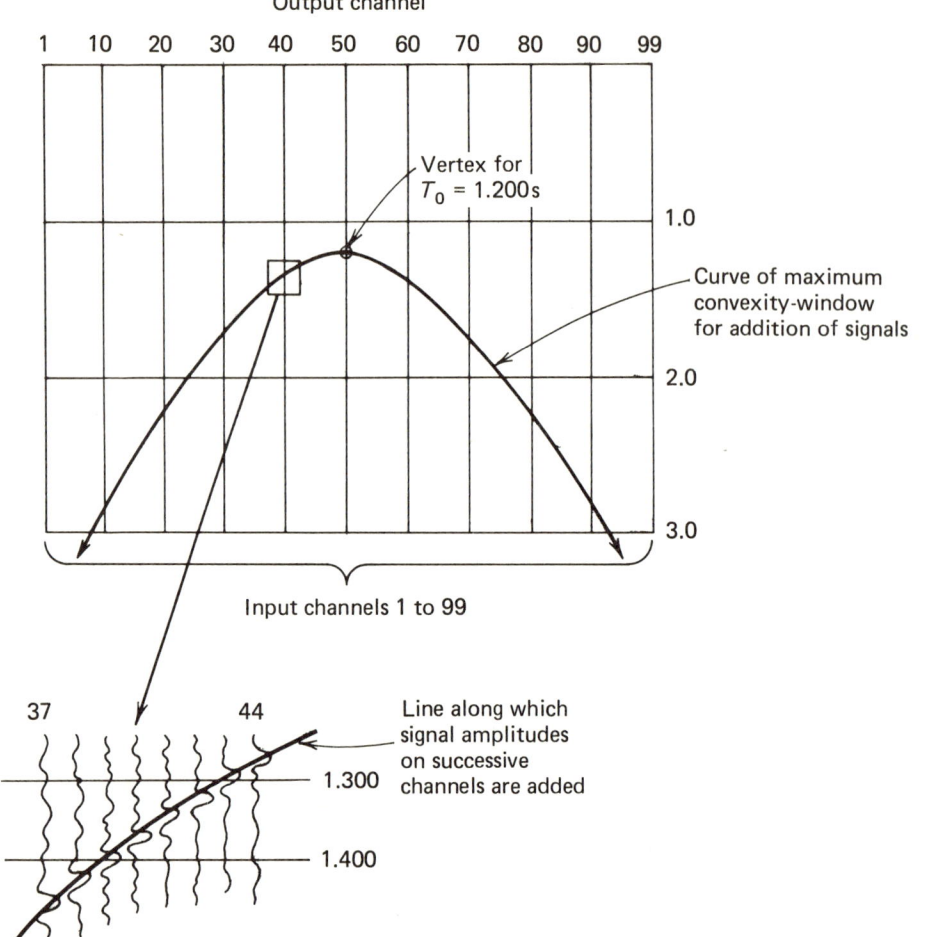

Figure 2. Principle of computer migration based on Fresnel diffraction. Signals on each trace are added at times determined by intersection with trace of hyperbolic curve. If source is at vertex, sum will have observable strength when plotted at that position on output trace. If not, signal amplitudes cancel and no event appears (Dobrin, 1976).

other sample times (every .002 sec) on channel 50 and then for all samples on channel 51, for which the window now extends from channel 2 to channel 100. This process is repeated one output trace at a time until all channels are swept. The record section thus plotted is the migrated section.

Figure 3 illustrates the results of Fresnel migration. It compares the conventional and the migrated sections for synclinal structure. The "bow-ties" are observed when the curvature of each synclinal reflection is greater than that of the wave-front at the same depth, leading to a crossover of the rays on their way to the surface.

A more recently introduced technique of automatic migration involves the introduction of the observed data into the wave equation using finite-differences instead of derivatives (Claerbout, 1976). Conceptually, the technique follows the reflected waves observed at the surface downward into the earth for a time equal to one-half the two-way reflection time. At this instant the wave should have reached the reflecting surface and its position should be equivalent to that of the reflector.

Figure 4 illustrates the principle. Assume a geologic section with three conformable synclinal surfaces at different depths, each generating a reflection. With source and receiver pairs evenly spaced along a profile at the surface, the resultant seismic record shows that the shape of the shallowest synclinal pattern on the time-horizontal-distance display is similar to that of the synclinal surface in space but that the resemblance gets poorer as the depth of the structure increases. The "bow tie" pattern previously noted becoming more pronounced as the syncline gets deeper.

We could reproduce the shape of the depth section on the time section if we conceptually or mathematically determined what the geophone would record if it and the shot were lowered a distance into the earth corresponding to the wave travel for half the reflection time. Figure 5 shows how the time section as recorded at the surface is transformed to true structure if brought down to the slanting surface. The events displayed on this curved surface when projected upward to the x-t plane constitute the migrated time section.

Figure 6 shows how a complex section can be made more meaningful geologically if it is transformed by wave-equation migration. The "bow ties" become synclines as the diffraction patterns are collapsed. Irregularities remain, due to sampling errors and lack of complete information on velocities, but there is no question that the migration gives a more realistic representation of true structure.

Figure 3. Results of computer migration using diffraction method. Unmigrated section (above) shows "bow ties" at greater depths. Migrated section (below) shows true geometry of synclinal structure (Prakla-Seismos).

SEISMIC REFLECTION PROSPECTING 153

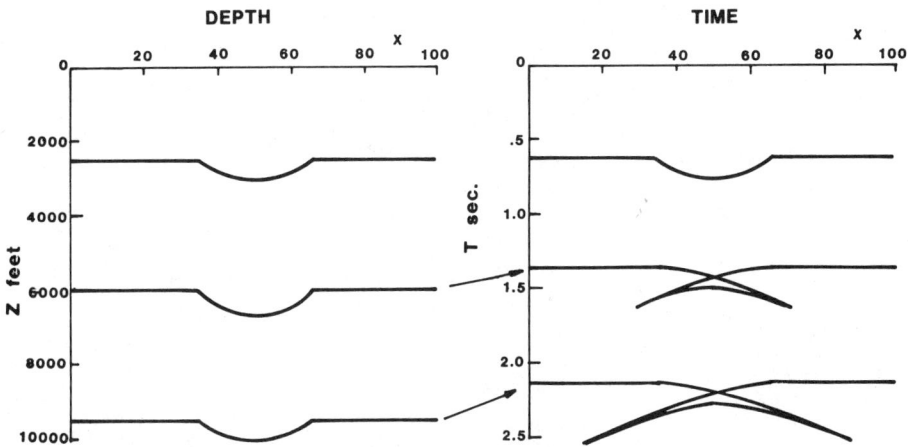

Figure 4. Synclinal structure shown at these levels on depth section is observed with increasing distortion on time section (unmigrated record) as depth gets greater.

A recent variation upon wave-equation migration converts the pattern on the record section from the space domain into the spatial-frequency domain by Fourier transformation, carries out the migration in the frequency domain, and then transforms the results back into spatial coordinates (Stolt, 1978). One advantage of this technique is that it allows accurate migration of steeper dips than conventional wave-equation techniques can handle.

Three-Dimensional Presentation

Conventional seismic surveys involve shooting and recording along lines and each record section thus produced shows structures which are presumed to be in the vertical plane of the recording line. If there are substantial dips in directions that cross the line of the profile it would be difficult to present the true structure on conventional vertical sections. A more accurate recording system for structurally disturbed areas involves mapping on regular two-dimensional grids instead of along linear profiles. A simple operational technique for producing such a grid is to shoot at closely spaced intervals along one line and to receive at comparable intervals on a line perpendicular to the shooting profile. This arrangement gives a uniform grid of reflection points which makes a rectangle with sides

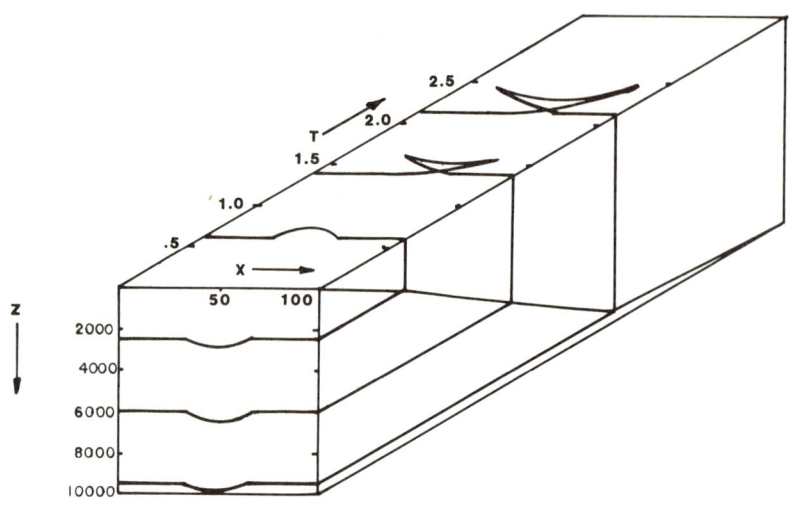

Figure 5. Principle of wave-equation migration. Wave field recorded in time is continued downward into earth until position corresponding to one-half of reflection time is reached. Along sloping surface having trace on $t_1 x$ plane indicated by inclined line, time structure has same shape as structure in depth. Projection of this configuration yields migrated time section.

one-half as long as the respective shooting and receiving lines. Another approach is to shoot a large number of parallel conventional profiles more closely spaced than in regular mapping. The latter technique is the only practical one for offshore three-dimensional surveys.

In order to realize the full potential of three-dimensional recording, it is necessary to migrate the reflections observed in the field to their true positions in space. This involves the same principles as two-dimensional migration, but it is considerably more tedious and requires a computer having a large storage capacity.

All reflection points are put into computer storage at their correct positions in the x, y plane and their amplitudes for a given reflection time are presented as variable-density or variable-area patterns in the horizontal plane corresponding to that time. Comparing the

SEISMIC REFLECTION PROSPECTING 155

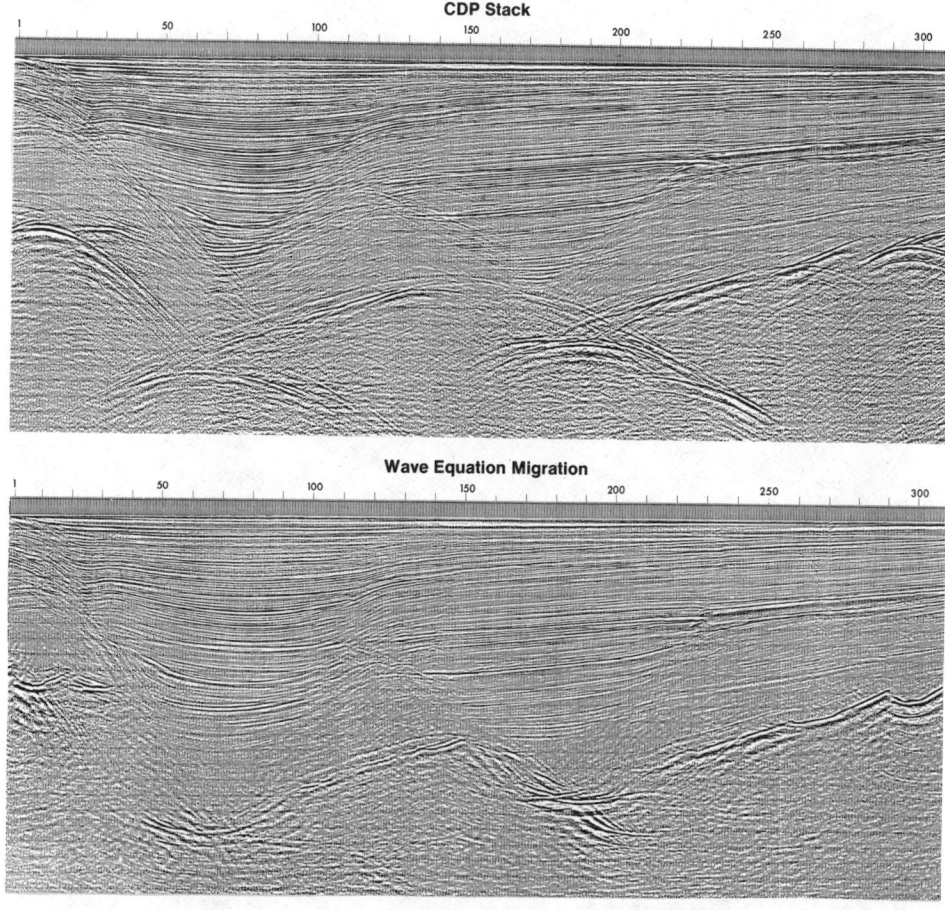

Figure 6. Example of wave-equation migration. Unmigrated section (above) shows diffraction patterns and "bow-ties" that are collapsed in migrated section (below) (Seiscom-Delta).

patterns on these horizontal "slices" for times that are close together, it is possible to follow faults and other structural features downward into the earth. Figure 7 shows a typical "slice" presentation (both unmarked and interpreted) for a three-dimensional survey over the Gulf of Thailand (Dahm and Graebner, 1979). It shows the reflection pattern over the horizontal plane corresponding to a reflection time (two-way) of 1.384 sec. The fault pattern can be followed upward and downward by comparison with other slices for nearby depths.

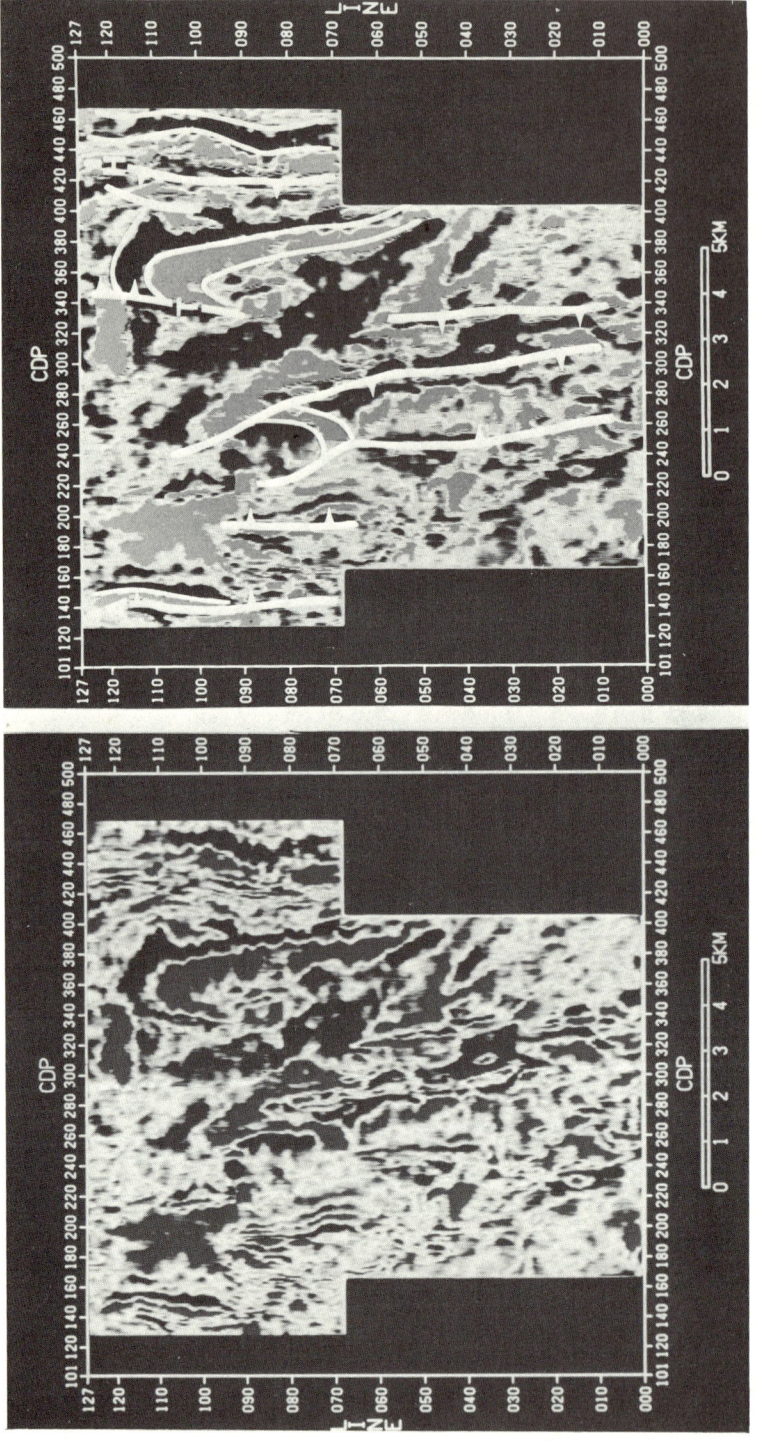

Figure 7. Presentation of three-dimensional migrated data. Variable density image is "slice" for time of 1.384 sec plotted from seismic coverage in Gulf of Thailand. Slice on left is uninterpreted. Faults and folds are superimposed on right (Dahm and Graebner, 1979).

In profile shooting over three-dimensional structures, the interpretation may be facilitated by modeling, either physical or mathematical. Figure 8 shows a series of sections obtained by mathematical modeling along six lines over a hypothetized structure that could be a reef or similar build-up. Note how indications of the feature can be observed on lines laterally displaced from it by substantial distances. In such modeling, the seismic data obtained over a hypothesized subsurface configuration is compared with that observed in the actual field measurements. The structure of the model is changed progressively until the best possible agreement is obtained between the two. The positions of the receiving lines must have the same geometrical relation to the subsurface feature in the two examples.

USE OF SEISMIC REFLECTION FOR LITHOLOGIC AND STRATIGRAPHIC STUDIES

When only analog techniques were available for recording reflection signals, the dynamic range of the ground-motion velocities to be recorded was so much greater than that of the recording systems that it was not possible to reproduce true relative amplitudes, and a great amount of compression was necessary to register all events without distortions due to saturation. Digital recording can accommodate such large dynamic ranges that it is possible to register the true relative amplitudes of most reflection signals. This new capability has turned out to be valuable in ways that were not visualized at the time that digital recording first appeared on the scene. It has made it possible to measure *reflectivities* of subsurface geological boundaries, a type of information that has led to the development of practical techniques for identifying lithology and also for detecting hydrocarbons.

Direct Detection of Hydrocarbons

The fact that the contrast in acoustic impedance (velocity times density) between an impervious shale and a gas-filled sand underlying it is greater than if the contact were with a water- or oil-filled sand should make the reflection from the top of the gas sand stronger than that from the top of a sand containing water or oil. This is the basis for the "bright spot," an anomalously strong reflection on a record section which is diagnostic of the presence of gas.

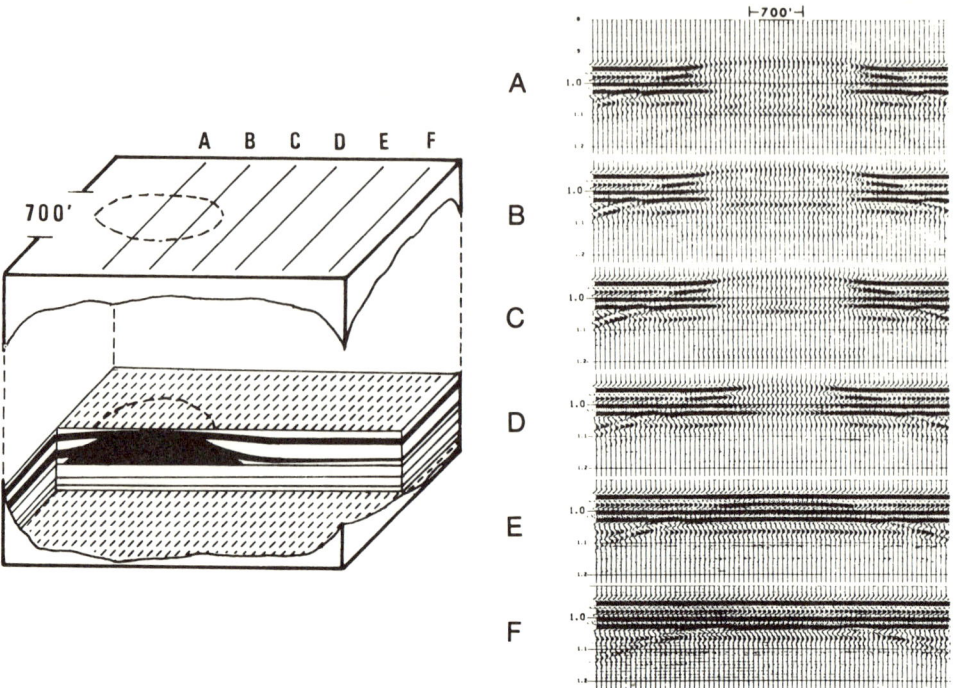

Figure 8. Seismograms obtained by computer modeling along lines crossing buried structure at various lateral distances. Note how indications of the subsurface features can be observed on lines displaced from it farther than its diameter.

It is evident from the reflectivities shown on the schematic cross-section in Figure 9 that the amplitude of a reflection from the top of a gas sand should be about two-and-a-half times that at the top of an oil sand and nearly three times that at the top of a water sand. The principle has been known for a long time and there are numerous papers in the Soviet literature, dating from the early 1960's, proposing that it be applied for detecting hydrocarbons directly.

In spite of the simplicity of this concept, it could not be put to use on a practical basis until computer programs could be developed that would separate out reflectivity from many other factors, such as distance from the source (spherical spreading) and attenuation. Figure 10 shows a conventionally processed section over a known gas

SEISMIC REFLECTION PROSPECTING

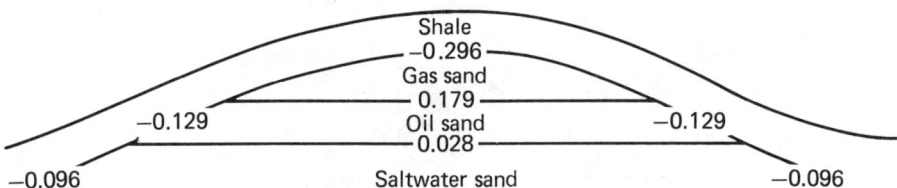

Figure 9. Reflection coefficients at top and bottom surfaces of gas sand and oil sand and at contact between shale and salt-water sand (Dobrin, 1976).

accumulation, whereas Figure 11 illustrates the section obtained when true amplitudes are extracted. The "bright spot", the high-amplitude reflection which shows the position of the gas sand, is obvious after the special processing but it hardly stands out at all on the conventionally processed section.

Wavelet Processing

In working with reflection amplitudes, it is desirable to show every reflection signal as a simple symmetrical pulse rather than as a multicycle train of peaks and troughs. "Wavelet processing" (Neidell and Poggiagliolmi, 1977) is an automatic filtering technique for converting the complex reflection pulse actually recorded at the surface to a symmetrical wavelet on which only a central peak and two lower amplitude troughs on opposite sides of it are observable. The added resolution this procedure allows is valuable particularly in mapping stratigraphic features.

Figure 12 illustrates some of the stages in the process of extracting a simple symmetrical wavelet from a broad, complex signal observed from a marine source. Phase and amplitude spectra are determined first. Then the phases are shifted to zero at all frequencies, yielding a symmetrical but multicycle wavelet. The final step is a filtering operation that will concentrate the energy in the center pulse, reducing amplitudes of side lobes to the point that they fall below noise level. The resultant signal has no frequencies outside the range encompassed by the input pulse. The amplitude of such an extracted wavelet should be proportional to the reflection coefficient at the boundary from which the reflection originates.

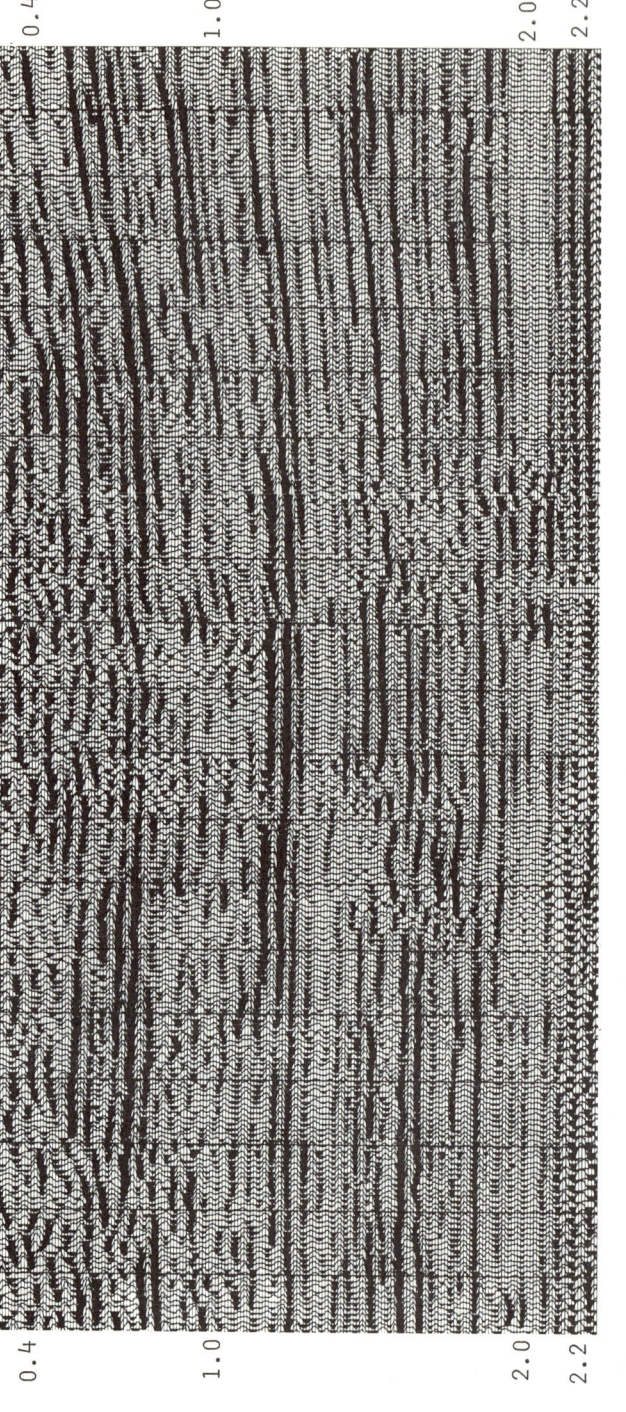

Figure 10. Conventionally processed section shot over gas accumulation. Amplitude equaliation makes it impossible to distinguish reflection originating at surface of gas sand at 1.4 sec from other reflections (Western Geophysical Co. of America).

SEISMIC REFLECTION PROSPECTING

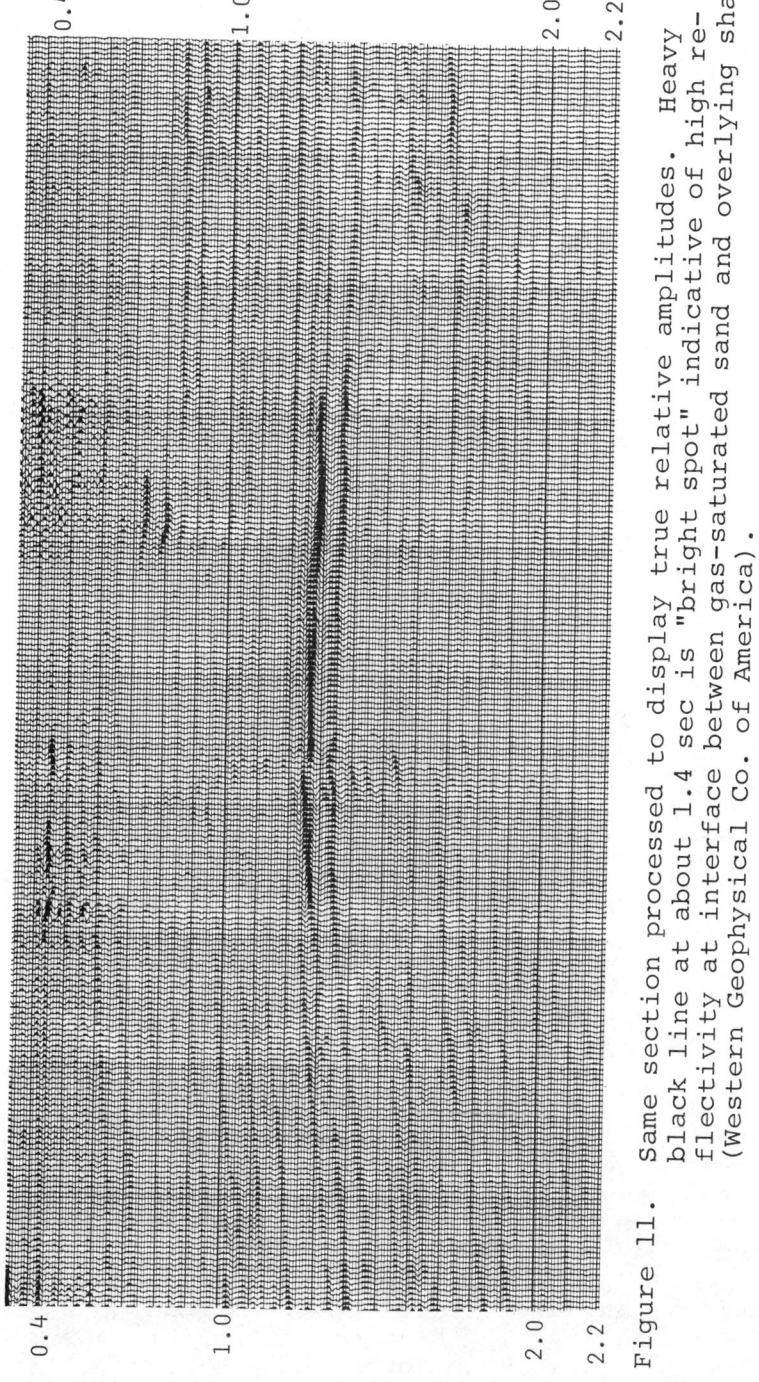

Figure 11. Same section processed to display true relative amplitudes. Heavy black line at about 1.4 sec is "bright spot" indicative of high reflectivity at interface between gas-saturated sand and overlying shale (Western Geophysical Co. of America).

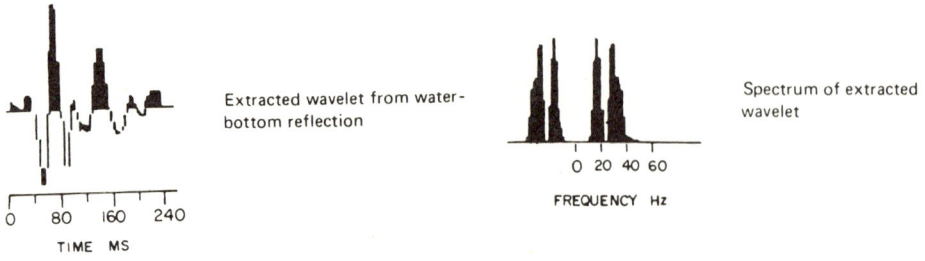

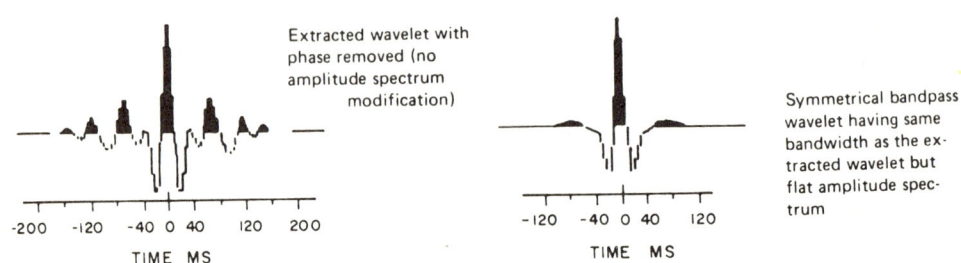

Figure 12. Conversion of complex wavelet from marine source into symmetrical wavelet by appropriate phase correction and filter operator (Neidell and Poggiagliomi, 1977).

Synthetic Sonic Logs

The capability of recording amplitudes of reflections thus processed and relating them to acoustic-impedance contrasts at interfaces makes it possible to determine velocity as a function of depth from reflection signals alone because the reflectivity (proportional to the amplitude observed) is equal to the difference of the acoustic impedances (velocity times density) across the interface divided by their sum. We can solve for the velocity below each reflector if we know the reflectivity and the velocity above it and assume an empirical velocity-density relation. The resulting information when plotted as velocity vs time or vs depth could be looked upon in the same light as a velocity log and may be referred to as an "inverse velocity log" or sometimes as a "pseudovelocity log."

An important difference between a real sonic log and a synthetic one lies in the fact that the frequency of the synthetic log will be lower because of the filtering

SEISMIC REFLECTION PROSPECTING

effect of the earth upon the reflection signals from which the latter is made (Lindseth, 1979). Figure 13 shows a comparison between a sonic log and a synthetic velocity log from a record shot over the logged well. The maximum frequency of the sonic log is obviously higher. The agreement is good but there are differences that show some of the limitations of the method.

If a series of adjacent seismic traces constituting a record section is converted to a corresponding series of synthetic velocity logs a new type of geologic section can be constructed which shows seismic velocity vs position along the profile and also vs time on the record (which is convertible to depth). The velocities are represented on the sections by a color code (e.g., the low velocities are blue, intermediate ones yellow, higher ones orange, and highest velocities red.) Such sections are available commercially under a number of trademarks such as "Seislog." Where well ties are available it is possible to associate the color bands corresponding to different ranges of velocity with types of lithology, such as sand or shale, or with porosity variations which may be correlative with velocity changes in limestones.

Figure 14 shows an example of a synthetic seismic log where a number of stratigraphic units are characterized on the basis of velocity. Each "wiggly line" is a plot of velocity vs depth. The color patterns here are replaced by different types of cross-hatching which represents various types of lithology. The advantage of having such information in stratigraphic studies hardly can be overstated. There are enough hazards to this approach, however, that one should have restricted confidence in any such lithologic projections of indicated velocities in the absence of well ties.

Complex Analysis as Source of Geologic Information

A new seismic technique which facilitates stratigraphic interpretation has been described recently by Taner, Koehler, and Sheriff (1979). It involves complex analysis of the seismic waveform with calculation of a quadrature (imaginary) trace for every real trace observed on a seismogram. The transformation is carried out by computing the Hilbert transform of the real signal.

By comparing the amplitudes of the real trace and imaginary trace at a given time, it is possible to determine reflection strength, phase, and frequency as functions of time on the record. These attributes are

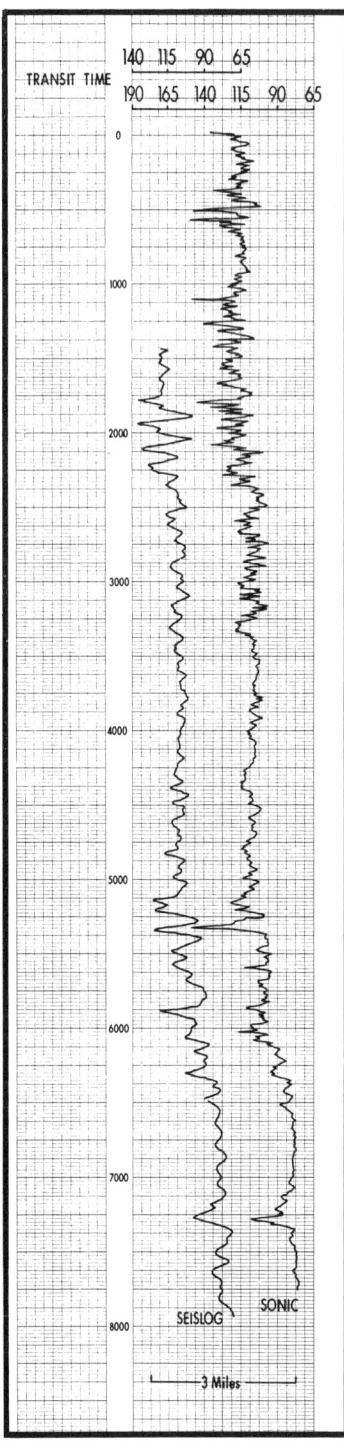

Figure 13. Comparison of sonic log and Seislog from reflection record shot over well in which log was made (Teknica, Ltd.).

SEISMIC REFLECTION PROSPECTING

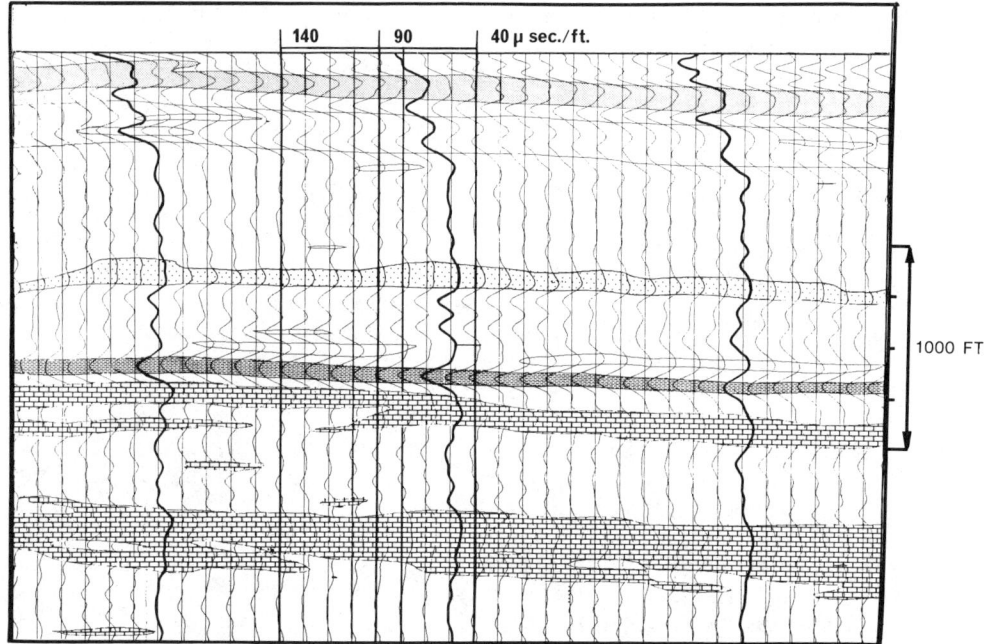

Figure 14. Typical Seislog made from record section. Lithology indicated by cross-hatching based on velocity indications from Seislog traces (Teknica, Ltd.).

represented by a color code and bands with various colors corresponding to various ranges of the attribute values are superimposed on the conventional variable-area section in black and white. The reflection strength gives information on reflectivity and might be used to locate hydrocarbons by the "bright-spot" approach. Instantaneous phase when superimposed on a record shows the degree of continuity of reflection. Record sections showing phase are good particularly for pointing up faults as well as depositional patterns. Hydrocarbon accumulations may be associated with anomalously low frequencies (although the reason is not understood) and color superposition based on frequency can show up deposits of oil or gas under favorable circumstances.

SEISMIC PROSPECTING IN THE 1980'S

Finally, let us try to project the developments in seismic reflection in the 1980's in geophysics for the past decade as far as we dare into the next decade.

What might we predict about the state of the geophysical art by the end of the 1980's? In making such predictions there is no better way to look into the future than to examine present trends and speculate where they may be leading us.

One new development that may have a great impact on future technology is the use of shear waves instead of compressional waves for reflection work. A comparison of shear and compressional reflections could provide valuable information not otherwise obtainable on lithology and fluid content. Another is the introduction of many more recording channels in the field than are used conventionally in current operations. This increase (up to 1024 channels in one system now operating commercially) allows more effective three-dimensional coverage and also enhances resolution by allowing correction of individual signals before compositing groups of them for noise cancellation. Many of the systems designed for recording a large number of channels at once use a smaller number of bits than are employed conventionally in recording. This allows transmission of more multiplexed signals on a single conductor in the field than would be possible with the full complement of bits. The most extreme level of such reduction is to confine each recorded sample to a single bit indicating the sing of the signal. Amplitude is built up by compositing of many identical signals, generally from Vibroseis sources.

A major limitation upon the number of channels that can be recorded at the same time without drastic reduction in the number of bits is the large capacity that would be needed to put all the information into storage. A 500-channel full-bit record of standard length would require the capability of storing 10 million bits per second. Such a recording rate is not practical even with 1600 bit-per-inch tape, although a 2-inch video tape of the type used in color television recording may allow fast enough storage to meet these requirements. Even then the processing of such a concentration of data is not feasible presently because no existing computer will accept data at the necessary rates.

The increase in recording channels will improve the accuracy of structural mapping and should increase resolution to some extent but the principal limitation on resolution is in the earth, where absorption of seismic signals increases exponentially both with frequency and with distance traveled. As resolution of reflection signals is limited by the frequency that can be transmitted to the reflector, we must expect it to decrease rapidly as reflector depth increases. Digital filtering can only optimize the resolving power of the waves that earth materials will pass but cannot improve it further. Thus, we will never be able to map all productive pinchouts and truncations at depths of many thousands of meters.

In spite of these real limitations imposed by nature, we can look forward to substantial progress over the next decade toward the ultimate goal of geophysics, which is to determine, as was said before, exactly what the drill would find at any specified location and depth. This goal will never be attained fully but sophisticated computer technology as developed for geophysical applications has brought us much closer to it than anyone would have dreamed ten years ago.

The rapid rate of technological progress in exploration geophysics for the past decade can be attributed to two factors. One is the rapid growth during this period in the capabilities of digital computers. The other is the incentive and support resulting from the world-wide energy shortage to improve geophysical exploration techniques in ways that can take advantage of these capabilities. Although alternative forms of energy will have to provide a long-term solution to our present problems, the only practical short-term alleviation (i.e., through the next few decades) will be in more effective exploration for the undiscovered oil and gas yet in the earth. It is fortunate that the progress in geophysical technology, much of it computer-based, which has characterized the past decade gives every indication of continuing through the next decade. The result should be an increased effectiveness in locating new hydrocarbon supplies so greatly needed to maintain the world's economy until new types of energy resources can be developed in sufficient quantities to meet the world's needs.

REFERENCES

Claerbout, J., 1976, Fundamentals of geophysical data processing: McGraw-Hill Book Co., New York, 274 p.

Dahm, C.G., and Graebner, R.J., 1979, Field development with three-dimensional seismic methods in the Gulf of Thailand - a case history: Offshore Technology Conference, 11th Ann. Proc., Houston, Texas, v. 4, p. 2591-2606.

Dobrin, M.B., 1976, Introduction to geophysical prospecting (3rd ed.): McGraw-Hill Book Co., New York, 630 p.

Lindseth, R.O., 1974, Recent advances in digital processing of geophysical data: Teknica, Ltd., Calgary.

Lindseth, R.O., 1979, Synthetic sonic logs - a process for stratigraphic interpretation: Geophysics, v. 44, no. 1, p. 3-26.

Neidell, N., and Poggiagliolmi, E., 1977, Stratigraphic modeling and interpretation - geophysical principles and techniques, *in* Seismic Stratigraphy, Applications to Hydrocarbon Exploration: Am. Assoc. Petroleum Geologists Mem. 26, Tulsa, Oklahoma, p. 389-416.

Stolt, R.H., 1978, Migration by Fourier transform: Geophysics, v. 43, no. 1, p. 23-45.

Taner, M.T., Koehler, F., and Sheriff, R.E., 1979, Complex seismic trace analysis: Geophysics, v. 44, no. 6, p. 1041-1063.

REGIONAL MINERAL- AND FUEL-RESOURCE FORECASTING - A MAJOR CHALLENGE AND OPPORTUNITY FOR MATHEMATICAL GEOLOGISTS

John W. Harbaugh

Stanford University

ABSTRACT

Mathematical geologists need to exert a stronger influence on government policy with regard to forecasts of mineral- and energy-resource commodities. The problems of forecasting inherently are statistical, and they should be based on geological, geophysical, and mineral-production data insofar as possible. Thus, the problems are both geological and statistical, or in other words, strong "geostatistical".

There are three main categories of problems in applying and using geostatistical methods in resource forecasts, namely (1) problems of policy, (2) problems of philosophy, and (3) problems of direction. The problems of policy, for example, are centered principally on the failure of governmental policy makers to incorporate appropriately statistical uncertainty in energy-resource policies. If effective, long-term forecasts of the world's energy and other mineral resources are to be made, geomathematicians must take a leading role. Three major tasks lie ahead of them, namely: (1) prepare comprehensive inventories of present resources that incorporate geological considerations and which, in turn, may be analyzed from a statistical standpoint; (b) develop formal optimization methods to guide exploration; and (c) promote policies that will make available publicly much of the exploration data that now are "frozen" because they are kept confidential by industry and government.

INTRODUCTION

My purpose in this paper is to emphasize that mathematical geologists need to make themselves heard with regard to pressing national and global problems that involve forecasting of mineral and fuel resources. Governmental planning should be facilitated greatly if effective procedures for forecasting petroleum, coal, uranium, and metallic resources were applied on a systematic regional, national, and world-wide basis. Unfortunately, such procedures are developed either inadequately or are not being applied to any significant extent.

Assessing a mineral or energy resource on a regional or national basis requires that geology be incorporated in the assessment processes. Furthermore, assessment processes are inherently "probabilistic", and demand use of the statistician's tools because realistic forecasts necessarily incorporate statistical uncertainty in the estimates. The skills needed are both geological and statistical, and thus are "geostatistical" in the broad sense and lie much within the mathematical geologists' domain. Now, let us turn to some major problems.

THREE MAIN CATEGORIES OF PROBLEMS

Let us begin by discussing three main categories of problems, namely *Problems of Policy*, *Problems of Philosophy*, and *Problems of Direction*. It seems appropriate to focus on these classes of problems before considering some of the technical details.

Problems of Policy. If resource forecasts are to be used properly for public policy making, it is essential that the policy makers understand the nature of the forecasts. Unfortunately, few policy makers in government seem to have much appreciation for the fact that any realistic resource forecast necessarily must incorporate a substantial degree of uncertainty. In fact, forecasts of any of the major energy mineral resources - oil, natural gas, coal, and uranium - are beset with enormous uncertainties. Yet the policy makers at the Federal level in the United States seem reluctant to allow for uncertainty in their resource policy. For example, the U.S. Congress seems to have assumed that by simply establishing a large exploration program (with public funds) that they could establish firmly the petroleum-resource potential of the National Petroleum

Reserve in Alaska, which spans a large area in northwestern Alaska. There seems to have been no appreciation by the Congress of the immense uncertainty surrounding the undertaking, and that at best there would be a high probability of not finding any commercial oil or gas accumulations in spite of an investment of many hundreds of millions of dollars. We witness similar misunderstanding concerning the large uncertainty that surrounds the United States' uranium resource potential. Thus, a major problem of policy concerns allowence for uncertainty in national decisions that are affected by resource estimates.

A second major policy issue concerns the planning of Federal energy-resource programs. For example, how much debate and professional attention have we mathematical geologists directed toward two of the U.S.'s major current energy-resource forecasting programs, namely the assessment of the petroleum potential in the nation's outer continental shelves (OCS's), and the National Uranium Resource Evaluation (NURE) program. These programs have involved a number of highly talented geologists, engineers, and other professionals, but it cannot be said there has been appropriate input from the fraternity of mathematical geologists. In hindsight, it is clear that we as mathematical geologists should be involved in planning the acquisition, coordination, and analysis of data.

Problems of Philosophy. My next main concern is to point out persistent problems in philosophy that affect geologists (and policy makers, too). First, the geological fraternity as a whole seems to adhere more or less stubbornly to "cause-and-effect" philosophy. By this I mean the belief that decrees that we may understand ultimately any geological phenomenon if we are given sufficient information. Cause-and-effect philosophy, for example, seems to dominate oil exploration, where it may be argued that if we knew enough about the origin, migration, and entrapment of petroleum, that we could predict with perfect assurance the location of undiscovered oil and gas pools. The truth of it is that even with our highly advanced technology, we are incapable yet of making forecasts that are more or less "assured". Witness the disappointing results in the Atlantic OCS and the Gulf of Alaska. Most oil explorationists will admit that exploration remains a risky, "chancy" business, but exploration strategies and plays seem to be guided by mechanistic cause-and-effect procedures which do not incorporate uncertainty in a formal manner.

A second major problem of philosophy is the reluctance to accept that there may be some irreducible level of uncertainty in a resource forecast, and that this uncertainty may be large. Again, the view seems to be that given enough information, uncertainty can be eliminated.

The third problem of philosophy is unwillingness to treat uncertainty from a quantitative analytical standpoint. The counter argument is that if uncertainty exists, we should take steps to measure and quantify it and appreciate the fact that the analytical treatment of uncertainty is the business of statisticians and geostatisticians.

A fourth problem of philosophy is the reluctance of policy makers and business decision makers to incorporate formally uncertainty in their decision-making procedures. Even if the uncertainty is fully acknowledged, the traditional "seat-of-the-pants" decision-making procedures generally fail to use formalized, rigorous methods that are available. Part of the problem is that procedures to quantify uncertainty (such as expressing an exploratory well outcome or a resource forecast as an objectively obtained probability distribution) have been slow in materializing.

Problems of Direction. There should be concern with problems of direction. I see three main problems in this context. First, there is widespread confusion between "economic geology" and the "geology of economic deposits". The distinction between the two is important. For example, most technical papers labeled "economic geology" do not address questions of economics at all. There is nothing wrong with study solely of the geology of economic deposits as long as the objective is understood clearly. In general, however, the goal of the study of the geology of economic deposits is an understanding of the origin and occurrence. By contrast, the goal of economic geology is drastically different because it is both economic and geologic. In a profit-seeking enterprise, the ultimate role of the economic geologist is to provide information which will facilitate decision making in the quest to maximize profits. Thus, we can generalize that the ultimate role of economic geology is to facilitate decision making.

A second problem, which is related closely, is the general failure to focus on the manner in which geology is employed in making economic, financial, public policy decisions. How many papers can be cited that analyze critically the manner in which geology was used in making economic or financial decisions? The answer is that there are surprisingly few, and there are almost no papers that dissect the decision-making process from a rigorous, analytical standpoint. On a more general level, there are papers that are concerned with the role of geology in its affect upon decisions, but the actual decision processes themselves are not presented. We can argue that there is a strong need for research at this interface between geology and formalized decision analysis, it being a no-man's land between two seemingly disparate disciplines, bridged by only a few researchers.

A third main problem of direction concerns the need for research on geological "sample design" from an optimization standpoint. For example, how should the optimum spacing of boreholes be established to block out a commercial coal seam? For example, we might apply techniques of universal Kriging to estimate the variance of the coal seam's thickness between boreholes, and then space the holes close enough so that the maximum acceptable variance is not exceeded. However, the problem actually is one of financial optimization because we must compare the financial value of the information obtained from each borehole, with the cost of the borehole. The cost of the borehole is easy to forecast, but estimating the financial value of the information in the form of the reduced variance is more difficult. The problem of an optimum borehole spacing arises daily in many exploration contexts, but it remains essentially unstudied in many applications.

SOME ISSUES INVOLVING FORECASTING

To return to the theme of resource forecasting, let us consider three more important issues. First, there is a major difference between a resource forecast (on a regional basis) and a forecast of the result of an exploration campaign. If we were to design a program to forecast the resources of a region, it should differ substantially from a program to explore the region commercially. We can argue that a program to obtain a resource forecast in a frontier region should concentrate on gathering information throughout the region that has to do with the occurrence of the commodity being sought, for example petroleum. By contrast, an exploration campaign should concentrate on finding petroleum that may be be exploited profitably. In a regional petroleum forecast, it is as important to estimate where oil does not occur as it is to estimate where it does occur. By contrast, in exploration the objective is to focus on those localities where oil may be profitably extracted and ignore the rest.

The second issue concerns the magnitude of the effort that should be directed toward resource forecasting on a national basis (in the U.S., for example). We can argue that major decisions are being made (for example, the U.S.'s 20-billion dollar synthetic fuels program) with grossly inadequate information about supplies of both conventional and nonconventional hydrocarbon resources. Would a 100 million dollar resource forecasting study be adequate to confirm or deny the need for a synthetic fuels program? We do not know because we have little experience in forecasting the results of large- vs. small-scale resource forecasting programs. The issue itself is pertinent to the discipline of economic geology, however, because it is basically an economic issue.

The third issue is whether resources can be forecast effectively without incorporating exploration methodology. One point of view is that the resource should be estimated without regard to the procedures used to explore for it. An alternative view (to which I subscribe), is that a resource estimate that does not consider exploration methodology (and economics, in turn) is of little use. For example, if we are to forecast the hydrocarbon resources of the Atlantic OCS in terms of its commerically producible oil and gas, we can be assured that present exploration practices will require that an exploratory well be drilled only where there is an attractive seismic prospect present. It can be argued that a useful resource forecast of the Atlantic OCS's hydrocarbon potential depends heavily on knowledge of seismic prospects, plus the results of exploratory drilling to date. Thus, a resource forecast for the Atlantic OCS might consist of the estimated sizes of the fields discovered to date, plus the sum of the probability distributions assigned to each seismic prospect presently discernible, plus a probability distribution for the "wild card" factor for the rest of the region. The summation process would require Monte-Carlo methods. Such a definition of a resource forecast is obviously a restricted one that may not satisfy everyone. The fact that there are differences of opinion emphasizes the need for satisfactory, alternative definitions of resource potential; however, any definition is necessarily both semantic and statistical.

REGIONAL MINERAL- AND FUEL-RESOURCE FORECASTING

RESOURCE INVENTORIES

Forecasts of undiscovered resources require that we know what we have discovered already. In other words, we need inventories of our existing resources. Surprisingly, the United States lacks an effective inventory of its mineral and fuel resources. I suggest that there are five categories of inventories that should be created, namely, one for the nation's oil and gas resources, a second for its nonmetallic resources, a third for its metallic resources, a fourth for its uranium resources, and a fifth for its coal resources.

What should be included in a resource inventory? An inventory should be more than a listing of deposits. Ideally it should contain information pertaining to size, grade, quality, critical geological characteristics, and cumulative production and remaining reserves. Unfortunately, few existing resource inventories contain this much information. Let us now consider the information that should be incorporated in an inventory of the oil and gas fields of a region. For each field the inventory should contain the following:

(1) The field's cumulative production
(2) Area size
(3) Producing horizons, with cumulative production allocated to each horizon
(4) Type or types of trap
(5) Year of discovery
(6) Depth ranges
(7) Number of producing wells
(8) Well spacing
(9) Oil and gas initially in place (expressed as a probability distribution)
(10) Reserves, expressed as a series of probability distributions for different economic and engineering scenarios
(11) Reservoir characteristics segregated by producing horizon, including thickness, porosity, permeability, and relative permeability

Such an inventory would be immensely useful. For example, it would enable us to determine the form of the frequency distributions of field sizes in the region, either for the region as a whole, or for specific subregions, or for specific periods of discovery, or for specific producing horizons, and so forth. Obtaining frequency distributions is essential in resource forecasting, because knowledge of the form of field size frequency distributions is fundamental in forecasting the remaining, undiscovered resources of the region. Thus, the current lack of appropriate resource inventories seriously hampers our efforts in making effective forecasts.

Several regional resource-inventory efforts have been completed recently. One involves both the metallic and the nonmetallic resources of the California Desert Conservation Area (CDCA) in southeastern California. This effort has been undertaken on behalf of the U.S. Bureau of Land Management, and involves a compilation of all reported mineral occurrences in the CDCA. It is a major undertaking because the CDCA embraces about one fifth the area of California, and lies within a region that has been explored extensively for more than a century.

SOME BROAD RESEARCH AREAS

There are some general, broad areas for research that should provide challenges for mathematical geologists. I shall speak of four areas which are all interrelated and to which I have alluded previously. The first concerns methods for estimating exploration outcome probabilities, as for example, forecasting the outcome of an exploratory well drilled in search of oil and gas. The forecast should be expressed as a probability distribution, which provides for a dryhole probability, plus a spectrum of oil and gas field sizes. The probability statement produced should be conditional upon the geological and geophysical data which form the rationale for drilling the prospect. Such a procedure for estimating well-outcome probabilities would have immediate relevance for exploration. They also would be relevant in resource forecasting because resource forecasts necessarily are based on information supplied mainly by exploration. Furthermore, if there is agreement that resource forecasts should incorporate exploration methodology, then a regional resource forecast may be regarded in part as the Monte-Carlo sum of probability distributions attached to individual prospects.

The second general research area involves application of formalized optimization methods in exploration, particularly in the design of borehole drilling programs. The third and fourth involve, respectively, the interfaces between exploration geology and formalized decision analysis, and the interfaces between regional resource appraisal and public policy making. As I stated earlier, these are no-man's lands which have received little attention, perhaps as a consequence of their seeming tenuous nature.

A CAVEAT: "FROZEN" DATA

Before proceeding, let me touch on a sobering fact that affects the whole issue of regional resource appraisal. Many of the data are "frozen" in that they are inaccessible because of government or industry policies. The reasons for their inaccessibility are familiar and include the proprietary nature of information which may have large financial value. For example, most seismic data obtained in oil exploration are highly sensitive and are accessible only to their owners. Even seismic data which have been submitted to the United States government in OCS operations virtually are inaccessible, even to Federal agencies other than the agency receiving the data. There are historical reasons for such security policies, but it would seem that some of the data, particularly in relatively thoroughly explored areas, could be made accessible for inventory analysis purposes.

We encounter similar problems in the use of ore-deposit data. For example, data pertaining to uranium ore reserves on a property-by-property basis is totally inaccessible, yet much of the data reside in Federal files and many of the ore deposits are on public lands.

What can be done about this? We could argue that companies that explore and produce on public lands ought to make their data eventually accessible to the public. This would involve a compromise between the need to withhold public inspection of the data during an interim period, with the long-term public good provided by eventual accessibility of the data. Many in industry will argue that eventual release of the data is of greatest value to industry itself. Witness the value to the oil industry of the voluntary public exchange of electric logs of virtually all oil wells drilled in the United States. I have heard only praise for this system from persons in industry. Therefore, we can ask if well logs are that different from seismic data? It would be appropriate for a committee that includes IAMG members to prepare a proposed policy statement that would provide standards for release of exploration and exploitation data that pertain to the public lands.

A SPECIFIC RESEARCH OPPORTUNITY

Before concluding, let me outline a somewhat more specific research opportunity. There is a strong need to explore the statistical relationships that presumably exist between seismically perceived structures and hydrocarbon occurrences. The OCS regions would provide good opportunities to examine these relationships, particularly in offshore Louisiana and Texas. Onshore regions, however, need to be analyzed similarly for comparison purposes. Consider the problems of assessing the petroleum potential of the Atlantic OCS. There is a large body of seismic information available for the Atlantic OCS, but most of the seismically detected structures are untested. If these untested structures are to be incorporated in an appraisal of the Atlantic OCS's hydrocarbon potential, they should be compared with similar structures that have been perceived seismically elsewhere and have been tested by drilling. Such a comparison would provide a structure-by-structure forecast of the Atlantic OCS's hydrocarbon potential under present exploration technology. Although a forecast would be influenced strongly by drilling results in the Atlantic OCS to date, it also would be highly uncertain because we are not sure where the best analogs to the Atlantic OCS are. The Louisiana and Texas OCS is suitable partly because it contains many structures that are similar to those of the Atlantic OCS. But the Atlantic OCS is lithologically more similar to the rocks of the northern part of the Gulf Coastal Plain (northern Louisiana and southern Arkansas) where the oil-producing rocks (Cretaceous and Jurassic) also are more nearly equivalent in age to those of the Atlantic OCS.

RECOMMENDATION

Mathematical geologists need to involve themselves to a greater degree in resource forecasting by providing guidance on a sustained basis. There are many interesting technical challenges and there also is a strong social need. It is a unique opportunity for mathematical geologists.

COMPUTERS IN OCEANOGRAPHY

William W. Hay

Rosenstiel School of Marine and Atmospheric
  Science, University of Miami

ABSTRACT

   Oceanography is a broad topic encompassing not only
marine geology and geophysics but also physical, chemical, and biological oceanography. Computers are used for
data storage, analysis, and modeling. Data storage for
this field is centralized in the National Oceanographic
Data Center which archives a vast amount of data in a
variety of forms. Many of the oceanographic institutions
also archive sets of data of particular interest at that
institution. The Deep Sea Drilling Project (DSDP) archives measurements and observations on the cores that
have been recoved by drilling, and this constitutes the
largest body of homogenous computer-accessible data in
geology. The DSDP computer files have been transferred
by duplication to Germany and to the USSR. Computer
analysis ranges from complex processing problems, such
as multichannel seismics and satellite images, to relatively simpler current dynamics, paleoecologic interpretations, lithologic analysis, and construction of age-
depth curves in marine stratigraphic sequences. Modeling
is most sophisticated in physical oceanography and atmospheric science, where the general circulation models require large computing facilities. Process modeling is
developing rapidly in marine chemistry. Ecological modeling is an active area in biology, with much interesting work being carried out on analog computers. A unique
development in oceanographic computing is the use of
satellites to transmit data from ships and buoys to

shore laboratories for processing and return by satellite of the processed data to the ships in time to permit modification of experiments in process as conditions change.

COMPUTER APPLICATIONS IN EXPLORATION AND MINING GEOLOGY:
TEN YEARS OF PROGRESS

G.S. Koch, Jr.

University of Georgia

ABSTRACT

Although many applications of computers to exploration and mining geology had been developed prior to 1969, their use and acceptance in mining was limited, even by the technologically advanced countries. Today, most mining companies and governmental agencies concerned with natural resources apply computers extensively.

During the decade, Matheron's French geostatistics has undergone extensive theoretical expansion on the one hand, and reduction to understandable form, through several textbooks, on the other hand. Classical statistics for mining geology has been refined further as the result of many detailed applications with much feedback from industrial practice.

The manipulation by computer of large data bases has made possible a closer relationship between geology and mine systems analysis, and between geology and mining.

Whereas in 1969 the contribution of statistics and computers to geological exploration for mineral deposits was in its infancy, today these methods are used widely, although the most extensive applications are in the mining of known ore deposits. Particularly influential have been (1) the development of models for drillhole exploration and (2) operations research methods to conceptualize and organize exploration effort.

A significant step has been the development of many computer programs, some relatively large and complex, for the computer analysis of data from the mineral industry. Some of these programs are proprietary, but many have been published in technical journals.

Exploration geochemistry and geophysics grew at an explosive rate. Today, many computer applications have been made for the organization, display, and interpretation of these data.

INTRODUCTION

When I wrote "Computer applications in mining geology" in 1969 for the previous symposium volume, the application and acceptance of computers and statistics was limited, even in the technologically advanced countries; today, most mining companies and government agencies concerned with natural resources apply computers extensively. I am sure that this trend will continue; as data increase dependence on computers necessarily follows. Computers will become easier to use, and the results of their use will be easier to understand.

A decade ago (Koch, 1969), I could attempt a summary of the literature. Today it is impossible to summarize or even mention all of the significant publications, nor is there a book that integrates the advances made; excellent papers in "Computer methods for the 80's in the mineral industry," (Weiss, 1979a) cover many phases. The time-consuming work of developing models and running computers has occupied most of us during the decade; we have been more engrossed with "doing our own thing" than with sharing and clarifying our goals and our progress.

Moreover, words have never been as well-matched to our field as are graphic displays and data. This explains much of the current emphasis on computer graphics, data-base development, and manipulation, and man/machine interaction. Many new devices, including touch sensing, terminals, and minicomputers are going to make communication easier, quicker, and more accurate.

In this article, I mention some published and unpublished articles with which I am familiar; most of those cited are recent but they in turn refer to earlier ones. Sixteen International Symposia on the Applications of Computers and Operations Research in the Mineral

Industries (APCOM) have been held since 1961; many computer applications discussed in this paper were first published in the proceedings of these symposia (listed in Table 1). Because most are out of print, they may not be easy to find.

EXPLORATION

Planning of Exploration

Computers are used widely by mining companies and to some extent by governmental agencies for exploration planning, but I am not aware of much published work with a geological component. Helpful are papers by Weiss (1979b, 1979c), Griffiths (1974), Gabelman (1976), Guarascio and Turchi (1976), Mackenzie (1972), Mackenzie and Bilodeau (1977), and Gaucher and Gagnon (1973).

Champigny, Sanders, and Sinclair (1980) provide an interesting example of structured property exploration evaluated by measures of "relative information gain." Much of the work done in the 1970's was proprietary; I hope that some of this will be published soon.

Resource Evaluation

Shortages of petroleum and other commodities have added to the new emphasis on resource evaluation. Because of the large amounts of data and the multivariate statistical analyses required, computers have been essential for these investigations.

The proceedings of two recent meetings of the International Geological Correlation Program Project 98, "Standards for Computer Applications in Resource Studies" (Cargill and Clark, 1977, 1978) are sources for recent developments. Also valuable is a book by Harris (1977), one of the pioneers in the subject.

Harris continued his work of the 1960's on classification and evaluation of gridded areas; he applied his methods to several parts of North America to establish generality (Harris, 1973). Later, he and colleagues (Harris and Brock, 1973) introduced concepts of subjective probability. Harris' current work stresses probabilistic interpretatins of inferences made by geologists applying scientific methodology consistently in regional evaluation. Thus, the discipline imposed by

Table 1. Summary of symposia on applications of computers and operations research in the mineral industries (APCOM). Sponsors: AIMM, Australasian Institution of Mining and Metallurgy; CIM, Canadian Institution of Mining and Metallurgy; CSM, Colorado School of Mines; EP, Ecole Polytechnique (Montreal); McG, McGill University; SAIMM, South African Institution of Mining and Metallurgy; SME, Society of Mining Engineers of the American Institute of Mining Engineers; PSU, the Pennsylvania State University; SU, Stanford University; TUC, Technische Universitat Clausthal; UA, University of Arizona. For each symposia, the proceedings volume was published by the first-listed sponsor.

No.	Year	Sponsor(s)	Location	Name of proceedings volume	Editor(s) of proceedings volume
1	1961	UA	Tucson, Arizona	Short course on computers and computer applications in the mineral industry	J.C. Dotson
2	1962	UA, SU	Tucson, Arizona	Computer short course and symposium on mathematical techniques and computer applications in mining and exploration	J.C. Dotson
3	1963	SU, UA	Stanford, California	Computers in the mineral industry	George A. Parks

No.	Year	Sponsor(s)	Location	Name of proceedings volume	Editor(s) of proceedings volume
4	1964	CSM, SU, PSU, SME, UA	Golden, Colorado	International symposium on application of statistics, operations research, and computers in the mineral industry	Sherman W. Spear
5	1965	UA, SU, CSM, PSU, SME	Tucson, Arizona	Short course and symposium on computers and computer applications in mining and exploration	J.C. Dotson, W.C. Peters
6	1966	PSU, SU, CSM, SME, UA	University Park, Pennsylvania	Proceedings of the symposium and short course on computers and operations research in mineral industries	Pamela L. Slingluff, Theresa A. Fike
7	1968	CSM, SU, PSU, SME, UA	Golden, Colorado	Seventh international symposium on operations research and computer applications in the mineral industries	Charles O. Frush
8	1969	SME, SU, CSM, PSU, UA	Salt Lake City, Utah	A decade of digital computing in the mineral industry -- a review of the state-of-the-art	Alfred Weiss

No.	Year	Sponsor(s)	Location	Name of proceedings volume	Editor(s) of proceedings volume
9	1970	CIM, McG, EP, SU, CSM, PSU, SME, UA	Montreal, Canada	Decision-making in the mineral industry	J.I. McGerrigle
10	1972	SAIMM, SU, CSM, PSU, SME, UA	Johannesburg, South Africa	Application of computer methods in the mineral industry: proceedings of the tenth international symposium	M.D.G. Salamon, F.H. Lancaster
11	1973	UA, SU, CSM, PSU, SME	Tucson, Arizona	Eleventh symposium on computer applications in the minerals industry	John R. Sturgul
12	1974	CSM, SU, PSU, SME, UA	Golden, Colorado	Twelfth symposium on the applications of computers and mathematics in the minerals industry	Thys B. Johnson, Donald W. Gentry
13	1975	TUC, SU, CSM, PSU, SME, UA	Clausthal, West Germany	Thirteenth international symposium on the application of computers and mathematics for decision making in the mineral industries	F.L. Wilke

No.	Year	Sponsor(s)	Location	Name of proceedings volume	Editor(s) of proceedings volume
14	1976	SME, PSU, SU, CSM, UA	University Park, Pennsylvania	Application of computer methods in the mineral industry: proceedings of the fourteenth symposium	R.V. Ramani
15	1977	AIMM, SU, CSM, PSU, SME, UA	Brisbane, Australia	Apcom 77: papers presented at the fifteenth international symposium on the application of computers and operations research in the mineral industries	Alban Lynch
16	1979	SME, UA	Tucson, Arizona	Sixteenth application of computers and operations research in the mineral industry	T.J. O'Neil

the inflexibility of the computer has helped to emphasize the importance of systematic geological observation; the computer has made possible a consistency of application of scientific methodology.

The Geological Survey of Canada also was active in resource evaluation. Agterberg and others (1971) made a geomathematical evaluation of copper and zinc potential in the Abitibi area of Ontario and Quebec. In this study, they established a data base for geological and geophysical parameters measured in small cells. Through multivariate statistics, they then compared these measurements to metal content for cells containing ore bodies and predicted metal endowment in other cells. They showed that meaningful data could be obtained from geological maps, that these data were incomplete, and that effective statistical methods could be devised. Later, they (Agterberg, 1975; Chung, 1978) refined their models and established their generality by applying them to other geographic areas and to large blocks of the earth's crust. Recently, Agterberg and Divi (1978) modeled the frequency distribution of concentration values for a chemical element in blocks of constant weight sampled at random from a segment of the earth's crust and developed statistical models to describe clustering of mineral deposits of a given type in a region.

Griffiths and his students, at the Pennsylvania State University, continued work on resource studies (e.g., Griffiths and Singer, 1973; Menzie, Labovitz, and Griffiths, 1976). Griffiths (1978) further developed his idea of unit regional value, a concept that has influenced greatly other investigators.

At the U.S. Geological Survey, in Reston, Virginia, the Office of Resource Analysis emphasized petroleum rather than mineral-resource geology. However, much of their work is directly relevant to mineral deposits, and their papers on mineral deposits as such have been influential. One approach, the discovery process model (Drew, Schuenemeyer, and Root, 1980; Barouch and Kaufman, 1977) estimates undiscovered petroleum resources in a partially explored region based upon characteristics of the discovery process; another, is a classification technique (Botbol and others, 1978) named characteristic analysis. A recent one is a study of rock geochemical data related to possible concealed porphyry copper mineralization (McCammon and others, 1979). All of these models have been successful for various geological situations; further work will define more clearly their generality.

Other models developed in the 1970's include abundance estimation (Celenk and others, 1978), volumetric estimation (Kingston and others, 1978), and subjective probability, using the Delphi method (Baxter and others, 1978; Miller and others, 1975). In explored areas, deposit modeling (Sinding-Larsen and Vokes, 1978) has been used successfully for resource estimation. Beauchamp and others (1979) applied discriminant analysis to identify favorable areas for uranium. Singer and Overshine (1979) assessed resources in Alaska, providing an example of suitable methodology for a poorly explored area.

Data bases for resources first were developed extensively during the 1970's; typically, geologists and engineers wrote the specifications for the bases, which were completed by computer and information scientists. Examples of large systems are those of the Geological Survey of Canada (Robinson, 1972; Eckstrand, 1977) and the U.S. Geological Survey and Bureau of Mines (1976); the systems allow these organizations and other geologists to retrieve easily and use valuable information.

Gill and others (1977) discussed the design of geological data systems for developing nations. Thompson (1975) described a computer-oriented system to portray minerals availability for policy planners in the U.S. government. Cargill and Clark's (1977) previously cited publication also contains papers on data bases.

Sampling Designs for Exploration

Most sampling designs for exploration are drilling plans. Progress on two-dimensional models (begun by Slicher in the 1950's and by Griffiths and his students in the 1960's) continued with Singer's work (1975); with Drew's (1979) procedures for determining the areal influence of drillholes; and with Shurygin's (1976) mathematical analyses.

Three-dimensional models are less developed than the two-dimensional ones and apply to a narrower range of physical situations. Koch, Link, and Schuenemeyer (1974) use simulation to locate exploration drillholes for ore bodies that can be represented as ellipses. Malmqvist and Malmqvist's (1979) model explores for three-dimensional folded targets.

Exploration Geochemistry

Exploration geochemistry developed rapidly in both scope and in volume of data, as costs of chemical analysis decreased and extensive regional surveys were undertaken. Rose, Hawkes, and Webb (1979, p. 520) point out that "for both data handling and a great deal of statistical interpretation of the results of large-scale geochemical surveys, computer-processing methods are virtually mandatory. Furthermore, the field is developing rapidly with the appearance of new statistical concepts and with computer services becoming progressively less expensive." McCammon (1974) provides another excellent summary chapter on the subject.

Several workers developed sampling designs and methods of statistical analysis implemented by computers. Excellent papers that will lead the reader to other references are by Miesch (1976) and by Garrett (1979). Sinclair (1976) developed probability graphs that have been used widely for determining thresholds. Govett and coworkers (Govett, 1972; Govett and others, 1975) were active in statistical analyses of data sets from Cyprus, New Brunswick, and elsewhere.

Regional geochemical surveys required a systematic collection of samples, analytical work, and presentation of results. Webb and others (1978) present work done in Great Britain; Howarth and others (1980) describe some of the techniques that were used for this atlas, and also for study of exploration data collected by the U.S. Department of Energy in the search for uranium.

MINING GEOLOGY

Geostatistics

Geostatistics provides a way to analyze variables that are distributed in space (or time); it was devised by G. Matheron, who published a book in French in 1962. Geostatistics could be considered an extension of the weighting schemes such as inverse distance (long used in ore-reserve calculations) and of trend-surface analysis (developed in the United States and South Africa in the 1950's and 1960's). Because Matheron's book required intensive study of a complicated system, it was difficult reading, even for those with a good knowledge of French.

In the 1970's the theory and practice of Geostatistics became accessible to English-speaking geologists and engineers through many short courses. Four books published in English were the following: "Geostatistical Ore Reserve Estimation," by Michel David (1977) explains the subject using a relatively small amount of mathematics; more mathematical, "Mining Statistics" by Andre G. Journel and C.J. Huijbregts (1978, is comprehensive and suitable for reference as well as a textbook; "An Introduction to the Geostatistical Methods of Mineral Evaluation," is a summary by J.M. Rendu (1978); "Practical Geostatistics" by Isobel Clark (1979) provides a short, relatively nonmathematical approach. The first three are by Matheron's students; Clark presents an independent view.

Many of the articles written on geostatistics during the last decade are thoroughly referenced in the extensive bibliographies in these books.

Other Developments in Ore-Reserve Estimation

During the 1970's, D.G. Krige and H.S. Sichel of South Africa continued the work they began in the 1950's on statistical and computer methods of ore reserve estimation. Krige summarized his results in a 1978 book entitled "Lognormal-de Wijsian geostatistics for ore evaluation." Sichel (1972) wrote on the statistical valuation of diamondiferous deposits, a difficult subject because of the extremely low grade of these deposits and their high variability. De Wijs, of the Netherlands, wrote (1972) on the method of successive difference applied to mine sampling.

In two papers, Parker and coauthors (Parker and Switzer, 1975; Parker, Journel, and Dixon, 1979) discuss the use of conditional probability distributions in ore-reserve estimation; the second is a case study for estimation of open-pit reserves in stratabound uranium deposits.

CONCLUSIONS

During the next decade, I anticipate continued rapid growth. Among the authors who have commented on the technological and human factors in this future development are Agterberg (1979), Krige (1977), and Merriam (1980). The role of Matheron's geostatistics and classical statistics will become more clearly defined. Many additional methods of data analysis, particularly those of the exploratory data-analysis school of Tukey (1977), will be employed widely and familiar to all practitioners. Enough information will be at hand from completed exploration studies in which statistics and computers played a part to allow appraisal of the models and their refinement. Planning for geochemical and geophysical surveys will be a necessity and methods of data interpretation will be better understood and more straightforward than at present. Methods for the routine development and handling of large data bases will be available and used, and they will be efficient. I believe that the application of computers in exploration and mining geology will be as beneficial in the years to come as it has been in the past.

REFERENCES

Many of the references in this list come from one of the sixteen symposia on the Application of Computer Methods in the Mineral Industry. For brevity, these references are identified in the following list by the number of the symposium and the acronym APCOM. Table 1 gives the bibliographic details necessary to identify fully the editors and publishers of these symposia volumes.

Agterberg, F.P., 1975, Statistical models for the regional occurrence of mineral deposits: 13th APCOM, p. CI1-CI15.

Agterberg, F.P., 1979, Statistics applied to facts and concepts in geoscience: Geol. en Mijnbouw, v. 58, no. 2, p. 201-208.

Agterberg, F.P., Chung, C.F., Fabbri, A.G., Kelly, A.M., and Springer, J.S., 1971, Geomathematical evaluation of copper and zinc potential of the Abitibi area, Ontario and Quebec: Geol. Survey of Canada Paper 71-41, 55 p.

Agterberg, F.P., and Divi, S.R., 1978, A statistical model for the distribution of copper, lead, and zinc in the Canadian Appalachian region: Econ. Geology, v. 73, no. 2, p. 230-245.

Barouch, E., and Kaufman, G.M., 1977, Estimation of undiscovered oil and gas in mathematical aspects of production and distribution of energy: Proceedings of Symposium in Applied Mathematics, v. 21, Am. Math. Soc., p. 77-81.

Baxter, G.G., Cargill, S.M., Chidester, A.H., Hart, P.E., Kaufman, G.M., and Urquidi-Barrau, F., 1978, Workshop on the Delphi method: Jour. Math. Geology, v. 10, no. 5, p. 581-588.

Beauchamp, J.J., Begovich, C.L., Kane, V.E., and Wolf, D.A., 1979, Application of discriminant analysis and generalized distance measures to uranium exploration: Union Carbide Technical Report K/UR-28, Oak Ridge, Tennessee, 58 p.

Bilodeau, M.L., and MacKenzie, B.W., 1977, The drilling investment decision in mineral exploration: 14th APCOM, p, 932-949.

Botbol, J.M., Sinding-Larsen, R., McCammon, R.B., and Gott, G.B., 1978, A regionalized multivariate approach to target selection in geochemical exploration: Econ. Geology, v. 73, no. 4, p. 534-546.

Cargill, S.M., and Clark, A.L., eds., 1977, Standards for computer applications in resource studies: Jour. Math. Geology, v. 9, no. 3, p. 205-337.

Cargill, S.M., and Clark, A.L., 1978, "Standards for computer applications in resource studies," Project 98: Jour. Math. Geology, v. 10, no. 5, p. 405-642.

Celenk, O., Clark, A.L., de Vletter, D.R., Garrett, R.G., and van Staaldvinen, C., 1978, Workshop on abundance estimation: Jour. Math. Geology, v. 10, no. 5, p. 473-480.

Champigny, N., Sanders, K.G., and Sinclair, A.J., 1980, Sepcogna gold deposit of Consolidated Cinola Mines - an example of structured property exploration: Western Miner, June, p. 35-44.

Chung, C.F., 1978, Computer program for the logistic model to estimate the probability of occurrence of discrete events: Geol. Survey of Canada Paper 78-11, 23 p.

Clark, I., 1979, Practical geostatistics: Applied Sci. Publ., Ltd., London, 129 p.

David, M., 1977, Geostatistical ore reserve estimation: Elsevier Sci. Publ. Co., Amsterdam, 364 p.

de Wijs, H.J., 1972, Method of successive differences applied to mine sampling: Trans. I.M.M., v. 81, p. A78-A81.

Drew, L.J., 1979, Pattern drilling exploration: optimum pattern types and hole spacings when searching for elliptical shaped targets: Jour. Math. Geology, v. 11, no. 2, p. 223-254.

Drew, L.J., Schuenemeyer, J.H., and Root, D.H., 1980, Resource appraisal and discovery rate forecasting in partially explored regions: Part A, an application to the Denver Basin: U.S. Geol. Survey Prof. Paper 1138, 13 p.

Eckstrand, O.R., 1977, Mineral resource appraisal and mineral deposits computer files in the Geological Survey of Canada: Jour. Math. Geology, v. 9, no. 3, p. 235-244.

Gabelman, J.W., 1976, Expectations from uranium exploration: Am. Assoc. Petroleum Geologists, v. 60, no. 11, p. 1993-2004.

Garrett, R.G., 1979, Sampling considerations for regional geochemical surveys, *in* Current research, part A, Geol. Survey of Canada Paper 79-1a, p. 197-205.

Gaucher, E., and Gagnon, D.C., 1973, Compilation and quantification of exploration data for computer studies in exploration strategy: Can. Inst. Min. Met. Bull., Sept., p. 113-117.

Gill, D., Beylin, J., Boehm, S., Frendel, Y., and Rosenthal, E., 1977, Design of geological data systems for developing nations: Jour. Math. Geology, v. 9, no. 2, p. 145-158.

Govett, G.J.S., 1972, Interpretation of a rock geochemical exploration survey in Cyprus - statistical and graphical techniques: Jour. Geochem Expl., v. 1, no. 1, p. 77-102.

Govett, G.J.S., Goodfellow, W.D., Chapman, R.P., and Chork, C.Y., 1975, Exploration geochemistry - distribution of elements and recognition of anomalies: Jour. Math. Geology, v. 7, no. 5/6, p. 415-446.

Griffiths, J.C., 1974, Quantification and the future of geosciences: Syracuse Univ. Geology Contr. 2, p. 51-66.

Griffiths, J.C., 1978, Mineral resource assessment using the unit regional value concept: Jour. Math. Geology, v. 10, no. 5, p. 441-472.

Griffiths, J.C., and Singer, D.A., 1973, Size, shape and arrangement of some uranium ore bodies: 11th APCOM, p. B82-B1112.

Guarascio, M., and Turchi, A., 1976, Exploration data management and evaluation techniques for uranium mining projects: 14th APCOM, p. 451-464.

Harris, D.P., 1973, A subjective probability appraisal of metal endowment of Northern Sonora, Mexico: Econ. Geology, v. 68, no. 2, p. 222-242.

Harris, D.P., 1977, Mineral endowment, resources, and potential supply: theory, methods for appraisal, and case studies: MINRESCO, TUSCON, Arizona, various pages.

Harris, D.P., and Brock, T.N., 1973, A conceptual bayesian geostatistical model for metal endowment: 11th APCOM, p. B113-B184.

Howarth, R.J., Koch, G.S., Jr., Chork, C.Y., Carpenter, R.H., and Schuenemeyer, J.H., 1980, Statistical map analysis techniques applied to regional distribution of uranium in stream sediment samples from the southeastern United States for the National Uranium Resources Evaluation Program: Jour. Math. Geology, v. 12, in press.

Journel, A.G., and Huijbregts, C., 1978, Mining geostatistics: Academic Press, New York, 600 p.

Kingston, G.A., David, M., Meyer, R.F., Ovenshine, A.T., Slamet, S., and Schanz, J.J., 1978, Workshop on volumetric estimation: Jour. Math. Geology, v. 10, no. 5, p. 495-500.

Koch, G.S., Jr., 1969, Computer applications in mining geology, *in* Merriam, D.F., ed., Computer applications in the earth sciences: Plenum Press, New York, p. 121-140.

Koch, G.S., Jr., Link, R.F., and Schuenemeyer, J.H., 1974, A mathematical model to guide the discovery of ore bodies in a Coeur d'Alene lead-silver mine: U.S. Bureau of Mines, Rept. Invest. No. 7989, 43 p.

Krige, D.G., 1977, The human element in APCOM's development: 15th APCOM. p. 1-6.

Krige, D.G., 1978, Lognormal-de Wijsian geostatistics for ore evaluation: South African Institute of Mining and Metallurgy Monograph Series, Geostatistics 1, 50 p.

Mackenzie, B.W., 1972, Corporate exploration strategies: 10th APCOM, p. 1-8.

Mackenzie, B.W., 1972, and Bilodeau, M.L., 1977, A model to assess the economic characteristics of base metal investment in Canada: 15th APCOM, p. 463-470.

Malmqvist, K., and Malmqvist, L., 1979, A feasibility study of exploration for deep-seated sulphide ore bodies: 16th APCOM, p. 25-37.

Matheron, G., 1962, Traite de geostatistique appliquee: Editions Technip, Paris, Tome 1 (1962), 334 p.; Tome 2 (1963), 172 p.

McCammon, R.B., 1974, The statistical treatment of geochemical data, *in* Levinson, A.A., Introduction to exploration geochemistry, Applied Publishing, Ltd., Calgary, Canada, p. 469-508.

McCammon, R.B., Botbol, J.M., and McCarthy, J.H., 1979, Drill-site favorability for concealed porphyry copper prospect. Rowe Canyon, Nevada based on characteristic analysis of geochemical anomalies: Fall Meeting, Soc. Mining Engineers, Tucson, Arizona, preprint, 13 p.

Menzie, W.D., Labovitz, M.L., and Griffiths, J.C., 1976, Evaluation of mineral resources and the unit regional value concept: 14th APCOM, p. 322-339.

Merriam, D.F., 1980, Some future developments in geology: Nature and Resources, v. 16, no. 2, p. 2-5.

Miesch, A.T., 1976, Geochemical survey of Missouri - methods of sampling, laboratory analysis, and reproduction of data: U.S. Geol. Survey Prof. Paper 954-A, 39 p.

Miller, B.M., Thomsen, H.L., Dolton, G.L., Coury, A.B., Henricks, T.A., Lennartz, F.E., Powers, R.B., Sable, E.G., and Varnes, K.L., 1975, Geological estimates of undiscovered recoverable oil and gas resources in the United States: U.S. Geol. Survey Circ. 725, 78 p.

Parker, D.H., and Switzer, P., 1975, Use of conditional probability distributions in ore reserve estimation: 13th APCOM, p. MII1-MII16.

Parker, H.M., Journel, A.G., and Dixon, W.C., 1979, The use of conditional lognormal probability distribution for the estimation of open-pit ore reserves in stratabound uranium deposits - a case study: 16th APCOM, p. 133-148.

Rendu, J.M., 1978, An introduction to geostatistical methods for mineral evaluation: Monograph, South African Inst. Min. Metall., Johannesburg, 100 p.

Robinson, S.C., 1972, The role of a data base in modern geology, *in* Merriam, D.F., ed., The impact of quantification in geology: Syracuse Univ. Geology Contr. 2, p. 67-82.

Rose, A.W., Hawkes, H.E., and Webb, J.S., 1979, Geochemistry in mineral exploration: Academic Press, New York, 657 p.

Shurygin, A.M., 1976, The probability of finding deposits and some optimal search grids: Jour. Math. Geology, v. 8, no. 3, p. 323-330.

Sichel, H.S., 1972, Statistical valuation of diamondiferous deposits: 10th APCOM, p. 17-25.

Sinclair, A.J., 1976, Application of probability graphs in mineral exploration: Spec. Vol. no. 4, Association of Exploration Geochemists, 95 p.

Sinding-Larsen, R., and Vokes, F.M., 1978, The use of deposit modelling in the assessment of potential resources as exemplified by Caledonian stratabound sulfide deposits: Jour. Math. Geology, v. 10, no. 5, p. 565-580.

Singer, D.A., 1975, Relative efficiencies of square and triangular grids in the search for elliptically shaped resource targets: U.S. Geol. Survey, Jour. Research, v. 3, no. 2, p. 163-167.

Singer, D.A., and Ovenshine, A.T., Assessing metallic resources in Alaska: Am. Scientist, v. 67, no. 5, p. 582-589.

Thompson, G.G., 1975, A computer-oriented minerals availability system: 13th APCOM, p. GII1-GII14.

Tukey, J.W., 1977, Exploratory data analysis: Addison-Wesley, Reading, Massachusetts, 499 p.

U.S. Geol. Survey and U.S. Bureau of Mines, 1976, Principles of the mineral resource classification system of the U.S. Bureau of Mines and U.S. Geological Survey: U.S. Geol. Survey Bull. 1450-A, p. A1-A5.

Webb, J.S., Thornton, I., Thompson, M., Howarth, R.J., and Lowenstein, P.L., 1978, The Wolfson geochemical atlas of England and Wales: Clarendon Press, Oxford, 69 p.

Weiss, A., ed., 1979a, Computer methods for the 80's: Soc. Mining Engineers, New York, 975 p.

Weiss, A., 1979b, Mining information systems planning and project management, introductory review, *in* Weiss, A., ed., Computer methods for the 80's: Soc. Mining Engineers, New York, p. 1-2.

Weiss, A., 1979c, Project management in mineral information technology, *in* Weiss, A., ed., Computer methods for the 80's, Soc. Mining Engineers, New York, p. 15-30.

SOME DEVELOPMENTS IN COMPUTER APPLICATIONS IN PETROLOGY

R.W. Le Maitre

University of Melbourne

ABSTRACT

During the last decade most of the computer applications in petrology have been concerned with data bases, petrological mixing models, the use of multivariate statistical techniques and, more recently, the modeling of igneous differentiation trends.

The characteristics of three of the major data bases, that is RKNFSYS (approximately 16,000 analyses), CLAIR (approximately 26,000 analyses), and PETROS (approximately 35,000 analyses) are compared. As a result of the experience gained in building these bases, a new international *ig*neous data *ba*se (IGBA) is being organized, which will supercede eventually the others due to the inclusion of mineralogical and more textural information.

Recent developments in petrological mixing models include a better understanding of the principles behind the various models, in particular the fact that for constant sum major element data, a constrained model should be used. A completely generalized model capable of dealing with metamorphic reactions of the type $A + B + C + .. = D + E + F + G..$ also has been developed.

The use of principal-components analysis, factor analysis, and discriminant analysis in the interpretation of the chemistry of igneous rocks and mineral groups has become increasingly popular, but unfortunately, either

through ignorance or the use of statistical packages, some
of the interpretations have been incorrect. In particular, principal-component analysis based on the correlation
matrix seems difficult to justify for major element geochemical data.

The modeling of igneous differentiation trends has
been approached in two different ways. One mainly empirical method predicts the path of crystallization of any
given anhydrous liquid composition at 1 atmosphere pressure. The inclusion of H_2O and pressure into the method
is only a question of time. The other method is
based on Rayleigh's Law of fractional crystallization.

INTRODUCTION

Possibly one of the most important developments in
computer applications in petrology in the last decade has
been the compilation of major petrological data bases,
for without these it is difficult to obtain unbiased results from the application of statistical methods, such
as principal-components analysis, factor analysis, and
discriminant analysis. The lack of availability of such
data bases also may have been responsible partly for the
slow development in the use of some of these multivariate
statistical methods.

One method which has gained wide acceptance and now
is used almost as routinely as a norm calculation, is the
least-squares fitting of compositional data, usually
known as mixing models. The reason for its popularity is
probably the fact that the results are in a form readily
understandable to petrologists and require no specialized
interpretation. However, perhaps the most exciting, and
potentially the most important, development of recent
years is the application of modeling methods to igneous
differentiation processes. Some aspects of some of these
developments now will be discussed in further detail.

PETROLOGICAL DATA BASES

The last decade has seen the development of three
major petrological data bases. In order of appearance
they are RKNFSYS (Chayes, 1971), CLAIR (Le Maitre, 1973,
1976a), and PETROS (Mutschler and others, 1976, 1978).
These files consist essentially of chemical analyses of
igneous rocks from all over the world, plus additional
information which differs slightly from file to file.

RKNFSYS contains approximately 16,000 analyses of Cenozoic volcanic rocks only. Stored with each analysis is the Troger number of the rock type and its location. The data base is accessed by machine-dependent software and uses random read/write facilities. Although the data base has never been distributed, data retrievals always have been available on request.

CLAIR (approximately 26,000 analyses) and PETROS (approximately 35,000 analyses) are similar types of sequential files of data of all types of igneous rocks (including some metamorphosed ones) of all ages, together with latitude and longitude and, where available, age, and trace-element data. The main difference between the two files is the manner in which the rock names, as given in the original publications, are recorded. In CLAIR they are coded in fixed format, which makes retireval by rock name extremely easy. In PETROS, however, they are in free format, which makes retrieval by rock name difficult, especially as abbreviations also are included. For example, to locate basalts one not only has to search for the term "basalt" or "bas" etc., but also to exclude "trachybasalt", "basaltic andesite", etc. The PETROS file is available freely at the cost of reproduction and a simple interactive operating system KEYBAM (Barr, Mutschler, and Lavin, 1977) also is available for data reductions. The CLAIR file has not been made available freely, although copies are in use at the C.R.P.G. Nancy, France and in the Department of Geology at the University of Leicester, England. To process the data file, a generalized data retrieval, storage, and processing system, termed the CLAIR DATA SYSTEM, was written (Le Maitre and Ferguson, 1978). In use, this open-ended system has proved to be extremely flexible and capable of processing many types of data, mainly due to the design philosophy of the system, which is described adequately by Le Maitre and Ferguson (1978).

One of the interesting things about these three files is the manner in which they differ. All three authors of the data files believed that they had collected a considerable amount of the data available, yet comparisons of the files revealed that only 15 percent of the total number of analyses in RKNFSYS and CLAIR were common to both files (Chayes and Le Maitre, 1972). Similarly, the overlap between CLAIR and PETROS was only 18 percent, indicating that probably well over 100,000 analyses already are available in the literature. Published maps of the geographical distribution of data from CLAIR

(Le Maitre, 1973) and PETROS (Mutschler and others, 1978) reveal that even these differ. Compared to CLAIR, PETROS contains little data from Europe or Australasia, but far more from the ocean floors and northwestern U.S.A.

One major deficiency of these three files is that they contain no mineralogical information apart from what may be in the rock name. In order to overcome this, and to avoid duplication of future effort, a new international data base called IGBA (for *ig*neous *ba*se) came into being. This was launched as IGCP Project 163 and considerable interest was shown in the concept at the inaugural meeting at Bochum in 1977 under the chairmanship of Felix Chayes. Since then some 12 working groups have been set up in various parts of the world to contribute data to the file. Eventually, IGBA will supercede the other three files both in number of analyses and in the type of information recorded.

The uses to which these files can be put is almost limitless and many examples of their application can be cited from the literature. However, their most important contribution to date probably is in the area of the classification of igneous rocks (Chayes, 1976, 1979; Le Maitre, 1976a, 1976b, 1976c; Streckeisen and Le Maitre, 1979) where work is in progress to aid the IUGS Subcommission on Systematics in Igneous Petrology.

PETROLOGICAL MIXING MODELS

The least-squares fitting of compositional data (usually termed mixing models) was applied first a decade ago and is one of the few statistical methods to have achieved wide acceptance and use amongst petrologists. The method has been described many times elsewhere (Bryan, 1969; Bryan, Finger, and Chayes, 1969a, 1969b; Wright and Doherty, 1970; Albarede and Provost, 1977; Stormer and Nicholls, 1978) so will not be repeated here. Minor differences in method have arisen, one of which is the choice of whether to constrain the estimated proportions to sum to unity, or to leave them unconstrained. Recently, the geometric interpretation of the differences between the two methods has indicated that, for compositional data with a constant sum, it is more logical to constrain the proportions to sum to unity (Le Maitre, 1979). This geometric approach also has led to a generalized mixing model that can be applied to metamorphic reactions of the type:

$$A + B + C \ldots = D + E + F + G \ldots$$

When the proportions on both sides of the equation are constrained to sum to unity, the problem is simply to locate the points of closest approach of two hyperplanes in n-dimensional composition space, one passing through compositions A, B, C, etc., and the other passing through compositions D, E, F, G, etc. The residual sum of squares then is the square of the shortest distance between the two hyperplanes. Prior to this such reactions had to be solved by proposing a model of the type:

$$A = B + C \ldots + D + E + F + G \ldots$$

and hoping that the proportions of B, C, etc. come out negative. However, such a solution, although giving proportions that are of the right order of magnitude, gives residual sums of squares that can be misleading. Used on actual analytical data, this generalized mixing model showed that the metamorphic reaction

Biotite + sillimanite + quartz = garnet + K-feldspar + water

is far from isochemical (Le Maitre, 1979).

PRINCIPAL-COMPONENTS ANALYSIS, FACTOR ANALYSIS, AND DISCRIMINANT ANALYSIS

Although not used as extensively as the mixing modles, these three techniques have become increasingly popular in the last decade. Considerable confusion exists in the literature about the difference between principal-components analysis and factor analysis with many authors considering principal-components analysis to be a part of factor analysis (Joreskog, Klovan, and Reyment, 1976). However, I, like many statisticians, consider principal-components analysis to be a method in its own right and distinct from true factor analysis. Thus, many papers using what is described as factor analysis only are using principal-components analysis.

One of the reasons for using principal-components analysis is to produce an optimal visual representation of complex multivariate data, by projecting the original data into the space of the first few eigenvectors. Fortunately, for most petrological systems, the first three eigenvalues derived from the variance-covariance matrix usually account for over 90 percent of the total variance, so that excellent projections can be produced (Le Maitre, 1968). There remains, however, a choice of whether to extract the eigenvectors and eigenvalues from the variance-covariance or the correlation matrix. In the latter situation, all the variables are scaled to have equal variance and, therefore, are given equal weight so that, for example, a change of 0.2 percent in MnO may be given as much importance as a change of 20 percent in MgO. Although this may be acceptable for variables measured in different units (for which this method of scaling was intended originally, there seems to be little justification for scaling of this type with major element data. There would seem to be some justification in using the correlation matrix if the variables are mixed major and trace elements (i.e. wt% and ppm), but as they are measured basically in the same type of unit (i.e. weight per unit weight), a more logical approach would be to convert all the values to wt% or ppm and to take logarithms. In this manner a major element that doubled in value would be given the same weight as a trace element that doubled in value.

A pitfall that some authors seem to have fallen into (e.g. Till and Colley, 1973; Saxena and Walter, 1974) is to use the correlation matrix to determine the eigenvectors and then to form the cross-product sum of the *raw* data and the eigenvectors in order to produce the plots. Processed in this manner, however, the data are not projected onto the eigenvectors (which is the object of the exercise) but are projected onto a set of axes which are rotated with respect to the original eigenvectors. The correct procedure is to form the cross-product sum of the *standardized* data and the eigenvectors.

Factor analysis has been used successfully by many authors to investigate the chemical variation within mineral and rock groups (e.g. Middleton, 1964; Mottana, Sutterlin, and May, 1971; Shaw, 1974), but the previous comments regarding the disadvantage of using a correlation matrix (the starting point for most factor-analysis procedures) yet apply. More recently, Miesch (1976) used factor analysis in a novel manner as an alternative to conventional mixing models and has achieved similar results.

Because of some of the earlier examples of discriminant analysis (Chayes, 1964, 1968; Le Maitre, 1968), the method has been accepted generally but has not been used as widely as it should have been, as many examples can be cited in the literature of groups being separated on a variety of simple scatter diagrams, usually constructed around some preconveived idea. Whether these diagrams are the only or best way of distinguishing the groups, seems to be of little interest to the authors. Most examples of the use of discriminant analysis continue to be two group examples (e.g. Chayes, 1975) although a few examples of multiple discriminant analyses also have been published (Gleadow and others, 1974; Le Maitre, 1978). The generalization of the classical two group discriminant analysis into multiple discriminant analysis, where more than two groups are involved, has the advantage that scatter diagrams can be produced showing the clustering of the groups in question. Although potentially useful for delimiting rock types from each other, the general use of discriminant analysis for classification purposes, unfortunately, has some disadvantages as pointed out by Le Maitre (1976c).

MODELING IGNEOUS DIFFERENTIATION TRENDS

The main difficulty with modeling differentiation trends, given a starting composition, is the problem of deciding what phases will crystallize from a particular liquid composition. Theoretically, this problem can be solved using thermodynamics, but unfortunately the required constants have not been determined yet. Taimre (1977) successfully modeled crystallization within the simple system Di-An-Fo using thermodynamic principles with constants determined empirically from experimental data. A slightly different method is used by Nathan and van Kirk (1978) in presenting a generalized model of magmatic crystallization of dry melts at 1 atmosphere pressure. By assuming that the crystallization temperature of each phase is a smooth function of the composition of the liquid, they use experimental data and multiple regression to determine the constants of these "mineral temperature equations". Only two of the "variables" in these equations, however, are based on "thermodynamics", the remainder being oxide fractions. With this method they successfully reproduce the trends of several natural rock series and the further extension of similar equations to anhydrous melts under pressure (to simulate plutonic conditions) is only a matter of awaiting relevant experimental results.

Another rather different approach, based on Rayleigh's Law of fractional crystallization, is used by Maaløe (1976) to determine the compositional trends within the simple Ab-An and Ab-An-Di systems. Presenting the solutions of the differential equations graphically, he demonstrates, for example, that with certain starting compositions, marked differences in trends may be expected depending upon whether perfect or partial fractional crystallization is involved. Similarly, by applying the same principles, Allegre and others (1977) and Minster and others (1977) further develop the theory to explain the behavior of trace elements during igneous processes with the object of deducing the main features of the differentiation process, including the initial liquid composition and the phases involved, given a suitable set of trace-element determinations on the suite of rocks. In this approach the differentiation trend is defined, and the possible mechanism of differentiation, which, of course, is the inverse of the previous applications.

REFERENCES

Albarede, F., and Provost, A., 1977, Petrological and geochemical mass-balance equations: an algorithm for least-squares fitting and general error analysis: Computers & Geosciences, v. 3, no. 2, p. 309-326.

Allegre, C.J., Treuil, M., Minster, J.F., Minster, B., and Albarede, F., 1977, Systematic use of trace elements in igneous process. Part 1: Fractional crystallisation processes in volcanic suites: Contrib. Mineral. Petrol., v. 60, no. 1, p. 57-75.

Barr, D.L., Mutschler, F.E., and Lavin, O.P., 1977, KEYBAM: a system of interactive computer programs for use with the PETROS petrochemical data bank: Computers & Geosciences, v. 3, no. 3, p. 489-496.

Bryan, W.B., 1969, Materials balance in igneous rock suites: Ann. Rept. Dir. Geophys. Lab. Carnegie Inst. Washington, v. 67, p. 241-243.

Bryan, W.B., Finger, L.W., and Chayes, F., 1969a, A least-squares approximation for estimating the composition of a mixture: Ann. Rept. Dir. Geophys. Lab. Carnegie Inst. Washington, v. 67, p. 243-244.

Bryan, W.B., Finger, L.W., and Chayes, F., 1969b, Estimating proportions in petrographic mixing equations by least-squares approximation: Science, v. 163, no. 3870, p. 926-927.

Chayes, F., 1964, A petrographic distinction between Cenozoic volcanics in and around the open oceans: Jour. Geophys. Res., v. 69, no. 8, p. 1573-1588.

Chayes, F., 1968, On locating field boundaries in simple phase diagrams by means of discriminant functions: Am. Mineral., v. 53, nos. 3 and 4, p. 359-371.

Chayes, F., 1971, Electronic storage, retrieval, and reduction of data about the chemical composition of common rock: Ann. Rept. Dir. Geophys. Lab. Carnegie Inst. Washington, v. 70, p. 197-201.

Chayes, F., 1975, On distinguishing alkaline from other basalts: Ann. Rept. Dir. Geophys. Lab. Carnegie Inst. Washington, v. 74, p. 546-547.

Chayes, F., 1976, Characterizing the consistency of current usage of rock names by means of discriminant functions: Ann. Rept. Dir. Geophys. Lab. Carnegie Inst. Washington, v. 75, p. 782-784.

Chayes, F., 1979, Partitioning by discriminant analysis: a measure of consistency in the nomenclature and classification of volcanic rocks, *in* The evolution of igneous rocks: Princeton Univ. Press, Princeton, New Jersey, p. 521-532.

Chayes, F., and Le Maitre, R.W., 1972, The number of published analyses of igneous rocks: Ann. Rept. Dir. Geophys. Lab. Carnegie Inst. Washington, v. 71, p. 493-495.

Gleadow, A.J.W., Le Maitre, R.W., Sewell, D.K.B., and Lovering, J.F., 1974, Chemical discrimination of petrographically defined clast groups in Apollo 14 and 15 lunar breccias: Chem. Geology, v. 14, no. 1, p. 39-61.

Joreskog, K.G., Klovan, J.E., and Reyment, R.A., 1976, Geological factor analysis, *in* Methods in Geomathematics 1: Elsevier Sci. Publ. Co., Amsterdam, 178 p.

Le Maitre, R.W., 1968, Chemical variation within and between volcanic rock series - a statistical approach: Jour. Petrology, v. 9, no. 2, p. 220-252.

Le Maitre, R.W., 1973, Experiences with CLAIR: a computerised library of analysed igneous rocks: Chem. Geology, v. 12, no. 4, p. 301-308.

Le Maitre, R.W., 1976a, Chemical variability of some common igneous rocks: Jour. Petrology, v. 17, no. 4, p. 589-637.

Le Maitre, R.W., 1976b, Some problems of the projection of chemical data into mineralogical classifications: Contrib. Mineral. Petrol. v. 56, no. 2, p. 181-189.

Le Maitre, R.W., 1976c, A new approach to the classification of igneous rocks using the basalt-andesite-dacite-rhyolite suite as an example: Contrib. Mineral. Petrol. v. 56, no. 2, p. 191-203.

Le Maitre, R.W., 1978, Numerical petrology: Trans. Leicester Lit. Phil. Soc., v. 72, in press.

Le Maitre, R.W., 1979, A new generalised petrological mixing model: Contrib. Mineral. Petrol. v. 71, no. 2, p. 133-137.

Le Maitre, R.W., and Ferguson, A.K., 1978, The CLAIR data system: Computers & Geosciences, v. 4, no. 1, p. 65-76.

Maaloe, S., 1976, Quantitative aspects of fractional crystallisation of major elements: Jour. Geology, v. 84, no. 1, p. 81-96.

Middleton, G.V., 1964, Statistical studies on scapolites: Can. Jour. Earth Sci., v. 1, no. 1, p. 23-34

Miesch, A.T., 1976, Q-mode factor analysis of geochemical and petrologic data matrices with constant row-sums: U.S. Geol. Survey Prof. Paper 574-G, p. 1-47.

Minster, J.F., Minster, J.B., Treuil, M., and Allegre, C.J., 1977, Systematic use of trace elements in igneous processes: Part II. Inverse problem the fractional crystallisation process in volcanic suites: Contrib. Mineral. Petrol., v. 61, no. 1, p. 49-77.

Mottana, A., Sutterlin, P.G., and May, R.W., 1971, Factor analysis of garnets and omphacites: a contribution to the geochemical classification of eclogites: Contrib. Mineral. Petrol. v. 31, no. 3, p. 238-250.

Mutschler, F.E., Rougon, D.J., and Lavin, O.P., 1976, PETROS - a data bank of major-element chemical analyses of igneous rocks for research and teaching: Computers & Geosciences, v. 2, no. 1, p. 51-57.

Mutschler, F.E., Rougon, D.J., Lavin, O.P., and Hughes, R.D., 1978, PETROS a data bank of major element chemical analyses of igneous rocks: U.S. Dept. Commerce, Nat. Oceanic Atmos. Admin., Envir. Data Serv., Pamphlet 1978(W), 4 p.

Nathan, H.D., and Van Kirk, C.K., 1978, A model of magmatic crystallisation: Jour. Petrology, v. 19, no. 1, p. 66-94.

Saxena, S.K., and Walter, L.S., 1974, A statistical-chemical and thermodynamic approach to the study of lunar mineralogy: Geochim. Cosmochim. Acta, v. 38, no. 1, p. 79-95.

Shaw, D.M., 1974, R-mode factor analysis on enstatite chondrite analyses: Geochim. Cosmochim. Acta, v. 38, no. 10, p. 1607-1613.

Stormer, J.C., and Nicholls, J., 1978, XLFRAC: a program for the interactive testing of magmatic differentiation models: Computers & Geosciences, v. 4, no. 2, p. 143-159.

Streckeisen, A., and Le Maitre, R.W., 1979, A chemical approximation to the modal OAPF classification of the the igneous rocks: Neues Jahrb. Mineral. Abh. Band 136, Heft 2, p. 169-206.

Taimre, T., 1977, Theoretical modelling of liquidus relation in silicate systems: unpubl. honours project, Univ. Melbourne, 132 p.

Till, R., and Colley, H., 1973, Thoughts on the use of principal component analysis in petrogenetic problems: Jour. Math. Geology, v. 5, no. 4, p. 341-350.

Wright, T.L., and Doherty, P.C., 1970, A linear programming and least squares computer method for solving petrologic mixing problems: Geol. Soc. America Bull., v. 81, no. 7, p. 1995-2008.

STRATIGRAPHIC ANALYSIS: DECADES OF REVOLUTION (1970-1979) AND REFINEMENT (1980-1989)

C. John Mann

University of Illinois

ABSTRACT

Significant progress in stratigraphic analysis has occurred during the 70s in several areas of stratigraphy. Multiband seismic data systems with greater resolution of detail have revolutionized our ability to interpret subsurface stratigraphic units, distributions, lithologies, physical properties, and contained fluids. For the first time, we are able to predict and map with reasonable accuracy what subsurface stratigraphic sequences will be encountered in poorly explored sedimentary basins.

Progress has occurred in development of quantitative stratigraphic correlation methods ranging from probablistic approaches in biostratigraphy to Fourier analyses in lithostratigraphy. Automated analyses of electrical-geophysical logs from boreholes have improved geological interpretations regarding strata. Multivariate approaches have enhanced our paleoenvironmental and paleoecological interpretations of earth's history. Advances in data representation, mapping methods, geographic analysis, and understanding of trend surfaces have resulted in improved regional stratigraphic analyses and interpretations. Interactive computer technology has permitted the stratigrapher to determine more quickly alternative methods of data display and analysis which leads to better understanding of relations and more rapid solution of problems. All of these advances have been accompanied by, and partially are, a result of improved stratigraphic data bases and information banks. Finally, the first significant step toward a more rigorous, theoretical stratigraphy occurred during the past decade with application of set theory principles to stratigraphic terminology, definitions, and concepts.

The next ten years predictably will give us improved and greater usage of existing quantitative methodologies. Better methods of displaying and depicting stratigraphic data, relationships, and interpretations will arise through greater use of three-dimensional displays, color coding, and improved resolution of output. Better data bases will provide broader coverage and permit more comprehensive analyses; nonetheless, much progress will remain to be accomplished because the final availability of all stratigraphic data in computer data banks seemingly is more distant than one decade. Similarly, data standardization will be greater but, unfortunately, probably will not be universal. Better stratigraphic analyses will result from more sophisticated applications of existing techniques to multiple data sets. All of this will become available more readily to geologists by development and marketing of smaller, cheaper, and faster computers having multiple processors and larger memories. More field usage of computers by geologists during the next ten years can be anticipated.

No methodological revolutions can be foreseen for the next decade; only improved usage of existing methodologies is predictable. This does not indicate that sudden advances will not occur, for who would have predicted ten years ago that seismic technology would revolutionize subsurface stratigraphy in less than a decade? One can say only that no revolutionary methodology or technology has been recognized to be lurking in our world, ready to surge forth in the 80s to alter some aspect of stratigraphic analysis as did occur in the 70s.

INTRODUCTION

The 70s were an active and interesting decade in stratigraphic analysis. With increasing availability generally of larger computers and increasing appreciation by geologists of benefits to be derived from computers, greater application of computers to the solution of stratigraphic problems was made. Both were major factors in seismic stratigraphy and quantitative stratigraphic correlation, two areas which advanced most significantly during this interval.

This review attempts to identify areas of major progress, trends, and growth in stratigraphy during the 70s. The bibliography is not exhaustive but hopefully it is comprehensive in areas of major quantification. Continued use and numerous papers in applications of older methods generally have not been noted individually. This does not indicate that these contributions are being ignored nor does it imply that they have been unimportant because obviously they are when applications have been frequent.

SEISMIC STRATIGRAPHY

Undoubtedly the most significant change and advance in stratigraphy during the past decade has been the rapid development of seismic stratigraphy. This has been made possible primarily through more complete integration of theory with practice in seismic exploration and data enhancement by digital computer (Payton, 1977). Seismic stratigraphic interpretations are possible now because resolution of seismic data has been increased and destructive signals have been reduced. Although some improvement in seismic data has been a result of improved seismic equipment, new and improved techniques, and electronic filtering of seismic signals, the vast majority of improvement has arisen through digital enhancement of raw data by various computer correction programs (Dobrin, 1975; Flowers, 1976; Sheriff, 1977).

A basic general improvement in seismic data comes about merely by repositioning the data to reflect more accurately subsurface geology in spatially correct relations (Flowers, 1976; Sheriff, 1977). These migrations are removal of horizontal components of seismic-wave travel paths which may have significant effects even in flat-lying strata. Although migration of seismic signals had been a recognized improvement for seismic data for sometime, extensive computations necessary to make them routinely has been prohibitive until recently, about 1969, when larger and more economical computational facilities became available. More recently, three-dimensional migrations have become possible whenever gridded seismic data are available.

Useless signals in seismic data such as random noise and reflective multiples have been reduced by a variety of techniques and computational corrections (Sheriff, 1977). Static correction programs may eliminate differences arising in the vicinity of seismic sources and geophones at earth's surface. Signature-processing remedies and employment of sources of known signal input eliminate much noise originating from seismic sources. Divergence corrections can remove many differences created by subsurface strata. Velocity filtering (common-depth point stacking) and redundancy recording procedures permit attenuation of many types of coherent wavetrains and random noise. Predictive deconvolution and common-depth point stacking reduce undesirable multiples. Deconvolution helps remove near-surface reverberation noise and broadens the frequency spectrum to improve or sharpen seismic wavelets. Wave-equation migration clarifies stratigraphic evidence arising from dipping strata even if bedding generally is flat.

Seismic resolution is a function primarily of signal wavelength that is reflected from subsurface interfaces (Sheriff, 1977). Generally, the greatest vertical resolution of events or features will be 1/8 to 1/4 the wavelength and greatest horizontal resolution will be proportional to the square root of the product of wavelength and depth to interface. Wavelengths normally increase with depth because velocity increases downward generally and frequency diminishes with depth; therefore resolution normally decreases with increasing depth. Because

$$\text{wavelength} = \text{velocity} \cdot \text{period} = \text{velocity}/\text{frequency}$$

and velocity is a function of geologic material and depth, the only manner that seismic resolution can be controlled is by altering input frequency and by recording a wide band of frequencies on signal return. Larger frequencies give greater resolution than do smaller frequencies. Previously most high frequencies were neither recorded nor retained in processing seismic data. Today, deliberate procedures are undertaken to generate, record, and retain during processing a broad band of frequencies. Nonetheless, earth attenuates higher frequencies (Sheriff, 1976; Dobrin, 1977) and enhanced resolution solely by generation of higher input frequencies is limited. Resolution also is improved by data enhancement, such as zero-phase wavelet extraction, analysis of amplitudes, and noise reduction (Dobrin, 1977).

Mathematical modeling (Flowers, 1976; Sheriff, 1976, 1977; Meckel and Nath, 1977; Schramm, Dedman, and Lindsey, 1977; Neidell and Poggiagliolmi, 1977; Farr, 1979; Ruotsala, 1979; Rice, Bakker, and Weinberg, 1979) also is introduced to improve seismic interpretations of stratigraphy. Many useless signals arise in near-surface geology which generally is well known and can be modeled effectively in order to reduce these unwanted signals in seismic data. Synthetic seismograms constructed from well data are effective in identifying interfaces from which reflections are arising. Synthetic stratigraphic columns or pseudologs constructed from seismic data are helpful in predicting lithologies and bed geometries in lesser known subsurface areas and detection of possible stratigraphic traps. All these modeling procedures are directed toward correcting, refining, and defining more precisely subsurface geology from raw seismic data.

Additional improvements in seismic data interpretations have been realized by considering seismic signals to be a component of a complex signal (Bracewell, 1965; Farnback, 1975; Taner and Sheriff, 1977; Taner, Koehler, and Sheriff, 1978). This transformation allows seismic signals to be examined from a local significance standpoint thereby providing insights that previously were not available for geologic interpretations. These transformations are comparable to Fourier transformations which have long been used in time-series and seismic analyses but which conversely provide an averaged value for a large portion of the signal trace for various wave properties. The real portion of a complex signal is that portion which is recorded by geophones whereas the imaginary or quadrature portion is not recorded; both real and imaginary traces are identical except for a 90° phase shift.

As a result of complex trace analysis, attributes may be defined which give point values and which when incorporated with normal seismic amplitude sections permit enhanced interpretations.

Reflection strength or amplitude of the complex trace is independent of phase and is a function of interface reflection quality. Good reflection strength is associated with major lithic changes and gas accumulations. Constancy indicates a single reflecting horizon or persistent composite reflectors. It forms a good reference for time-interval measurements in seismic stratigraphy and clearly reveals differential compaction, local or regional thinning, facies changes, and velocity variations. Local sharp changes of reflection strength may be indicative of faulting, hydrocarbon accumulations (bright spots), and rapid facies changes.

Instantaneous phase of the complex signal emphasizes continuity of subsurface events. It is good for recognizing pinchouts, angularities in deposition or structure, discontinuities, faults, prograding sediments, and offlap.

Instantaneous frequency, the time derivative of instantaneous phase, is helpful in recognizing pinchouts and edges of hydrocarbon-water interfaces. Low-frequency shadows may develop in strata below gas, condensate, and oil accumulations. Shadows also may occur below fracture zones in brittle rocks. Usually frequency is smoothed through some finite interval by moving windows so that an *averaged weighted frequency* actually is used rather than instantaneous frequency.

Apparent polarity is the sign of the seismic trace when reflection strength has a maximum value. It is sensitive to data quality and may distinguish between different types of bright spots.

Attributes aid significantly in stratigraphic interpretation of normal seismic amplitude sections.

A final contribution of improved seismic data enhancement during the past decade has been the effective use of color to convey complex types of multiple data to geological and geophysical interpreters of attributes for standard amplitude traces (Balch, 1971). Colors aid more accurate assimilation and digestion of seismic data by adding effectively another dimension; color results in better comprehension and geologic translations. Normally, attribute values are zoned arbitrarily and coded numerically in computer output so that a user may assign any color he wishes to various interval values. Colors normally are superimposed over conventional migrated, enhanced seismic amplitude displays.

Seismic stratigraphy originally was concerned only with improved depiction of subsurface geology to identify more accurately and faithfully possible traps of hydrocarbons. However, with increased seismic resolution and complex trace analysis, not only have stratigraphic relations become clearer in the subsurface but lithologies, facies changes, porosities, and fluid content of strata usually can be recognized. Discordances reveal erosional surfaces and depositional cycles. Coupling this knowledge with simple depositional configurations (Vail, Mitchum, and Thompson, 1977) and an assumption that persistently strong regional reflective interfaces are approximate time horizons, extensive inferences may be made concerning depositional histories, marine transgressions and regressions, structural movements, and geologic history of large regions which because of a lack of subsurface drilling exploration previously were unknown geologically. Seismic data from continental shelves have been interpreted in terms of relative sea-level fluctuations (Vail, Mitchum, and Thompson, 1977; Vail and Mitchum, 1979).

Continued improvements in seismic stratigraphy in the future may be expected and correspondingly will increase probabilities of detection of petroleum accumulations in stratigraphic traps by seismic methods. With ever increasing scarcity of hydrocarbons, more and more petroleum effort will be directed toward discovery of stratigraphic accumulations that are independent of geologic structures. Because stratigraphic accumulations remain our last great hope for significant quantities of undiscovered reserves in earth's crust, seismic stratigraphy will play an increasingly important role in the world's economic future.

QUANTITATIVE STRATIGRAPHIC CORRELATION

Considerable progress has been realized during the past decade in efforts to correlate automatically stratigraphic sequences, either biostratigraphically or lithostratigraphically, by computer from data files. Although a fully operational system yet is to be established, numerous methods have been proposed and demonstrated to be capable of quantitative correlations in certain situations. But so far, all have limitations and none has been demonstrated to be general and adequate for handling all situations.

These methods have ranged in biostratigraphy from two-dimensional graphic approximations (Miller, 1977; Edwards, 1978) through seriation (Scott, 1974; Davaud and Geux, 1978) and multivariate analyses of various types (Hazel, 1970, 1977; Hohn, 1978) to probabilistic approaches (Hay, 1972; Southam, Hay, and Worsley, 1975; Rubel, 1976; Worsley and Jorgens, 1977; Edwards and Beaver, 1978). In lithostratigraphy, methods have ranged from successive pairwise point comparisons (Dienes, 1974a) through graphic procedures (Leont'ev, 1972; Kemp, 1977; Shaw, 1978), assumed sedimentation rates (Dienes, 1974a), slotting (Gordon and Reyment, 1979), cross correlation (Rudman and Lankston, 1973), zonation (Hawkins and Merriam, 1973, 1974; Webster, 1973; Shaw, 1978; Shaw and Cubitt, 1979) and pattern recognition (Vincent, Gartner, and Attali, 1979) to Fourier analysis (Dowell, 1972; Henderson, 1973; Rudman, Blakely, and Henderson, 1975; Mann and Dowell, 1978).

Efforts of a great number of workers were coordinated with some modest organization in 1977 when, Quantitative Stratigraphic Correlation, Project 148 of the International Geological Correlation Program was established. The first meeting and organization of Project 148 was held at Syracuse University during the 6th Geochautauqua when a one-half day session was devoted to quantitative correlation (Computers & Geosciences, v. 4, no. 3). The first international meeting and report was held last year in Jerusalem in connection with the International Sedimentological Association. Numerous national meetings have been held locally in the past two years. One technical session is planned for the International Geological Congress in Paris to report on advances and accomplishments of Project 148.

PROBABILITIES

Applications of probabilities in stratigraphic analyses were more evident during the 70s than previously. Probabilistic stratigraphy of Hay (1972) and subsequent workers (Hay and Steinmetz, 1973; Worsley and others, 1973; Southam, Hay, and Worsley, 1975; Rubel, 1976; Edwards and Beaver, 1978) has been a major contribution to ordering and evaluating dependability of biostratigraphic data. Originally, the probablistic approach was a simple application of binomial probability density functions to the practical problems of determining the biostratigraphic value of numerous microfaunal forms encountered in deep-sea drilling data. By assuming that no two biostratigraphic events occurred simultaneously, pairwise comparisons of biostratigraphic events can be made and probabilities of their ordering being a random occurrence can be calculated. A matrix of probabilities then may be ordered to establish the most likely sequence of events. Recent work (Edwards and Beaver, 1978) has generalized this approach to a trinomial probability density function which recognizes synchroneity of events as well as earlier than and later than occurrences. Probabilistic stratigraphy also forms an important approach to biostratigraphic correlation (Hay, 1974; Hay and Southam, 1978).

Probabilities were utilized by McCammon (1970; Harris and McCammon, 1971) to estimate lithic components in stratigraphic sequences with simultaneous evaluation of sonic, density, and neutron logs. A probabilistic approach is necessary whenever the number of lithologies in a sequence exceeds the number of response equations available for their prediction. The method involves determining a relative entropy function because a maximum entropy estimate is the least biased solution for estimating lithologic components under uncertainty. An example by McCammon (1972), demonstrates that the probabilistic approach is consistent with actual stratigraphic sequences.

A probabilistic method of paleobiogeographic analysis was employed by Henderson and Heron (1977) in a study of Cretaceous ammonites. Inadequacies of existing binary coefficients in paleobiogeography analyses are replaced by a new technique which relates number of shared taxa to sample size and inferred population size for areas. The probability frequency function proposed assumes random selection of taxanomic samples in the stratigraphic record and provides a quantitative assessment of population diversity and error estimates.

Probabilities also are being employed increasingly in petroleum exploration (Harbaugh, Doveton, and Davis, 1977; Harbaugh, 1979) in both evaluation of geologic, stratigraphic, and economic questions and in decision-making processes. Undoubtedly this trend in stratigraphic analyses will continue.

STOCHASTIC-PROCESS MODELS

Applications of random-process models in stratigraphic studies have increased slowly during the past ten years. Markovian chains, matrices, and transitional probabilities continue to attract many workers (Dacey and Krumbein, 1970; Schwarzacher, 1972; Miall, 1973; Ethier, 1975; Read, 1976; Hattori, 1973, 1976; Smyth and Cook, 1976). These have been used widely in sequential analyses, especially in cyclic sequences. More importantly, however, has been an increasing interest in and application of other forms of stochastic modeling for solution of geologic problems (Krumbein, 1972, 1976; Merriam, 1976). These stochastic studies have ranged in content from general aspects of stratigraphic applications (Krumbein, 1972, 1976; Schwarzacher, 1976, 1978), to specific problems (Mizutani and Hattori, 1972; Switzer, 1976), and exploratory excursions in various geologic processes (Jacod and Joathon, 1971, 1972).

STRATIGRAPHIC DATA SYSTEMS

After considerable inertia during the preceding decade, geologists began the 70s with cautious evaluations of general data systems (Robinson, 1970; Hubaux, 1972b; Burk, 1973), philosophical examinations (Dixon, 1970; Hubaux, 1970, 1972a), and practical concern such as how to formulate geologic data into digits (Morgan and McNellis, 1971; Hubaux, 1971; Conley and Hea, 1972; Greisemer and Costello, 1972). Generalized geologic data systems were developed initially (Jeffery and Gill, 1973; Cubitt, 1976) but were followed quickly by more specialized systems that dominately were oriented stratigraphically (David and Lebuis, 1976; Odell, 1976, 1977a; Shaw and Simms, 1977). A necessary subsequent development was a method to interchange data between two or more systems of dissimilar formating (Sutterlin, Jeffery, and Gill, 1977) and to utilize easily and efficiently available data files in normal geologic fashions (Burns and Remfry, 1976; Odell, 1977b; Baer, 1979; Talley, 1979).

Now at the end of the decade, we see some progress but, except for individual data systems within corporate entities, commercial groups, and governmental agencies, overall progress is disappointing. Stratigraphic data systems are limited. The task of quantification and codification of existing stratigraphic records is monumental; it requires considerable manpower, time, and money to accomplish. The Illinois Geological Survey is perhaps a typical example in this regard of the data-system situation. Although starting in 1973 to establish a state data system consisting of more than one quarter million boreholes, only 37 items of perhaps 600 total stratigraphic and lithic items of geologic interest have been entered into the system for about 45 percent of the borings so far. Additional geological data will be added after this initial phase is completed for all wells in about 1982.

Along with establishment of numerous localized data systems, which are not necessarily compatible, a disappointing lack of standardization of data within data systems (Iglehart, 1979; Baer, 1979) has created additional problems.

Although continued improvement and greater general availability of data from data systems will occur during the next decade, an interval longer than ten years unfortunately will be necessary probably before stratigraphic data systems are fully satisfactory to the users.

QUANTITATIVE LITHIC ANALYSIS

Commercial well-logging groups and petroleum companies have pioneered computerized reduction of electric and geophysical log data (Konen and Helander, 1970; Poupon and others, 1970; Fertl and Hammack, 1971; Harris and McCammon, 1971) for obvious economic reasons. Nearly all commercial logging companies provide routine computerized analyses for more dependable interpretations of lithologies, porosities, permeabilities, fluid content, and other lithic features encountered in strata penetrated and logged (Watt, 1977; Schlumberger, n.d.) by various logging devices. Computerization of log analyses has improved lithic interpretations as well as providing analysis which incorporates all available data. These computer analyses and summary logs of lithic properties now are becoming available at the well site (Anonymous, 1975) and signal an improved and timely formation evaluation capability that is available to the wellsite geologist.

In addition to commercial investigations and systems of lithic analyses by computer in the petroleum and mineral industries, other geological workers also are exploring methods of quantitative lithic analysis (McCammon, 1970, 1972; Ruoff, 1976; Magara, 1979) for solution of surface and subsurface problems in stratigraphy. These range from probabilistic approaches to simple classificational procedures and interactive systems (Doveton and Cable, 1979).

PALEOENVIRONMENTAL-PALEOECOLOGICAL ANALYSIS

Computer applications continue to contribute heavily to paleoenvironmental and paleoecological analyses in stratigraphy as they did during the previous decade. Analyses during the 70s tended to use more than one form of quantitative analysis (Feldhausen and Ali, 1976a, 1976b; Warshauser and Smosna, 1979), more sophisticated forms of analyses (Price and Jorden, 1977; Cisne and Rabe, 1978), and new methods (Gordon and Birks, 1974). But largely, no significant changes or radical advances occurred in environmental reconstructions. Increasingly, applications were the most noteworthy aspect of computer applications in paleoenvironmetal and paleoecological analysis.

INTERACTIVE COMPUTER APPLICATIONS

Interactive computer applications in stratigraphy have grown during the past decade. So far, use seemingly resides primarily in two areas: petroleum exploration (Jones, Johnson, and Phillips, 1976; Abry, 1979; Rice, Bakker, and Weinberg, 1979; Doveton and Cable, 1979) and in education (Mann, 1976a, 1976b; Raffin, 1976; Brady, 1978). Although some applications also are being made by geological surveys in other geological areas (Miesch, 1975, p. 159; 1976, p. 181), stratigraphic applications seemingly are limited in scope. Advantages of interactive computer programs reside in the ability of a stratigrapher to quickly review available data, to select modes of presentation and scales, and to identify areas that have attractive data for further exploration. Actual programs and methods used by interactive computer applications range greatly.

SET THEORY

Applications of set theory have been made only slowly in geology. Yet many advantages in this mathematical area seem to exist when computer manipulation of stratigraphic data is being made. One area is stratigraphic mapping (Bouille, 1976) where graph theory permits simple and fast digitization of unit boundaries. Excessive data volume is reduced through gridding. From an oriented graph composed of arcs associated with stratigraphic boundaries, other graphs may be deduced one of which is used to structure the data and a second of which is used to summarize geologic properties of the map. These in turn are useful for graphic reconstructions, correlation, and other analyses. Selective reconstructions permit great flexibility of composition by a user.

Rubel (1976) described biostratigraphic ranges of taxa in set-theory notation for more efficient evaluation of biostratigraphic sequences. By using sets, rather than probabilities, some problems of the latter are avoided and maximum subdivision of stratigraphic sequences is accomplished. The method intuitively is suitable for geologists.

Formalized statement of traditional geologic concepts such as stratigraphic units, fossil zonations, temporal relationships, and stratigraphic order in set-theory notation (Dienes, 1974a, 1974b, 1977; Dienes, and Mann, 1977; Mann, 1977) can facilitate computer manipulation of stratigraphic data files for geologic correlations and graphic presentations. More importantly, however, than obvious advantages to be gained in computer operations is an opportunity for greater stratigraphic generalization and abstraction than presently is possible without the rules and rigor of mathematical set theory. Stratigraphic prediction also will be enhanced under a more theoretical approach to stratigraphic problems and associated uncertainties. Set theory provides the first practical basis for a rigorous theoretical approach to stratigraphy.

CORRESPONDENCE ANALYSIS

Correspondence analysis was introduced early during the 70s (Benzercri, 1970) and was quickly applied to geologic problems (Cazes, 1970; Cazes, Solety, and Vuillaume, 1970; David and Beauchemin, 1974; David, Campigilo, and Darling, 1974; Dagbert and David, 1974; David and Woussen, 1974; David and Dagbert, 1975; Dagbert and others, 1975; Teil and Cheminee, 1975; David, Dagbert, and Beauchemin, 1977). Beginning with factor analysis, it is a distribution-free technique which takes advantage of the duality between R- and Q-mode factor analysis. It detects associations and oppositions existing between variables and samples in any data set by measuring the contribution to total variance jointly for each factor. This reduces scaling problems between variables and samples in diagrams exhibiting relationships of variables and samples to factors. Unlike standard R- and Q-mode analyses which have asymmetrical transformations, scaling and weighting procedures of correspondence analysis involve a symmetrical transformation that results in both analyses being equivalent. Thus only one diagram is necessary to exhibit relationships revealed by the analysis. Subsets of both variables and samples are projected onto the same set of factoral axes thereby aiding in an interpretation of the analytical results. The validity of correspondence analysis as a viable variation of factor analysis has been questioned (Miesch, 1974, p. 139-140).

GRADIENT ANALYSIS

The term, gradient analysis, unfortunately is used in at least two specific, but different ways formally in geology. In addition to these formal meanings, gradient analysis is used informally when gradients of any type are being analyzed.

Gradient analysis from applications in structural geology (Loudon, 1964, 1967; Whitten, 1966, 1968) now has been extended to stratigraphic problems (Lahiri and Rao, 1974). Gradient analysis, in this sense, is based on concepts of principal axis and moving averages; it reveals the direction of dispersion in a data array. Vectorial decomposition of spatial scalar data in multivariate situations may aid in searching for patterns in many types of stratigraphic data and analyses. Increased uses of gradient analysis follows directly from greater accessibility to faster and more economical computing facilities.

Gradient analysis as used more recently in paleoecologic and paleoenvironmental studies is an entirely different mode of analysis. Here percentages of taxa similarities existing at several localities form a basis for a gradient or complex gradient of environmental variation Ordination of community samples provides information on abundance and distribution of taxa which permits construction of a continuous gradient (Cisne and Rabe, 1978). This gradient subsequently is useful in establishing co-encorrelations in stratigraphic sequences.

TREND SURFACES

Now in its third decade of applications in geology, trend surfaces are employed in routine fashion by workers in all areas of geology who would not term themselves mathematical geologists. The geological profession as a whole has embraced enthusiastically this computer method because they appreciate its value in daily geologic operations. Trend surfaces perhaps are most uniquely geological of all computer methodologies that we use today.

Because of the importance that trend-surface analyses have in geology, considerable activity is being directed yet toward improvement and greater understanding of these methods (Whitten, 1970; Rao and Rao, 1970; Watson, 1971, 1972; Jones, 1972; Krumbein and Watson, 1972; Rao, 1975; Lahiri and Rao, 1978). New variations and techniques were initiated during the 70s in the form of extension of orthogonal polynomials to irregularly spaced data (Whitten, 1970), Z-trends (Robinson and Merriam, 1971), and spline-surfaces comparisons (Whitten and Koelling, 1973).

THREE-DIMENSIONAL ANALYSIS AND PRESENTATION

Three-dimensional presentation and analysis of stratigraphic data continues to show slow progress (Tipper, 1976, 1977). However, stratigraphic applications generally have not kept pace with three-dimensional technologies that are available to stratigraphers from other disciplines. Most significant advances perhaps have been in interactive computer technology and seismic exploration. The latter area has demonstrated conclusively beneficial aspects of three-dimensional analysis (Bone, Graebner, and Brown, 1979) but nonetheless, its increased expense may prohibit extensive utilization except for the most complex geologic situations requiring detailed studies and stratigraphic correlations. Certainly more extensive utilization of three-dimensional analyses and presentations in stratigraphy can be expected in the future.

ANTICIPATED ADVANCES DURING NEXT DECADE

Predictions of future advances are easy to make for clearly developed trends in stratigraphic analysis; they are difficult to make in those recently emerged areas which have not yet established a clear direction. Generally, all the present methodologies may be expected to grow and improve with greater usage and better understanding by workers. Better methods of displaying and depicting stratigraphic data, relationships, analytical results, and interpretations are anticipated during the eighties. These advances will ensue from greater use of three-dimensional displays, colors, and greater resolution of analytical methods.

Improved data bases will provide broader stratigraphic coverage, permit more comprehensive analyses, and allow new forms of analyses which previously have not been possible. However, data bases will remain incomplete most likely and much progress will remain to be accomplished in the following decade. Similarly, data standardization, unfortunately, also will be incomplete by the end of next decade, although this shortcoming may be circumvented in many instances. Overall, better stratigraphic analyses will result because of improvement in methodologies, more sophisticated applications of available techniques, and better data bases; in general, a refinement in analyses.

Computer applications in stratigraphy also will grow during the 80s because computers will be smaller, faster, more powerful, and cheaper for geologists to use. Multiple processors and larger memories will facilitate solution of some presently difficult geologic problems due to their complexities or sheer volume of input data required for solution. Greater field usage of computers is anticipated for the next decade. Computer displays in color will accelerate utilization of color in stratigraphic analyses.

No revolution in stratigraphic analysis comparable to seismic stratigraphy is predictable for the 80s. By nature, these type of events are difficult to predict except for moments of sheer genius or soothsayers. This does not indicate that sudden advances will not occur. It indicates merely that a revolutionary methodology or technique has not been recognized to be lying undeveloped, ready to surge forth in the 80s and radically alter some aspect of stratigraphy.

CONCLUSIONS

The 70s have been a period of rapid growth in many areas of computer application in stratigraphic analysis.

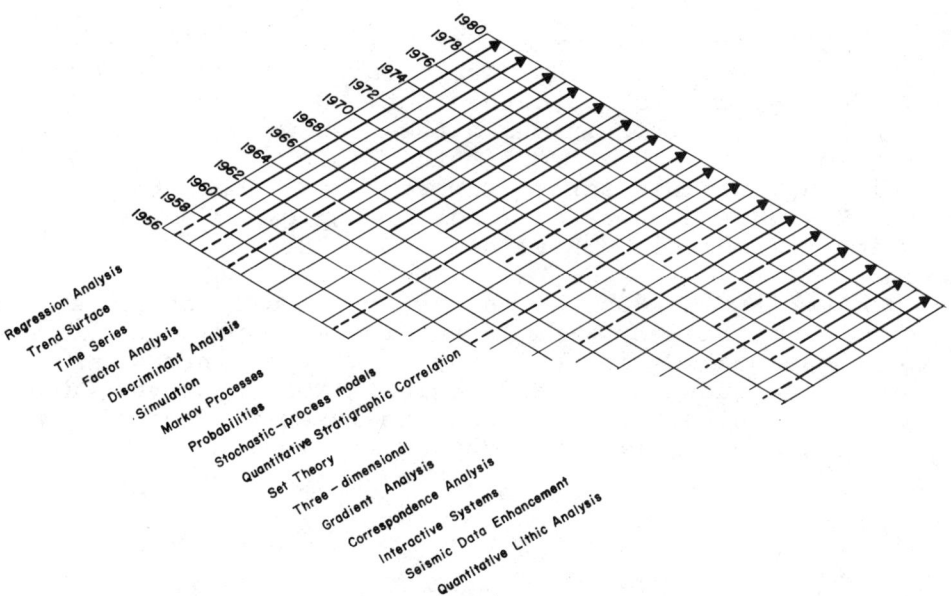

Figure 1. Quantitative methods used in stratigraphy and their approximate date of introduction (after Krumbein, 1969).

Seismic stratigraphy has been the most impressive advancement in both total computational effort required and increased stratigraphic capability which has been given to us. Quantitative stratigraphic correlation has grown from small exploratory excursions into considerations and evaluations of possible numerous and sometimes intricate and complex procedures suitable to either biostratigraphic or lithostratigraphic data. Probabilistic approaches to stratigraphy have radiated from curious examinations of Markovian matrices and transitional probabilities of stratigraphic sequences in the 60s into true stochastic modeling and simple but elegant applications of probability density functions. Applications of set theory have been introduced in stratigraphy to handle more efficiently various difficult stratigraphic concepts and classification problems; it also provides the first viable basis for a theoretically rigorous stratigraphy. New modes of stratigraphic analyses were introduced during the 70s in the form of correspondence analysis, gradient analysis, and lithic analysis. Continued growth and improvement were seen in three-dimensional analyses and representation of stratigraphic data, interactive computer systems, stratigraphic data files and all the older methods used prior to 1970.

Bill Krumbein summarized computer applications in geology through the 60s (Krumbein, 1969) in a symposium at Kansas comparable to this one. He presented a three-dimensional matrix in which the axes were time, geologic subject, and mathematical method (his figs. 1 and 2). Building upon his stratigraphic level, new methods of the 70s have been added (Fig. 1) and older methodologies which he recognized have been updated.

During the 80s, we may expect to see continued growth, refinement, and increasing widespread applications of these methods in use today as well as introduction of new techniques and methodologies. None, however, are likely to alter stratigraphic analyses as radically as did seismic stratigraphy during the 70s.

REFERENCES

Abry, C.G., 1979, Interactive graphics in subsurface stratigraphic methodology (abst.): 10th Ann. APG Symp., West Virginia Geol. Survey Circ. C-15, p. 1.

Anonymous, 1975, New digital logging system improves well data accuracy: World Oil, v. 181, no. 6, p. 79-82.

Baer, C.B., 1979, Computer-compatible handling of geologic data (abst.): Am. Assoc. Petroleum Geologists Bull., v. 63, no. 3, p. 412-413.

Balch, A.H., 1971, Color sonagrams: a new dimension in seismic data interpretation: Geophysics, v. 36, no. 6, p. 1074-1098.

Benzecri, J.P., 1970, Distance distributionnelle et metrique du Chi deux en analyse factorielle des correspondances (3rd ed.): Lab Statis. Math., Fac. Sci. Univ. de Paris, 173 p.

Bone, M.R., Graebner, R.J., and Brown, A.R., 1979, Three-dimensional seismic technology (abst.): 10th Ann. APG Symp., West Virginia Geol. Survey Circ. C-15, p. 2.

Bouille, F., 1976, Graph theory and digitization of geological maps: Jour. Math. Geology, v. 8, no. 4, p. 375-393.

Bracewell, R.N., 1965, The Fourier transform and its applications: McGraw-Hill Book Co., New York, 381 p.

Brady, J.B., 1978, Symmetry: an interactive graphics computer program to teach symmetry recognition: Computers & Geosciences, v. 4, no. 2, p. 179-187.

Burk, C.F., Jr., 1973, Computer-based storage and retrieval of geoscience information: Bibliography, 1970-1972: Geol. Sur. Canada Paper 73-14, 38 p.

Burns, K.L., and Remfry, J.G., 1976, A computer method of constructing geological histories from field surveys and maps: Computers & Geosciences, v. 2, no. 2, p. 141-162.

Cazes, P., 1970, Application de l'analyse de donnees an traitement de problemes geologique: These de 3eme cycle, Fac. Sci. Univ. de Paris, 132 p.

Cazes, P., Solety, P., and Vuillaume, Y., 1970, Exemple de traitement statistique de donnees hydrochimiques: Fr. Bur. Rech. Geol. Minieres Bull. (Ser. 2), Sect. 3, no. 4, p. 75-90.

Cisne, J.L., and Rabe, B.D., 1978, Coenocorrelation: gradient analysis of fossil communities and its applications in stratigraphy: Lethaia, v. 11, no. 4, p. 341-364.

Conley, C.D., and Hea, J.P., 1972, A lithologic data-recording form for a computer-based well-data system: Jour. Math. Geology. v. 4, no. 1, p. 61-72.

Cubitt, J.M., 1976, An analysis and management system suitable for sedimentological data, *in* Merriam, D.F., ed., Quantitative techniques for the analysis of sediments: Pergamon Press, Oxford, p. 1-9.

Dacey, M.F., and Krumbein, W.C., 1970, Markovian models in stratigraphic analysis: Jour. Math. Geol., v. 2, no. 2, p. 175-191.

Dagbert, M., and David, M., 1974, Pattern recognition and geochemical data: an application to Monteregian Hills: Can. Jour. Earth Sci., v. 11, no. 11, p. 1577-1585.

Dagbert, M., Pertsowsky, R., David, M., and Perrault, G., 1975, Agpaicity revisited: pattern recognition in the chemistry of nepheline syenite rocks: Geochim. et Cosmochim. Acta., v. 39, no. 11, p. 1499-1504.

Davaud, E., and Geux, J., 1978, Traitement analgique <<manuel>> et algorithmique de problemes complexes de correlations biochronologiques: Eclogae Geol. Helvetiae, v. 71, no. 3, p. 581-610.

David, M., and Beauchemin, Y., 1974, The correspondence analysis method and a FORTRAN IV program: GEOCOM Program 10, 14 p.

David, M., Campiglio, C., and Darling, R., 1974, Progress in R- and Q-mode analyses: correspondence analysis and its application to the study of geological processes: Can. Jour. Earth Sci., v. 11, no. 1, p. 131-146.

David, M., and Dagbert, M., 1975, Lakeview revisited: variograms and correspondence analysis. New tools for the understanding of geochemical data: Proc. Intern. Geochem. Symposium, Vancouver, p. 163-181.

David, M., Dagbert, M., and Beauchemin, Y., 1977, Statistical analysis in geology: correspondence analysis method: Colorado Sch. Mines Quart., v. 72, no. 1, 60 p.

David, M., and Woussen, G., 1974, Correspondence analysis, a new tool for geologists: Proc. Mining Pribram, v. 1, p. 41-65.

David, P.P., and Lebuis, J., 1976, LEDA: a flexible codification system for computer-based files of geological field data: Computers & Geosciences, v. 1, no. 4, p. 265-278.

Dienes, I., 1974a, General formulation of the correlation problem and its solution in two special situations: Jour. Math. Geology, v. 6, no. 1, p. 73-81.

Dienes, I., 1974b, Subdivision of geological bodies into ordered parts, *in* Mathemathika es szamitastecknika a nyersanyagbutatosban, Proceedings of a conference held at Budapest, v. 1: Hungarian Geol. Soc., p. 138-147.

Dienes, I., 1977, Formalized stratigraphy: basic notions and advantages, *in* Merriam, D.F., ed., Recent advances in geomathematics: Pergamon Press, Oxford, p. 81-87.

Dienes, I., and Mann, C.J., 1977, Mathematical formalization of stratigraphic terminology: Jour. Math. Geology, v. 9, no. 6, p. 587-603.

Dixon, C.J., 1970, Semantic symbols: Jour. Math. Geology, v. 2, no. 1, p. 81-87.

Dobrin, M.B., ed., 1975, Continuing education course on principles of seismic stratigraphy: Am. Assoc. Petroleum Geologists, Dallas, Texas, 87 p.

Dobrin, M.B., 1977, Seismic exploration for stratigraphic traps: Am. Assoc. Petroleum Geologists Mem. 26, p. 329-351.

Doveton, J.H., and Cable, H.W., 1979, KOALA - minicomputer log analysis system for geologists (abst.): Am. Assoc. Petroleum Geologists Bull., v. 63, no. 3, p. 441.

Dowell, T.P.L., Jr., 1972, An automated approach to subsurface correlation: unpubl. master's thesis, Univ. Illinois, Urbana, 34 p.

Edwards, L.E., 1978, Range charts and no-space graphs: Computers & Geosciences, v. 4, no. 3, p. 247-255.

Edwards, L.E., and Beaver, R.J., 1978, The use of a paired comparison model in ordering stratigraphic events: Jour. Math. Geology, v. 10, no. 3, p. 261-272.

Ethier, V.G., 1975, Application of Markov analysis to the Baniff Formation (Mississippian), Alberta: Jour. Math. Geology, v. 7, no. 1, p. 47-61.

Farnback, J.S., 1975, The complex envelope in seismic signal analysis: Seis. Soc. America Bull., v. 65, no. 4, p. 951-962.

Farr, J.B., 1979, High-resolution seismic work as tool to locate stratigraphic traps (abst.): Am. Assoc. Petroleum Geologist Bull., v. 63, no. 3, p. 448.

Feldhausen, P.H., and Ali, S.A., 1976a, Sedimentary environmental analysis of Long Island Sound, USA, with multivariate statistics, *in* Merriam, D.F., ed., Quantitative techniques for the analysis of sediments: Pergamon Press, Oxford, p. 73-98.

Feldhausen, P.H and Ali, S.A., 1976b, A multivariate statistical approach to sedimentary environmental analysis: Trans. Gulf Coast Assoc. Geol. Soc., v. 24, p. 314-320.

Fertl, W.H., and Hammack, G.W., 1971, A comparative look at water saturation computations in shaly pay sands: SPWLA Logging Symp.

Flowers, B.S., 1976, Overview of exploration geophysics - recent breakthrough and challenging new problems: Am. Assoc. Petroleum Geologists Bull., v. 60, no. 1, p. 3-11.

Gordon, A.D., and Birks, H.J.B., 1974, Numerical methods in Quaternary paleoecology, II. Comparison of pollen diagrams: New Phytol., v. 73, p. 221-249.

Gordon, A.D., and Reyment, R.A., 1979, Slotting of borehole sequences: Jour. Math. Geology, v. 11, no. 3, p. 309-327.

Griesemer, A.D., and Costello, D.F., 1972, An attempt at an unambiguous scheme to code the geometry of bedding planes: Jour. Math. Geology, v. 4, no. 4, p. 345-351.

Harbaugh, J.W., 1979, Computer-based oil exploration decision systems (abst.): 10th Ann. APG Symp., West Virginia Geol. Survey Circ. C-15, p. 15-16.

Harbaugh, J.W., Doveton, J.H., and Davis, J.C., 1977, Probability methods in oil exploration: John Wiley & Sons, New York, 261 p.

Harris, M.H., and McCammon, R.B., 1971, A computer-oriented generalized porosity-lithology interpretation of neutron, density and sonic logs: Jour. Petr. Tech., v. 23, no. 2, p. 239-248.

Hattori, I., 1973, Mathematical analysis to discriminate two types of sandstone-shale alternations: Sedimentology, v. 20, no. 3, p. 331-345.

Hattori, I., 1976, Entropy in Markov chains and discrimination of cyclic patterns in lithologic successions: Jour. Math. Geology, v. 8, no. 4, p. 477-497.

Hawkins, D.M., and Merriam, D.F., 1973, Optimal zonation of digitized sequential data: Jour. Math. Geology, v. 5, no. 4, p. 389-395.

Hawkins, D.M., and Merriam, D.F., 1974, Zonation of multivariate sequences of digitized geologic data: Jour. Math. Geology, v. 6, no. 3, p. 263-269.

Hay, W.W., 1972, Probabilistic stratigraphy: Eclogae Geol. Helvetiae, v. 65, no. 2, p. 255-266.

Hay, W.W., 1974, Implications of probabilistic stratigraphy for chronostratigraphy: Verhandl. Naturf. Ges. Basel, v. 84, p. 164-171.

Hay, W.W., and Steinmetz, J.C., 1973, Probabilistic analysis of distribution of Late Paleocene - Early Eoecene calcareous nannofossils: Soc. Econ. Paleon. and Min., Calc. Nannofossil Symp., Houston, Texas, p. 58-70.

Hay, W.W., and Southam, J.R., 1978, Quantifying biostratigraphic correlation: Ann. Rev. Earth Planet. Sci., v. 6, p. 353-375.

Hazel, J.E., 1970, Binary coefficients and clustering in biostratigraphy: Geol. Soc. America Bull., v. 81, no. 11, p. 3237-3252.

Hazel, J.E., 1977, Use of certain multivariate and other techniques in assemblage zonal biostratigraphy: examples utilizing Cambrian, Cretaceous, and Tertiary benthic invertebrates, *in* Kauffman, E.G., and Hazel, J.E., eds., Concepts and methods of biostratigraphy: Dowden, Hutchinson & Ross, Inc., Stroudsburg, Pennsylvania, p. 187-212.

Henderson, G.J., 1973, Correlation and analysis of geologic time series: unpubl. doctoral dissertation, Univ. Indiana, Bloomington, 289 p.

Henderson, R.A., and Heron, M.L., 1977, A probabilistic method of paleobiogeographic analysis: Lethaia, v. 10, no. 1, p. 1-15.

Hohn, M.E., 1978, Stratigraphic correlation by principal components: effects of missing data: Jour. Geology, v. 86, no. 4, p. 524-532.

Hubaux, A., 1970, Description of geological objects: Jour. Math. Geology, v. 2, no. 1, p. 89-95.

Hubaux, A., 1971, Scheme for a quick description of rocks: Jour. Math. Geology, v. 3, no. 3, p. 317-322.

Hubaux, A., 1972a, Dissecting geological concepts: Jour. Math. Geology, v. 4, no. 1, p. 77-80.

Hubaux, A., ed., 1972b, Geological data files: CODATA Bull., v. 8, 30 p.

Iglehart, C.F., 1979, The API well number: what is it? Where is it available? (abst.): 10th Ann. APG Symp., West Virginia Geol. Sur., Circ. C-15, p. 18-19.

Jacod, J., and Joathon, P., 1971, Use of random-genetic models in the study of sedimentary processes: Jour. Math. Geology, v. 3, no. 3, p. 265-279.

Jacod, J., and Joathon, P., 1972, Conditional simulation of sedimentary cycles in three dimension, *in* Merriam, D.F., ed., Mathematical models of sedimentary processes: Plenum Press, New York, p. 139-165.

Jeffery, K.G., and Gill, E.M., 1973, G-EXEC: A generalized FORTRAN system for data handling: Geol. Sur. Canada Paper 74-63, p. 59-61.

Jones, T.A., 1972, Multiple regression with correlated independent variables: Jour. Math. Geology, v. 4, no. 3, p. 203-218.

Jones, T.A., Johnson, C.R., and Phillips, D.C., 1976, Interactive computer graphics and petroleum exploration (abst.): Geol. Soc. America, v. 8, p. 484-485.

Kemp, L.F., Jr., 1977, An algorithm for the stratigraphic correlation of well logs: Amoco Prod. Co., Research Dept. Rept. Cr 77-3, 21 p.

Konen, C.E., and Helander, D.P., 1970, A computer analysis of shaly sands using multiple porosity logging devices: The Log Analyst, v. 11, no. 1, p. 3-12.

Krumbein, W.C., 1969, The computer in geological perspective, *in* Merriam, D.F., ed., Computer applications in the earth sciences: Plenum Press, New York, p. 251-275.

Krumbein, W.C., 1972, Probabilistic models and the quantification process in geology: Geol. Soc. America Sp. Paper 146, p. 1-10.

Krumbein, W.C., 1976, Probabilistic modeling in geology, *in* Merriam, D.F., ed., Random processes in geology: Springer-Verlag, New York, p. 39-54.

Krumbein, W.C., and Watson, G.S., 1972, Effects of trends on correlation in open and closed three-component systems: Jour. Math. Geology, v. 4, no. 4, p. 317-330.

Lahiri, A., and Rao, S.V.L.N., 1974, Gradient analysis: a technique for the study of spatial variation: Modern Geology, v. 5, no. 1, p. 33-45.

Lahiri, A., and Rao, S.V.L.N., 1978, A choice between polynomial and Fourier trend surfaces: Modern Geology, v. 6, no. 3, p. 153-162.

Leont'ev, G.I., 1972, An attempt at a synchronization of old cyclically bedded sediment by the method of graphic connections: Lithology and Mineral Resources, v. 7, p. 103-113.

Loudon, T.V., 1964, Computer analysis of orientation data in structural geology: Ofc. Naval Research, Geography Branch, Tech. Rept. no. 13, ONR Task No. 339-135, 129 p.

Loudon, T.V., 1967, The use of eigenvector methods in describing surfaces: Kansas Geol. Survey Computer Contr. 12, p. 12-15.

Magara, K., 1979, Identification of sandstone body types by computer method: Jour. Math. Geology, v. 11, no. 3, p. 269-283.

Mann, C.J., 1976a, The PLATO system, its language, assets, and disadvantages: Computers & Geosciences, v. 2, no. 1, p. 41-50.

Mann, C.J., 1976b, Geology lessons on PLATO (abst.): Geol. Geol. Soc. America, v. 8, p. 492.

Mann, C.J., 1977, Toward a theoretical stratigraphy: Jour. Math. Geol., v. 9, no. 6, p. 649-652.

Mann, C.J., and Dowell, T.P.L., Jr., 1978, Quantitative lithostratigraphic correlation of subsurface sequences: Computers & Geosciences, v. 4, no. 3, p. 295-306.

McCammon, R.B., 1970, Component estimation under uncertainty, in Merriam, D.F., ed., Geostatistics, Plenum Press, New York, p. 45-61.

McCammon, R.B., 1972, Estimating lithologic components in stratigraphic sequences under uncertainty: Geol. Soc. America Sp. Paper 146, p. 11-24.

Meckel, L.D., Jr., and Nath, A.K., 1977, Geologic considerations for stratigraphic modelling and interpretation: Am. Assoc. Petroleum Geologists Mem. 26, p. 417-438.

Merriam, D.F., ed., 1976, Random processes in geology: Springer-Verlag, New York, 161 p.

Miall, A.D., 1973, Markov chain analysis applied to an ancient alluvial plain succession: Sedimentology, v. 20, no. 3, p. 347-364.

Miesch, A.T., 1974, Q-mode factor analysis: U.S. Geol. Survey Prof. Paper 900, p. 139-140.

Miesch, A.T., 1975, Simulation of sampling problems: U.S. Geol. Survey Prof. Paper 975, p. 158-159.

Miesch, A.T., 1976, Statistical geochemistry and petrology: U.S. Geol. Survey Prof. Paper 1000, p. 181-182.

Miller, F.X., 1977, The graphic correlation method in biostratigraphy, in Kauffman, E.G., and Hazel, J.E., eds., Concepts and Methods of biostratigraphy: Dowden, Hutchinson & Ross, Inc., Stroudsburg, Pennsylvania, p. 165-186.

Mizutani, S., and Hattori, I., 1972, Stochastic analysis of bed-thickness distribution of sediments: Jour. Math. Geology, v. 4, no. 2, p. 123-146.

Morgan, C.O., and McNellis, J.M., 1971, Reduction of lithologic-log data to numbers for use in the digital computer: Jour. Math. Geology. v. 3, no. 1, p. 79-86.

Neidell, N., and Poggiagliolmi, E., 1977, Stratigraphic modelling and interpretation - geophysical principles and techniques: Am. Assoc. Petroleum Geologists Mem. 26, p. 389-416.

Odell, J., 1976, An introduction to the LSDO2 system for rock description: Computers & Geosciences, v. 2, no. 4, p. 501-505.

Odell, J., 1977a, Description in the geological sciences and the lithostratigraphic description system, LSDO2: Geol. Mag., v. 114, no. 2, p. 81-163.

Odell, J., 1977b, LOGGER, a package which assists in the construction and rapid display of stratigraphic columns from field data: Computers & Geosciences, v. 3, no. 2, p. 347-379.

Payton, C.E., ed., 1977, Seismic stratigraphy - applications to hydrocarbon exploration: Am. Assoc. Petroleum Geologists Mem. 26, 516 p.

Poupon, A., Clavier, C., Dumanoir, J., Gaymard, R., and Misk, A., 1970, Log analysis of sand-shale sequences - a systematic approach: Jour. Petr. Tech., v. 22, no. 7, p. 867-881.

Price, R.J., and Jorden, P.R., 1979, A FORTRAN IV program for foraminiferid stratigraphic correlation and paleoenvironmental interpretation: Computers & Geosciences, v. 3, no. 4, p. 601-615.

Raffin, T.G., 1976, Oilfield, A PLATO lesson of the stratigraphy and economics related to oil drilling (abst.): Geol. Soc. America, v. 8, no. 4, p. 505.

Rao, M.S., 1975, Study of trend models: Modern Geology, v. 5, no. 2, p. 75-93.

Rao, S.V.L.N., and Rao, M.S., 1970, Geometric properties of hypersurfaces (trend surfaces) in three-dimensional space: Jour. Math. Geology, v. 2, no. 2, p. 203-205.

Read, W.A., 1976, An assessment of some quantitative methods of comparing lithological succession data, *in* Merriam, D.F., ed., Quantitative techniques for the analysis of sediments: Pergamon Press, Oxford, p. 33-51.

Rice, G.W., Bakker, M.L., and Weinberg, D.M., 1979, Geoseismic modelling - an interactive computer approach to stratigraphic and structural interpretation (abst.): Am. Assoc. Petroleum Geologists Bull., v. 63, no. 3, p. 515.

Robinson, J.E., and Merriam, D.F., 1971, Z-trend maps for quick recognition of geologic patterns: Jour. Math. Geology, v. 3, no. 2, p. 171-181.

Robinson, S.C., 1970, A review of data processing in the earth sciences in Canada: Jour. Math. Geology, v. 2, no. 4, p. 377-397.

Rubel, M., 1976, On biological construction of time in geology: Eesti NSV Tead. Akad Toim Keem Geol., v. 25, p. 136-144.

Rudman, A.J., Blakely, R.F., and Henderson, G.J., 1975, Frequency domain methods of stratigraphic correlation: Offshore Tech. Conf., v. 2, p. 265-277.

Rudman, A.J., and Lankston, R.W., 1973, Stratigraphic correlation of well logs by computer techniques: Am. Assoc. Petroleum Geologists Bull., v. 57, no. 3, p. 577-588.

Ruoff, W.A., 1976, A technique for interpreting depositional environments of sandstones from the SP log utilizing the computer: Log Analyst, v. 17, no. 4, p. 3-10.

Ruotsala, J.E., 1979, Seismic modelling (abst.): 10th Ann. APG Symp. West Virginia Geol. Survey Circ. C-15, p. 23-24.

Schlumberger, no date, Schlumberger engineered open hole services: Schlumberger, Houston, Texas, 40 p.

Schramm, M.W., Jr., Dedman, E.V., and Lindsey, J.P., 1977, Practical stratigraphic modelling and interpretation: Am. Assoc. Petroleum Geologists Mem. 26, p. 477-502.

Schwarzacher, W., 1972, The semi-Markov process as a general sedimentation model, *in* Merriam, D.F., ed., Mathematical models of sedimentary processes: Plenum Press, New York, p. 247-267.

Schwarzacher, W., 1976, Stratigraphic implications of random sedimentation, *in* Merriam, D.F., ed., Random processes in geology: Springer-Verlag, New York, p. 96-111.

Schwarzacher, W., 1978, Mathematical geology and sedimentary stratigraphy, *in* Merriam, D.F., ed., Geomathematics: past, present, and prospects: Syracuse Univ. Geol. Contr. 5, p. 65-71.

Scott, G.H., 1974, Essay review: stratigraphy and seriation: Newsletters of Stratigraphy, v. 3, p. 93-100.

Shaw, B.R., 1978, Quantitative lithostratigraphic correlation of digitized borehole-log records: Upper Glen Rose Formation, Northeast Texas: unpubl. doctoral dissertation, Syracuse Univ., 168 p.

Shaw, B.R., and Cubitt, J.M., 1979, Stratigraphic correlation of well logs: an automated approach, *in* Gill, D., and Merriam, D.F., eds., Geomathematical and petrophysical studies in sedimentology: Pergamon Press, Oxford, p. 127-148.

Shaw, B.R., and Simms, R., 1977, Stratigraphic analysis system, SAS: Computers & Geosciences, v. 3, no. 3, p. 395-427.

Sheriff, R.E., 1976, Inferring stratigraphy from seismic data: Am. Assoc. Petroleum Geologists Bull., v. 60, no. 4, p. 528-542.

Sheriff, R.E., 1977, Limitations on resolution of seismic reflections and geologic detail derivable from them: Am. Assoc. Petroleum Geologists Mem. 26, p. 3-14.

Smyth, M., and Cook, A.C., 1976, Sequence in Australian coal seams: Jour. Math. Geology, v. 8, no. 5, p. 529-547.

Southam, J.R., Hay, W.W., and Worsley, T.R., 1975, Quantitative formulation of reliability in stratigraphic correlation: Science, v. 188, no. 4186, p. 357-359.

Sutterlin, P.G., Jeffery, K.G., and Gill, E.M., 1977, FILEMATCH: a format for the interchange of computer-based files of structured data: Computers & Geosciences, v. 3, no. 3, p. 429-441.

Switzer, P., 1976, Applications of random process models to the description of spatial distributions of qualitative geological variables, *in* Merriam, D.F., ed., Random processes in geology: Springer-Verlag, New York, p. 124-134.

Talley, B.J., 1979, Lithology data systems - rocks to applications (abst.): Am. Assoc. Petroleum Geologists Bull., v. 63, no. 3, p. 537.

Taner, M.T., Koehler, F., and Sheriff, R.E., 1978, The computation and interpretation of seismic attributes by complex trace analysis: Seiscom Delta, Inc., Calgary, Alberta, Canada, 29 p.

Taner, M.T., and Sheriff, R.E., 1977, Application of amplitude, frequency, and other attributes to stratigraphic and hydrocarbon determinations: Am. Assoc. Petroleum Geologists Mem. 26, p. 301-327.

Teil, H., and Cheminee, J.L., 1975, Application of correspondence factor analysis to the study of major and trace elements in the Erta Ale Chain (Afar, Ethiopia): Jour. Math. Geology, v. 7, no. 1, p. 13-30.

Tipper, J.C., 1976, The study of geological objects in three dimensions by the computerized reconstruction of serial sections: Jour. Geology, v. 84, no. 4, p. 476-484.

Tipper, J.C., 1977, Three-dimensional analysis of geological forms: Jour. Geology, v. 85, no. 5, p. 591-611.

Vail, P.R., Mitchum, R.M., Jr., and Thompson, S., III, 1977, Seismic stratigraphy and global changes of sea level, part 3: relative changes of sea level from coastal onlap: Am. Assoc. Petroleum Geologists Mem. 26, p. 63-97.

Vail, P.R., and Mitchum, R.M., Jr., 1979, Global cycles of sea-level change and their role in exploration: Preprint for "Tenth World Petroleum Congress", September 9-14, Bucharest, Romania, 28 p.

Vincent, Ph., Gartner, J.-E., and Attali, G., 1979, An approach to detailed dip determination using correlation by pattern recognition: Jour. Petr. Tech., v. 31, no. 2, p. 232-240.

Warshauser, S.M., and Smosna, R.A., 1979, Multivariate analysis of carbonate data for paleoenvironmental interpretation (abst.): 10th Ann. APG Symp., West Virginia Geol. Survey Circ. C-15, p. 29.

Watson, G.S., 1971, Trend-surface analysis: Jour. Math. Geology, v. 3, no. 3, p. 215-226.

Watson, G.S., 1972, Trend-surface analysis and spatial correlation: Geol. Soc. America Sp. Paper 146, p. 39-46.

Watt, H.B., 1977, A complete analysis of complex and sandstone reservoirs: Dresser Atlas Tech. Memo, v. 6, no. 1, 8 p.

Webster, R., 1973, Automatic soil-boundary location from transect data: Jour. Math. Geology, v. 5, no. 1, p. 27-37.

Whitten, E.H.T., 1966, Sequential multivariate regression methods and scalars in the study of fold-geometry variability: Jour. Geology, v. 74, no. 5, pt. 2, p. p. 744-763.

Whitten, E.H.T., 1968, FORTRAN IV CDC 6400 computer program to analyze subsurface fold geometry: Kansas Geol. Survey Computer Contr. 25, 46 p.

Whitten, E.H.T. 1970, Orthogonal polynomial trend surfaces for irregularly spaced data: Jour. Math. Geology, v. 2, no. 2, p. 141-152.

Whitten, E.H.T., and Koelling, M.E.V., 1973, Spline-surface interpolation, spatial filtering, and trend surfaces for geological mapped variables: Jour. Math. Geology, v. 5, no. 2, p. 111-126.

Worsley, T.R., Blechschmidt, G., Ralston, S., and Snow, B., 1973, Probability-based analysis of the area-time distribution of Oligocene calcareous nannofossils: Soc. Econ. Paleon. and Min., Calc. Nannofossil Symp., Houston, Texas, p. 71-79.

Worsley, T.F., and Jorgens, M., 1977, Automated biostratigraphy, *in* Ramsay, A.T.S., ed., Oceanic micropaleontology: Academic Press, London, p. 1201-1229.

COMPUTER METHODS FOR GEOCHEMICAL AND PETROLOGIC MIXING PROBLEMS

A.T. Miesch

U.S. Geological Survey

ABSTRACT

Mixing problems arise frequently in examinations of compositional variations in rock bodies, particularly in studies of magmatic differentiation. At one time the problems were examined graphically, but ten years ago geochemists and petrologists were shown the matrix algebra that could be used to obtain least-squares solutions. Computer programs based on the fundamental matrix operations, and variations of them, have been circulated broadly and used widely by petrologists.

More recently, it has been determined that the methods of Q-mode factor analysis, extended for treating compositional data, are well suited for all types of chemical and mineralogic mixing problems. The methods can be used not only to estimate the mixing proportions, but also to determine the number of end members required in a given problem and to aid in determination of end-member compositions. The first step is to derive a matrix of recomputed data. The recomputed data matrix can approximate closely the original data matrix even though it may be of lower rank. The rank equals the number of end members required in the mixing model. Possible end-member compositions are represented by vectors in the same space as the row vectors in the matrix of recomputed data. Various methods can be used to select end-member vectors that might represent the compositions of materials involved in the mixing process. Also, selected compositions can be tested individually for mathematical suitability and modified accordingly. Interactive computer programs allow one to test geochemical hypotheses by trying various sets of end-member compositions until the derived mixing proportions are compatible with all that is known about the samples and the geologic environment from which they were collected.

INTRODUCTION

Mixing problems are abundant in geochemistry and petrology because most rocks have formed by processes of mixing or unmixing. Mixing occurs, for example, when sediments from different sources are deposited together or when magmas incorporate foreign materials. Unmixing occurs when minerals precipitate from aqueous solutions or silicate melts or when constitutents are removed from rocks by chemical or physical processes of alteration. However, the concept of mixing can be important even where mixing did not actually occur. Most rocks, for example, are regarded as mixtures of minerals although all the minerals may have crystallized in place from a common solution with no mixing at all. Also, many minerals are regarded as mixtures of theoretical end members even though the end members actually might not ever have existed in a pure state.

The basic mathematical model assumed in mixing problems is:

$$X_{NM} \simeq P_{Nm} C_{mM} \qquad (1)$$

X_{NM} is the matrix of compositional data for M constituents in N samples of rocks, sediments, soils, or water, and is approximated by the product of matrices P_{Nm} and C_{mM}. Matrix P_{Nm} contains the mixing proportions for the m end members in each of the N samples, and matrix C_{mM} contains the compositions of the m end members. The elements of matrices X_{NM} and C_{mM} are generally in units of percent concentration and all the rows of both matrices sum to 100. Most of the factor-analysis methods discussed here require this constant row-sum. Some of the elements in P_{Nm} may be negative, but the m values for each of the N rows must sum to plus one.

Ten years ago Bryan, Finger, and Chayes (1969) showed that a least-squares solution for the P_{Nm} matrix

GEOCHEMICAL AND PETROLOGIC MIXING PROBLEMS   245

may be derived from:

$$P_{Nm} = X_{NM} C'_{Mm} (C_{mM} C'_{Mm})^{-1} \qquad (2)$$

and variations of this method have been proposed and used by Wright and Doherty (1970) and by Stormer and Nicholls (1978). The method allows one to estimate mixing proportions given a particular set of end-member compositions. However, depending on the end-member compositions used, the matrix derived as the product of the derived mixing proportions and the given end-member compositions may or may not approximate the matrix of original data. If it does not, either alternative end-member compositions must be used or others must be added to those used previously. The method provides no information about the number of end-member compositions required for a given problem, although it is generally known that m need be no greater than M in order to account perfectly, in a mathematical sense, for any observed data. There are a great many mixing problems in geochemistry and petrology where the C_{mM} matrix to be used in equation (2) will be obvious, as, for example when determining the proportions of albite and anorthite in a series of plagioclase specimens: or in estimating the proportions of known minerals of known composition in a suite of rock samples. However, there also are a great many problems where determination of a mathematically adequate and geologically plausible C_{mM} matrix is as much as or more of a problem than determination of P_{Nm}. In this type of situation, methods of factor analysis and vector geometry can be useful.

FACTOR ANALYSIS

The methods of Q-mode factor analysis described by Klovan and Imbrie (1971) lead to a matrix of principle components or varimax factor loadings, A_{Nm}, and a matrix of principal components or varimax factor scores, F_{mM}. Each of the N rows of the loadings matrix can be taken as m coordinates of a vector that represents the corresponding sample. Each row of the scores matrix pertains to one of the m reference axes (principal components or varimax) for the vector system. If the extended Q-mode methods (Miesch, 1976a) are used, one can use A_{Nm} and F_{mM} to derive an approximate data matrix, $\hat{X}_{NM}$. A brief description of the procedures is given in the Appendix. The column means of $\hat{X}_{NM}$ are the same as those of the actual data matrix,

X_{NM}; the principal difference between the two matrices is that $\hat{X}_{NM}$ is of lower rank. The rank of X_{NM} is always M if it consists of actual chemical or mineralogic determinations. The rank of $\hat{X}_{NM}$ may be much less than M without the two matrices differing by an appreciable amount. Thus, a mixing model with relatively few end members (m) may be derived to account for $\hat{X}_{NM}$, whereas M end members are required to account for X_{NM}. If the two matrices are substantially the same, as may be the situation, the extra end members are superfluous and may cause development of the mixing model to be unncessarily difficult.

VECTOR REPRESENTATION OF COMPOSITIONS

The compositions represented by the vectors defined by the rows of matrix A_{Nm} (that is, the sample vectors) are contained in the corresponding rows of matrix $\hat{X}_{NM}$. The compositions represented by other vectors may be determined by the following procedure. Determine the scores for the vector from:

$$G_M = B_m F_{mM} \qquad (3)$$

where B_m contains the coordinates (loadings) of the vector with respect to either the principal components or varimax axes represented by F_{mM}. A scale factor for the kth set of scores then is derived from:

$$s_k = \frac{K - \sum_j b_j}{\sum_j (g_j(a_j - b_j))} \qquad (4)$$

where K is the constant row sum in X_{NM} and $\hat{X}_{NM}$ (generally 100), g_j is an element of the score vector G_M, and a_j and b_j are constants used in the initial scaling of the jth variable (see equation 1a of Appendix). The composition represented by the vector then is given by:

$$c_j = s_k g_j(a_j - b_j) + b_j \qquad 1 \geq j \geq M \qquad (5)$$

On the other hand, to derive the coordinates of a vector that best represents a given composition, the composition is first scaled by:

$$w_j = \frac{c_j-b_j}{a_j-b_j} \qquad 1 \geq j \geq M \qquad (6)$$

where c_j is the original compositional value (in the same units as X_{NM}), w_j is the scaled value, and a_j and b_j are the same constants used to scale the original data (see equation 1a of Appendix). The scaled data then are row-normalized by:

$$y_j = w_j/(\Sigma_j w_j^2)^{1/2} \qquad 1 \geq j \geq M \qquad (7)$$

and the coordinates are obtained from:

$$B_m = Y_M F'_{Mm} \qquad (8)$$

where Y_M is a row-vector containing y_j. If m in equation (8) is equal to M, the sum of squares of the elements in B_m (the vector communality) will equal unity. Otherwise, the sum of squares generally will be less than unity indicating that the composition does not fit perfectly into the m-dimensional vector space. Use of the vector B_m derived with equation (8) in equation (3) followed by use of equations (4) and (5), however, will give the composition represented by the vector after it has been projected into the m-dimensional compositional system. The differences between the c_j values used in equation (6) and those later derived with equation (5) will indicate the degree of departure of the composition tested from the compositional system represented by $\hat{X}_{NM}$.

MIXING MODELS FROM THE RESULTS OF FACTOR ANALYSIS

The basic mixing model is:

$$\hat{X}_{NM} = P_{Nm} C_{mM} \qquad (9)$$

where $\hat{X}_{NM}$ is an approximation of X_{NM} and P_{Nm} and C_{mM} are as defined for equation (1). Each of the m rows of the C_{mM} matrix contains a set of c_j values from equation (5), and the elements of the P_{Nm} matrix are derived from an A_{Nm} matrix by:

$$p_{ik} = a_{ik}/(s_k \Sigma_k (a_{ik}/s_k)) \qquad 1 \leq i \leq N; \ 1 \leq k \leq m \qquad (10)$$

where s_k is the scale factor derived from equation (4) corresponding to the kth row in C_{mM} and the kth column in A_{Nm}.

The principal task in the development of a mixing model by Q-mode factor methods consists of determining reference vectors within the vector space represented by A_{Nm} that lead to matrices of P_{Nm} and C_{mM} that are acceptable on geologic grounds. The compositions contained in C_{mM} must be those of geologic materials that were or, at least, could have been involved in the mixing process, or if the model is conceptual only, the compositions in C_{mM} must be consistent with the model's purpose. Also, the mixing proportions in P_{Nm} must be acceptable for each sample in both sign and magnitude. Once the principal-components or varimax coordinates of m reference vectors have been determined, they are used to form the matrix B_{mm}. Multiplication of the matrix A_{Nm} by the inverse of B_{mm} yields a new A_{Nm} matrix containing the coordinates of the sample vectors with respect to the new reference axes (Imbrie, 1963). Each row of matrix B_{mm} can be used in equation (3) to derive a score vector G_M. Equations (4) and (5) then are used to derive the matrix of end-member compositions, C_{mM}, and equation (10) applied to the new A_{Nm} matrix gives the mixing proportions P_{Nm}. The product of the matrices of mixing proportions and end-member compositions will equal the same matrix of $\hat{X}_{NM}$ as determined, for the corresponding value of m, from the principal components or varimax axes by the procedures outlined in the Appendix.

SEARCHING FOR END-MEMBER COMPOSITIONS

As noted previously, the principal task in the development of a geologic mixing model generally is to determine m end-member compositions that both are satisfactory mathematically and plausible geologically in view of all available geochemical and geologic data. The task is reduced considerably if the matrix $\hat{X}_{NM}$ is accepted as a satisfactory approximation of X_{NM} because, in this situation, the number of end-member compositions necessary is known and all possible end-member compositions that are suitable mathematically are represented by hypothetical vectors contained in the space defined in matrix A_{Nm}. In other words, the m end-member compositions are all linear combinations of

GEOCHEMICAL AND PETROLOGIC MIXING PROBLEMS   249

the compositions contained in the matrix $\hat{X}_{NM}$.

The compositions represented by the vectors within the space defined by A_{Nm} change systematically across the space in all directions. Moving outward from the approximate center of the space, away from the first principal-components axis, the compositional value for some constitutent will eventually become zero and then negative. If this operation is repeated a large number of times, moving outward in various directions, the points where various constitutents turn negative will define the margin of what can be termed the positive subspace. Because negative compositional values are impossible, vectors representing possible end-member compositions occur not only witnin the space defined by A_{Nm} but within the positive subspace of A_{Nm}.

All vectors within the positive subspace of A_{Nm} represent end-member compositions that are satisfactory mathematically in the sense that any m such vectors represent nonnegative compositions and will yield P_{Nm} and C_{mM} matrices whose product equals $\hat{X}_{NM}$ exactly. Searches for a set of geologically satisfactory end members can be conducted by selecting m vectors from the positive subspace, and then examining the resultant P_{Nm} and C_{mM} matrices. If the end-member compositions contained in the rows of C_{mM} are at least similar to those of geologic materials that might have been involved in the mixing (or unmixing) process, and if the mixing proportions contained in the rows of P_{Nm} are reasonable in both sign and magnitude, the search is ended. Most of these P_{Nm} and C_{mM} matrices, however, will be objectionable on geologic grounds. In this situation, either alternative end-member vectors must be selected, or one or more of the methods discussed next may be tried.

Alternative to trying various sets of vectors, a more direct approach is to begin with compositions that one would expect, from knowledge of the geology, to have taken part in the mixing (unmixing) process. For example, if the mineral magnetite is a prominant constituent in a series of differentiated lavas, it is likely that the separation (unmixing) of magnetite contributed to the differentiation process. Equations (6) to (8) can be used to determine the coordinates of the vector within the space defined by A_{Nm} that most closely represents the composition of magnetite. If the vector communality is less than one, the composition actually represented by the vector then must be determined from equations (3) to (5), and if it happens that the composition is partly negative (that is, not within the positive subspace), or if the composition is different substantially from the composition of magnetite, pure magnetite can be rejected as one of the end members for the mixing problem. This will not indicate that magnetite was not involved in the differentiation process, only that magnetite did not separate independently of other minerals. It is possible yet that magnetite separated along with other minerals such as, for example, olivine, plagioclase, or other iron oxides.

In many geologic environments it is to be expected that the addition or separation of one mineral will be accompanied by the addition or separation of one or more others. Thus, even though the correlations among the amounts added or separated may not be perfect, the correlated behavior of the various minerals may cause them to have the effect of a single end member in the determination of m. It is possible to anticipate groups of minerals that might behave in this manner in various geologic environments. For example, olivine separating from a magma might be expected to be accompanied by small amounts of magnetite and, perhaps, a calcic plagioclase, and clay minerals being deposited in a stream bed generally contain minor particles of other minerals as impurities. The proportions of the various minerals in the assemblage can be estimated using computer programs that perform iterative computations. The programs form mathematical mixtures of the various minerals in the assemblage, progressing systematically from zero to 100 percent for each mineral, and test each mixture in the same manner used to test an individual mineral as described in the preceeding paragraph. The mixture with the highest vector communality is taken as the most likely mixture in the total assemblage to have been involved in the mixing process. Examples involving the assemblage magnetite-limenite and the solid-solution series for forsterite-fayalite are given in Miesch (1976a, fig. 10, 13, and 15). If a large number of the mixtures in the assemblage have high vector communalities, a range of compositions within the assemblage may have been involved in the mixing process, and the assemblage may contain more than a single end member for the mixing model. For an example of this situation, involving the assemblage hornblende-albite-anorthite-magnetite in granitic rocks, see Miesch and Reed (1979, fig. 11 and accompanying discussion). Another example, involving the solid-solution series albite-anorthite-orthoclase, is given in Miesch (1976a, fig. 12).

AN EXAMPLE OF THE PROCEDURES

The procedures discussed in the preceding sections will be illustrated using analyses of five orthopyroxenes from Deer, Howie, and Zussman (1963, table 2, analyses 1, 6, 10, 17, and 19). The analyses used are for SiO_2, Fe_2O_3, FeO, and MgO, and CaO, but Fe_2O_3 was recomputed and combined with FeO as is customary in many types of petrochemical calculations, and all four oxide values then were adjusted to sum to 100 percent. The adjusted analyses form the matrix of original data:

$$X_{NM} = \begin{matrix} SiO_2 & FeO & MgO & CaO \\ 59.84 & 0.38 & 39.46 & 0.32 \\ 57.05 & 9.13 & 33.32 & 0.50 \\ 54.25 & 18.74 & 24.29 & 2.72 \\ 51.95 & 32.42 & 14.15 & 1.48 \\ 48.23 & 43.11 & 7.14 & 1.52 \end{matrix} \quad (11)$$

Orthopyroxenes comprise a mineral group whose members range in composition mainly between enstatite ($MgSiO_3$) and ferrosilite ($FeSiO_3$), but small amounts of the wollastonite molecule ($CaSiO_3$) also may be present. The ideal compositions of these theoretical end members, as percents, are:

$$C_{mM} = \begin{matrix} SiO_2 & FeO & MgO & CaO & \\ 59.85 & 0 & 40.15 & 0 & \text{Enstatite} \\ 45.54 & 54.46 & 0 & 0 & \text{Ferrosilite} \\ 51.72 & 0 & 0 & 48.28 & \text{Wollastonite} \end{matrix} \quad (12)$$

Use of matrices (11) and (12) in equation (2) gives the following matrix of mixing proportions:

$$P_{Nm} = \begin{matrix} \text{Enstatite} & \text{Ferrosilite} & \text{Wollastonite} \\ 0.9854 & 0.0081 & 0.0082 \\ 0.8241 & 0.1653 & 0.0069 \\ 0.6010 & 0.3425 & 0.0540 \\ 0.3682 & 0.6018 & 0.0401 \\ 0.1772 & 0.7913 & 0.0311 \end{matrix} \quad (13)$$

And finally, multiplication of matrices (12) and (13) according to equation (9) gives the following approximation of the original data:

$$\hat{X}_{NM} = \begin{matrix} SiO_2 & FeO & MgO & CaO \\ 59.77 & 0.44 & 39.57 & 0.40 \\ 57.21 & 9.00 & 33.09 & 0.33 \\ 54.36 & 18.65 & 24.13 & 2.61 \\ 51.52 & 32.78 & 14.79 & 1.94 \\ 48.25 & 43.10 & 7.11 & 1.50 \end{matrix} \quad (14)$$

If each of the rows of matrix P_{Nm} is divided through by the row-sum prior to the computation of $\hat{X}_{NM}$, then the rows of P_{Nm} will sum to unity and those of $\hat{X}_{NM}$ will sum to 100 as is required for a mixing model based on percentages. However, in this example the P_{Nm} and $\hat{X}_{NM}$ matrices are changed only slightly by this procedure. Comparison of the estimated matrix in (14) with the original matrix in (11) is a test of the mathematical adequacy of the mixing model.

RESULTS FROM FACTOR ANALYSIS

Methods of extended Q-mode factor analysis may be used to arrive at a mixing model similar to that contained in matrices (12) and (13), but it is not necessary that the precise end-member compositions be known beforehand. The need for a priori knowledge of the end-member compositions is no problem in treating data of the type used in this example, but is the major obstacle in many other types of geologic mixing problems.

Application of the methods described in the Appendix to the data in matrix (11), but omitting the initial scaling according to equation (1a), leads to the following coefficients of determination between corresponding columns of X_{MN} and $\hat{X}_{NM}$ derived using m = 2 to m = 4 factors:

m	SiO_2	FeO	MgO	CaO	
2	0.9876	0.9997	0.9986	0.2673	
3	0.9879	1.0000	0.9997	0.9425	(15)
4	1.0000	1.0000	1.0000	1.0000	

The coefficients for CaO point clearly to the need for 3 end members in order to account for appreciable portions of the variances in all four oxide variables. The non-zero eigenvalues of the cosine-theta matrix derived from the data in matrix (11) are:

$$4.5690 \quad 0.4300 \quad 0.0008 \quad 0.0002$$

These values give no clear indication that three end members are required. The estimated data matrix derived with m = 3 is:

$$\hat{X}_{NM} = \begin{array}{cccc} SiO_2 & FeO & MgO & CaO \\ 59.52 & 0.48 & 39.53 & 0.46 \\ 57.37 & 9.04 & 33.23 & 0.35 \\ 54.60 & 18.68 & 24.16 & 2.57 \\ 51.23 & 32.44 & 14.51 & 1.82 \\ 48.59 & 43.14 & 6.92 & 1.34 \end{array} \quad (16)$$

The matrix of principal-components scores is (first 3 rows only):

$$F_{mM} = \begin{array}{cccc} SiO_2 & FeO & MgO & CaO \\ 0.8643 & 0.3368 & 0.3729 & 0.0217 \\ -0.0850 & 0.8287 & -0.5527 & 0.0240 \\ 0.2265 & -0.2409 & -0.3582 & 0.8731 \end{array} \quad (17)$$

and the matrix of principal-components loadings is (first 3 columns only):

$$A_{Nm} = \begin{array}{ccc} PC1 & PC2 & PC3 \\ 0.9287 & -0.3707 & -0.0055 \\ 0.9718 & -0.2352 & -0.0116 \\ 0.9989 & -0.0391 & 0.0232 \\ 0.9723 & 0.2333 & 0.0029 \\ 0.9048 & 0.4257 & -0.0106 \end{array} \quad (18)$$

Multiplication of matrices (18) and (17) yields a product matrix with rows that can be scaled, using equations (4) and (5), to yield the estimated data matrix in (16). Alternatively, the rows of matrix (17) can be scaled according to equations (4) and (5), and matrix (18) scaled by equation (10), so that the product of the two will yield matrix (16) directly. Regardless, the rows of matrix (18) each contain the coordinates of a vector that represents the composition given in the corresponding row of matrix (16). Because the vector system occupies only three dimensions, it can be represented on a stereogram as shown in Figure 1. The limits of the positive subspace also are shown on the stereogram as well as the positions of the principal components and varimax reference axes. The five sample vectors cluster about a plane that occurs near one margin of the positive subspace. The compositions represented by vectors at various locations within the positive subspace can be inferred from the contours in Figure 2. It will be seen that vectors near the upper-left corner of the subspace represent compositions close to that of enstatite, those near the upper-right corner represent compositions close to that of ferrosilite, and those near the lower corner represent compositions close to that of wollastonite. Data for drawing Figures 1 and 2 were derived from programs EQSPIN and EQSTER described in Miesch (1976b).

Four specific methods that can be used to explore for end-member compositions for the mixing model will be reviewed. Each of the methods is applicable theoretically in situations where the vector system occupies any number of dimensions, but available computer programs are restricted to ten or fewer dimensions (end members) and most of the successful applications have involved no more than four.

Method 1 is used in situations where there is a theoretical basis for assuming what one or more of the end-member compositions might have been. In the orthopyroxene example being used here the obvious end-member compositions are those given in matrix (12) - the ideal compositions of enstatite, ferrosilite, and wollastonite. Each of these compositions was normalized with equation (7) and corresponding principal-components vector coordinates were obtained with equation (8) using matrix (17). [Because the data were not scaled prior to row-normalization and derivation of the R_{MM} matrix - see Appendix - scaling according to equation (6) prior to the use of equation (7) can be omitted.] The computed principal-components coordinates and respective vector communalities are:

	(1)	(2)	(3)	Communality	
Enstatite	0.933748	0.357743	0.011441	0.999996	
Ferrosilite	0.194553	0.980077	0.039510	0.999963	(19)
Wollastonite	0.502661	0.409309	-0.761384	0.999908	

Although the communalities are high, all of them are less than one and the compositions actually

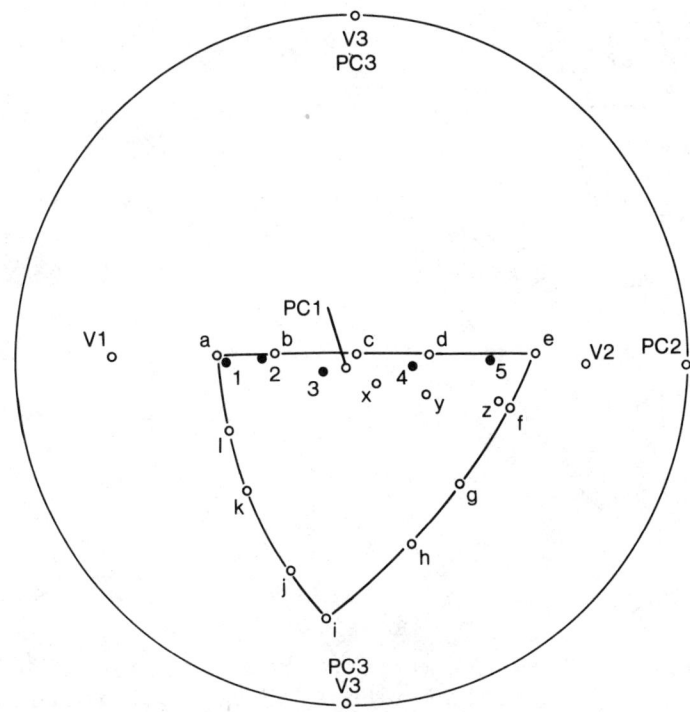

Figure 1. Stereogram showing relative positions of five vectors (solid dots) that represent compositions of orthopyroxenes. Open circles labeled V1, V2, and V3 represent varimax reference axes. Open circles labeled PC1, PC2, and PC3 represent principal-components reference axes. Other open circles represent vectors discussed in text. All vectors have been projected from upper hemisphere vertically onto plane of stereogram. Vectors outside of triangular area represent compositions that are partly negative.

represented by the three vectors, determined with equations (3), (4), and (5), are:

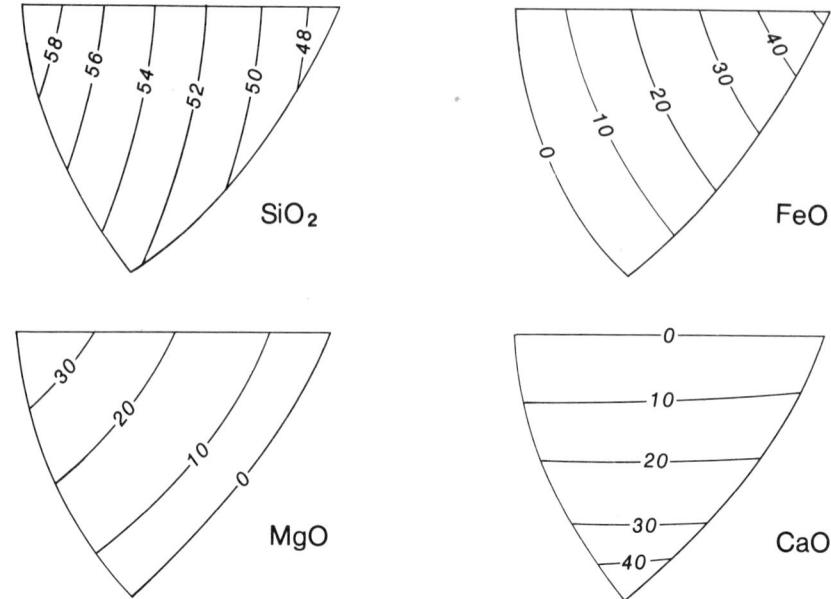

Figure 2. Contours over triangular area of Figure 1 showing nature of compositional variations.

	SiO$_2$	FeO	MgO	CaO	
Enstatite	59.69	0.05	40.18	0.07	
Ferrosilite	45.94	54.55	-0.28	-0.21	(20)
Wollastonite	52.41	-0.26	-0.45	48.30	

Therefore, although the ideal compositions of these minerals cannot be represented precisely in the 3-dimensional space of the sample vectors, vectors that represent closely similar compositions can be identified. Unfortunately, two of the three vectors lie outside of the positive subspace; the two recomputed compositions are partly negative. This difficulty could be overcome by selecting vectors near those listed in matrix (19), but within the positive subspace or on its margin. The computations required for method 1 are provided by program EQEXAM (Miesch, 1976b).

Method 2 is used in situations where one or more of the end-member compositions are thought to be real or conceptual mixtures of two or more other compositions. For example, if an end member is thought to be some type of feldspar, the end-member composition may be a conceptual mixture of albite ($NaAlSi_3O_8$), anorthite ($CaAl_2Si_2O_8$), and orthoclase ($KAlSi_3O_8$). Similarly, if we were unaware that our example data in matrix (11) pertained to orthopyroxenes but suspected that the end members might be iron and magnesium silicate minerals, we could consider the compositional system forsterite (Mg_2SiO_4)-fayalite (Fe_2SiO_4)-silica (SiO_2), as represented in Figure 3. The compositional system then could be examined throughout at small increments by testing each selected composition by the same procedures used in method 1. Using increments of 2 molar percent, 1,326 compositions within the system of Figure 3 were tested and 35 were determined to have communalities equal to or greater than the average of the sample vector communalities (0.9999 - the average row sum-of-squares for matrix 18). These compositions are identified on Figure 3 and range from ideal enstatite to ideal ferrosilite. With the exception of nine compositions close to but slightly outside this range (small row of dots in Fig. 3), no other compositions within the system met the test criterion. Therefore, the extremes in the range of compositions identified in Figure 3 could be taken as two of the end-member compositions for the mixing model. Alternatively, three end-member compositions could be identified by examination of the four-component system forsterite-fayalite-silica-orthosilicate (Ca_2SiO_4). The computational procedures for method 2 are contained in program EQSCAN (Miesch, 1976b).

Method 3 is used to determine an array of compositions that can be added to or subtracted from one composition to produce another. The initial and final compositions may be represented by two vectors in m-dimensional space, and all possible compositions that could produce one from the other are represented by vectors in the plane of these two vectors. For example, in order to change the composition represented by sample vector 2 on Figure 1 to the composition represented by sample vector 3, it is necessary to subtract a composition represented by a vector in the plane of vectors 2 and 3, but to the left of vector 2; or to add a composition represented by a vector in the same plane, but to the right of vector 3. Most vectors in that plane to the left of vector 2 are outside of the positive subspace and

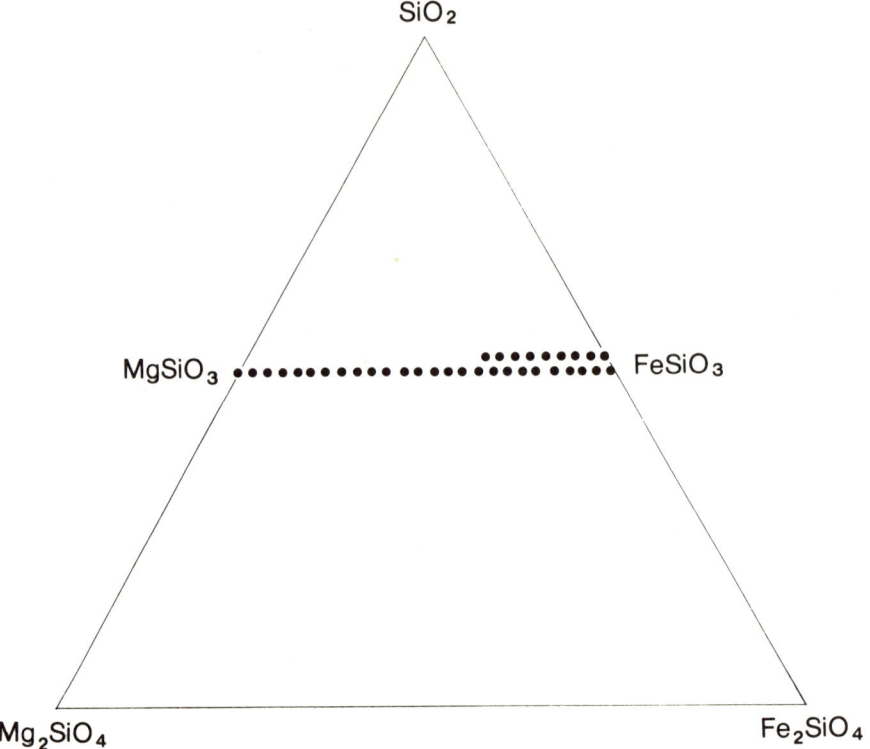

Figure 3. Triangular diagram representing compositional system enstatite (Mg_2SiO_4)- ferrosilite (Fe_2SiO_4)- silica (SiO_2) as molecular percentages. Solid dots identify compositions within system that can be represented in vector system for orthopyrexene data by vectors with communalities greater than 0.9999.

therefore, represent compositions that are partly negative. Some of those to the right of vector 3 are located on Figure 3 and represent the following compositions:

Vector	SiO_2	FeO	MgO	CaO	
x	53.58	22.20	20.85	3.37	
y	51.70	28.72	14.70	4.87	(21)
z	47.56	43.10	1.17	8.17	

Examples of the manner in which method 3 can be used to determine the end members for mixing models are given in Miesch and Reed (1979) and Miesch (1979). A FORTRAN computer program that performs the necessary computations and also derives CIPW norms for the determined compositions (EQFANN) is unpublished but available from the author (U.S. Geological Survey, Stop 925, Box 25046, Denver Federal Center, Denver, CO 80225).

Method 4 is based on the premise that many end-member compositions to be expected in certain petrologic systems include zero concentrations of one or more of the compositional variables. For example, a precipitate from basaltic magmas may be the mineral olivine which contains little or no Al_2O_3, CaO, Na_2O, or K_2O, variables usually represented in matrices of petrochemical data. Another precipitate from a broad range of magmas is plagioclase which contains little or no FeO or MgO. Because these compositions of interest contain zero values, the vectors represented by them in a system of sample vectors will occur on the margins of the positive subspace. Therefore, it is useful to search these margins for compositions that might be plausible geological end-member compositions. Twelve compositions represented by vectors at approximately regular intervals along the margins of the positive subspace represented in Figure 1 are as follows:

Vector	SiO_2	FeO	MgO	CaO	
a	59.72	0.00	40.28	0.00	
b	56.56	12.49	30.96	0.00	
c	53.46	24.71	21.82	0.00	
d	50.81	35.15	13.97	0.00	
e	46.07	53.93	0.00	0.00	
f	47.28	43.66	0.00	9.05	(22)
g	48.83	30.57	0.00	20.59	
h	50.00	20.76	0.00	29.24	
i	52.46	0.00	0.00	47.54	
j	54.39	0.00	10.75	34.86	
k	56.35	0.00	21.59	22.06	
l	57.74	0.00	29.32	12.92	

Compositions on the margins of the positive subspace may be determined with programs EQSPIN, INSECT, and EQSTER (Miesch, 1976b) if the sample vector system is 3-dimensional, or with program EQZERO (unpublished but available from the author) otherwise.

Compositions a, e, and i in matrix (22) are close to the compositions of ideal enstatite, ferrosilite, and wollastonite, respectively, as given in matrix (12). Therefore, in order to illustrate the computational methods for determining the mixing porportions, these three compositions will be taken as matrix C_{mM} containing the end-member compositions for the mixing model. The coordinates of the end-member vectors with respect to the principal-components axes may be determined using equations (7) and (8). These form the matrix B_{mm}:

$$B_{mm} = \begin{matrix} \text{PC1} & \text{PC2} & \text{PC3} & \\ 0.925064 & -0.379526 & -0.012518 & \text{(composition a)} \\ 0.817467 & 0.574884 & -0.036049 & \text{(composition e)} \\ 0.655018 & -0.046869 & 0.754127 & \text{(composition i)} \end{matrix} \quad (23)$$

The product of matrix (18) and the inverse of matrix (23) gives the coordinates of the five sample vectors with respect to vectors a, e, and i. These are:

$$A_{Nm} = \begin{matrix} (a) & (e) & (i) & \text{Sample} \\ 0.989205 & 0.009006 & 0.009557 & 1 \\ 0.887927 & 0.177705 & 0.007851 & 2 \\ 0.689883 & 0.392403 & 0.060973 & 3 \\ 0.415913 & 0.683940 & 0.043443 & 4 \\ 0.189385 & 0.868018 & 0.030580 & 5 \end{matrix} \quad (24)$$

Equations (3) and (4) then are used to derive scale factors for each of the three row-vectors in matrix (23) and the mixing proportions, matrix P_{Nm}, are derived by application of equation (10) to matrix (24). The final mixing model is:

$$X_{NM} \simeq \hat{X}_{NM} = P_{Nm} C_{mM} \quad (25)$$

where X_{NM} is as given in matrix (11), $\hat{X}_{NM}$ is as given in matrix (16), and the mixing proprotions and end-member compositions (from matrix 22) are:

$$P_{Nm} = \begin{matrix} \text{"Enstatite"} & \text{"Ferrosilite"} & \text{"Wollastonite"} & \text{Sample} \\ 0.9812 & 0.0091 & 0.0097 & 1 \\ 0.8249 & 0.1677 & 0.0074 & 2 \\ 0.5997 & 0.3464 & 0.0540 & 3 \\ 0.3602 & 0.6015 & 0.0383 & 4 \\ 0.1718 & 0.7999 & 0.0282 & 5 \end{matrix}$$

and

$$C_{mM} = \begin{matrix} \text{SiO}_2 & \text{FeO} & \text{MgO} & \text{CaO} & \text{End member} \\ 59.72 & 0.00 & 40.28 & 0.00 & \text{"Enstatite"} \\ 46.07 & 53.93 & 0.00 & 0.00 & \text{"Ferrosilite"} \\ 52.46 & 0.00 & 0.00 & 47.54 & \text{"Wollastonite"} \end{matrix}$$

DISCUSSION

Representation of the matrix X_{NM}, which consists of measurements of M compositional variables on N samples, as a vector system requires a vector space of M dimensions. Fortunately, correlation among the compositional variables is a common property of geochemical and petrologic data matrices so that, although the vectors occupy M dimensions, they usually cluster about a space of only m dimensions - as when vectors in three or more dimensions cluster about a plane. When this circumstance occurs the vectors can be projected into the smaller space and only small angles separate the original and projected vectors. Thus, the projected vectors represent compositions close to those represented by the original vectors; the matrix $\hat{X}_{NM}$ represented by the projected vectors is a close approximation of X_{NM}.

Transformation of a matrix of original data, X_{NM}, into a matrix of approximate or recomputed data, $\hat{X}_{NM}$, may be viewed as an attempt to filter the data by removing the effects of random errors in the laboratory measurements and of geologic processes that affected the sample compositions in a minor and random, or at least haphazard, manner. Removal of random effects certainly is possible in computer-simulation experiments and there is no reason to believe that the transformation cannot perform similarly with real data. A comparison of petrographic diagrams constructed from the original and recomputed data by Stuckless and others (1979) suggests that it does.

The recomputed data matrix may be derived directly from the principal-components loadings and scores as shown in the Appendix, or it may be derived from the varimax loadings and scores or from any other reference axes one chooses to use. These initial reference axes are merely devices that serve as a temporary basis for the system of sample vectors. The object of the modeling procedure is to determine a new basis that is interpretable geologically and in accord with whatever is known about the geologic origin of the samples or otherwise in accord with the purpose of the modeling exercise.

The Q-mode approach offers several important advantabes in mixing problems to the basic method introduced to geologists ten years ago. The most important of these are (a) that the number of end members needed for the model can be determined before the end members are known, and (b) one has the ability to test individual compositions (rather than just groups of compositions) for suitability as end-member compositions for the mixing model and to modify these when necessary so that they are suitable. It also is possible to prepare lists of suitable compositions, using various methods, so that the user may scan these for compositions that are of interest geologically. Finally, however, the Q-mode method, similar to other methods, requires that selected sets of end-member compositions be tested by computing the mixing proportions. The sets must be altered or replaced if the mixing proportions are determined to be unreasonable for any substantive reason. The usual reason for rejection is that the mixing proportions are negative when geologic evidence calls for addition of the end member, or positive when the geologic evidence or theory argues for subtraction.

The Q-mode computations for the orthopyroxene example used here for purpose of illustration were performed without initial scaling of the data. As a result, the compositional variables were weighted and affected the outcome of the analysis in proportion to the variances. Various methods of initial scaling can be used to avoid this situation if desired.

Use of the Q-mode method is most effective when available in an interactive mode on a time-sharing computer system. The methods described here, which are contained mostly in published FORTRAN programs (Miesch, 1976b), can be used to explore for mathematically suitable and geologically plausible end-member compositions that yield reasonable mixing proportions so that models can be developed that are both mathematically sound and in accord with all available geologic evidence that pertains to the mixing problem.

REFERENCES

Bryan, W.B., Finger, L.W., and Chayes, F., 1969, Estimating proportions in petrographic mixing equations by least-squares approximation: Science, v. 163, no. 3870, p. 926-927.

Deer, W.A., Howie, R.A., and Zussman, J., 1963, Rock-forming minerals, Vol. 2, Chain silicates: John Wiley & Sons, Inc., New York, 379 p.

Imbrie, J., 1963, Factor and vector analysis programs for analyzing geologic data: Office Naval Research, Geography Branch, Tech. Rept. 6 [ONR Task No. 380-135], 135 p.

Klovan, J.E., and Imbrie, J., 1971, An algorithm and FORTRAN-IV program for large-scale Q-mode factor analysis and calculation of factor scores: Jour. Math. Geology, v. 3, no. 1, p. 61-77.

Miesch, A.T., 1976a, Q-mode factor analysis of geochemical and petrologic data matrices with constant row-sums: U.S. Geol. Survey Prof. Paper 574-G, 47 p.

Miesch, A.T., 1976b, Interactive computer programs for petrologic modeling with extended Q-mode factor analysis: Computers & Geosciences, v. 2, no. 4, p. 439-492.

Miesch, A.T., 1979, Vector analysis of chemical variation in the lavas of Paricutin volcano, Mexico: Jour. Math. Geology, v. 11, no. 4, p. 345-371.

Miesch, A.T., and Reed, B.L., 1979, Compositional structures in two batholiths of circum-Pacific North America: U.S. Geol. Survey Prof. Paper 574-H, in press.

Stormer, J.C., and Nicholls, J., 1978, XLFRAC: a program for the interactive testing of magmatic differentiation models: Computers & Geosciences, v. 4, no. 2, p. 143-159.

Stuckless, J.S., Miesch, A.T., Goldich, S.S., and Weiblen, P.W., 1979, A Q-mode factor model for the petrogenesis of the volcanic rocks from Ross Island and vicinity, Antarctica: Am. Geophy. Union Mem., Antarctica Research Series, in press.

Wright, T.L., and Doherty, P.C., 1970, A linear programming and least squares computer method for solving petrologic mixing problems: Geol. Soc. America Bull., v. 81, no. 7, p. 1995-2008.

APPENDIX

Determination of the matrix $\hat{X}_{NM}$

The matrix of original data, X_{NM}, with constant row-sums (generally equal to 100 percent), is scaled initially according to:

$$w_{ij} = \frac{x_{ij} - b_j}{a_j - b_j} \qquad 1 \leq i \leq N; \; 1 \leq j \leq M \qquad (1a)$$

where x_{ij} is the value in the ith row and jth column of X_{NM}, w_{ij} is the scaled value, and a_j and b_j are constants for the jth variable. If the data are to be scaled to proportions of the variable ranges, a_j and b_j are set to the maximum and minimum values, respectively, in the jth column of X_{NM}. If the initial scaling is to be omitted, the constants a_j and b_j are effectively set to 1 and 0, respectively, in equation (1a) and in all other equations where they occur.

The scaled data then are row-normalized to produce a Z_{NM} matrix where each element is determined by:

$$z_{ij} = w_{ij} / (\sum_j w_{ij}^2)^{1/2} \qquad 1 \leq i \leq N; \; 1 \leq j \leq M \qquad (2a)$$

Following Klovan and Imbrie (1971), then, a cross-products matrix is formed by:

$$R_{MM} = Z'_{MN} Z_{NM} \qquad (3a)$$

The eigenvectors of R_{MM} corresponding to the m largest eigenvalues form the matrix of principal-components scores, F_{mM}, and the matrix of principal-components loadings is obtained from:

$$A_{Nm} = Z_{NM} F'_{Mm} \qquad (4a)$$

The rows of F_{mM} (f_j) are set equal to g_j and scale factors for each of the rows are determined with equation (4); the composition scores for each of the rows then are obtained with equation (5), yielding the matrix C_{mM}. The loading matrix A_{Nm} is converted to a matrix, P_{Nm}, of composition loadings (or mixing proportions) by use of equation (10). The matrix $\hat{X}_{NM}$ is obtained with equation (9).

Repetition of all the steps listed after equation (3a) for values of m from 2 to M will yield M - 1 matrices of $\hat{X}_{NM}$. The best $\hat{X}_{NM}$ matrix to use in a mixing problem generally will be the one most similar to X_{NM} based on the smallest value of m, in accordance with the principle of parsimony (Imbrie, 1963). Similarity is measured conveniently by coefficients of determination (squares of correlation coefficients) between corresponding columns of $\hat{X}_{NM}$ and X_{NM} which can be summarized effectly on factor-variance diagrams (Miesch, 1976a). The number of end-member compositions required to account for all of the variability in matrix $\hat{X}_{NM}$ is equal to its rank which is m, the same as the number of columns in A_{Nm} and P_{Nm} and the number of rows in F_{mM} and C_{mM}.

COMPUTER AS A RESEARCH TOOL IN PALEONTOLOGY

David M. Raup

Field Museum of Natural History

ABSTRACT

In many areas of paleontological research, the computer has taken its place with the microscope and handlens as a basic research tool. Most of the standard applications are digital and include various types of biometrical analysis, information processing (including collection management), computer graphics, and statistical testing of hypotheses.

There is some evidence that paleontology, especially evolutionary paleontology, is undergoing a major transformation: from a primarily idiographic science devoted to building a chronology of the history of life (who begat whom?) to a more nomothetic science in search of general statistical laws. The computer may not be responsible for this shift of emphasis but it may be the vital catalyst. In the nomothetic approach species are treated as particles and the group behavior of large numbers of species (or evolutionary events) is best treated by computer. This is true for two reasons: (a) testing models with real data requires massive data-processing capability, and (b) Monte-Carlo and numerical simulations may be necessary in the formulation and testing of theoretical models.

Along with the search for general statistical laws in the evolution of life, there is new emphasis on complex Markov processes. Any history and particularly evolutionary history contains Markovian elements. Whether the computer analysis is done analytically or by simulation, it is important to be able to treat Markov series (especially time series and branching processes) in a massive and rigorous fashion. For this, the computer is indispensable.

THE PAST TEN YEARS

The use of computers in paleontological research has developed on a broad front during the past ten years. For many workers, computers have been absorbed to the point where they are a routine part of the research arsenal - along with the microscope, camera, and calipers. One now rarely sees "by computer" or "computer generated" in the titles of papers - suggesting that the use of computers in paleontology has matured. In fact, it is nearly impossible to survey computer use from the literature alone because it may not state whether computers have been used in a given study.

It cannot be claimed, of course, that all appropriate applications have been explored or that all studies that could benefit from computers actually use computers. By far the most usual applications in paleontology are in aid of fairly standard biometrical analysis of morphologic variation among fossils. Where multivariate methods are used, studies can be done by computer that would be impossible by any other method. Biometrical work thus is no longer restricted to simple univariate and bivariate statistics. Multivariate methods also are being used increasingly in various distributional studies - especially in biofacies analysis and other aspects of community paleoecology.

Certain applications that showed promise ten years ago have developed little. Automatic image analysis in the study of morphology yet has tantalizing possibilities but relatively little actual progress has been made. This is due in part to the lack of suitably efficient and economical hardware and in part to the lack of explicit mathematical models of growth and form in most plant and animal groups. Development of the computerization of biostratigraphy also has been disappointing: much potential but little real progress.

The use of computers in the cataloging and management of large museum collections has not developed to the extent that many people had hoped. Although not strictly a research tool, computerization of data on systematic collections could have important research implications because of the host of interesting and important questions that can be asked of a good electronic file of museum data. In particular, studies of diversity and biogeography could be enhanced greatly in this manner. One problem that yet faces EDP projects in museum collections is the sheer enormity of the data entry problem (Jones, 1979).

Thus, the past ten years have seen important advances in computer applications in paleontology but the advances have been uneven.

THE NEXT TEN YEARS

Any predictions made now probably will be wrong because they will not anticipate technological innovations nor the influence of ideas generated by a few, key individuals. Nonetheless, some trends are suggestive. My own hunch is that the next ten years will see increasingly ambitious analyses of large data sets having to do with distribution of fossil taxa in time and space. There are several reasons for thinking this:

(1) There is a renewed interest in the search for statistical laws or generalizations that may be used to predict geographic or evolutionary phenomena.
(2) Published compilations of paleontologic data are more comprehensive and more available than ever before. The *Treatise on Invertebrate Paleontology* and the JOIDES reports provide a consistent and rigorous basis for synthetic studies that has not been available before.
(3) The improvement of text-editing systems (such as WYLBUR and SUPERWYLBUR) and of basic utility programs for sorting and retrieval of data has facilitated the building and processing of the large data files which are the necessary raw material for broad synthesis.

Careful analysis of large data sets is applicable to a wide range of paleoecological, biogeographic, biostratigraphic, and evolutionary problems. Interesting questions are being asked, especially by the younger research workers, which can be answered only by massive data-processing efforts. We thus may be seeing the convergence of good questions being asked, the availability of appropriate data, and the methods of analysis. The computer may provide the essential catalyst for significant advances in the science. With this in mind, I will devote the remainder of my space to some aspects of the search for statistical laws in evolution and to the contributions that computers are making or are likely to make to this search.

THE SEARCH FOR STATISTICAL LAWS

Paleontology has long been dominated by what has been termed an idiographic approach (Raup and Gould, 1974). That is, the fossil record has been subjected to painstaking and detailed description and chronicling of what happened. Emphasis has been on description of individual species and interpretation of their habitats and on ancestor-descendent relationships (who began whom? and similar questions). This has been necessary and desirable and nearly all fields of natural science have gone through this phase: it has been essential to establish the data base before making large-scale interpretations. This is not to say that paleontologists did not look for generalizations or laws during the idiographic phase. In fact, the literature is strewn with generalizations, particularly in the field of evolution. Examples include the Biogenetic Law, various Cope's Rules, orthogenesis, and many generalizations about evolutionary rates. Therefore it is not fair to say that paleontological research has been idiographic completely in the past and now is nomothetic (roughly the opposite of idiographic). Rather, it seems that the nomothetic aspect is becoming relatively more prominent as a research approach - which is entirely reasonable if we can assume that the data base has improved to the point where synthetic studies are more nearly justified.

At present, there are many paleontologists who feel that the time has come to treat fossil species as particles in a larger dynamic system. To use a rough analogy with the study of gases in physics, it can be argued that we are beyond the point where it is fruitful to describe and track each "molecule"; rather, we should be looking statistically at the group behavior of large numbers of "molecules" so that generalizations having predictive ability can be made (that is, "gas laws" for evolutionary systems). It is probably safe to assume that an evolutionary system is far more complex than most physical systems; it even may be that simplifying generalizations or laws are not possible. But the search is on!

MARKOV PROCESSES

Nearly all time sequences in geology and paleontology can be looked upon as Markov series. Whether the change is physical (as in a sedimentary sequence) or biological (as in evolutionary series), the state of the system at time = t is to some extent dependent upon the state at time = t-1. A brachiopod species, for example, may change from one stratigraphic level to the next but the descendent form is constrained to be at least similar to the ancestral form. This indicates that two succeeding forms in the series are more similar to each other than each is to a brachiopod drawn at random from a larger series or array. In other words, the evolutionary system has memory in the sense that a given step is constrained by the preceding condition. The state at time = t is not determined fully by the state at t-1, however - at least as far as we know. A given step is thus a combination of (a) the influence (legacy) of the preceding step and (b) an indeterminate array of factors influencing the change itself. The uncertainty of the change introduces a probabilistic element. The changes from step-to-step even may be said to contain a random element as long as one understands that random in this context simply refers to our inability to predict the changes in advance.

Time series in purely physical systems in geology can be treated in the same manner, as Krumbein (1969) and others have shown. In evolution, the two most abundant Markov processes are the random walk and the branching process.

RANDOM WALKS

A general exercise in evolutionary paleontology is to plot the state of a morphological character through time. Figure 1 shows such a plot for shell diameter in the classic *Kosmoceras* lineage analyzed by Brinkman (1929). This example satisfies the requirements of a Markov chain because the increments in shell diameter from one sample to the next are small relative to the differences in diameter between randomly chosen samples:

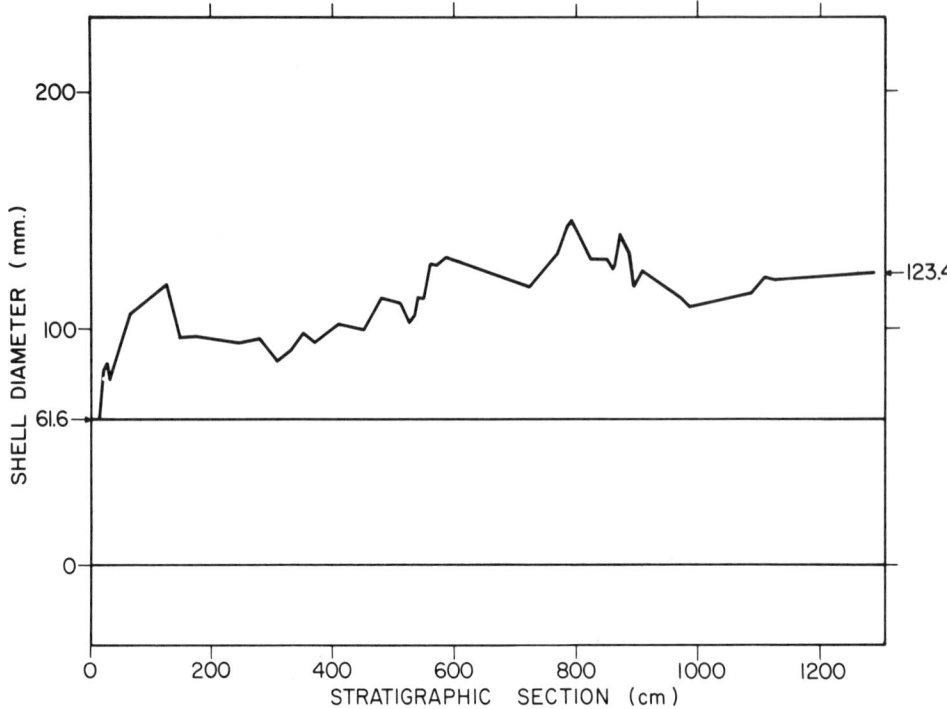

Figure 1. Change in mean shell diameter for assemblages of Jurassic ammonite *Kosmoceras* (*Zugokosmoceras*) through about 12 m of stratigraphic section. Data from Brinkmann (1929).

that is, the shell diameter at one horizon is constrained to be relatively close to that at the preceding horizon. And there is uncertainty in the direction and magnitude of any horizon-to-horizon change. This establishes that we are concerned with a Markov process and that the simple Markov chain or random walk is a reasonable conceptual and mathematical framework for analysis.

This does not indicate that the evolution of *Kosmoceras* is random in any literal sense - although it may be. In the simplest random walk, all changes in the dependent variable (shell diameter, in this situation) should be equal and the probabilities of upward and downward movement should be the same. It is clear from Figure 1 that the increments are not constant. Figure 2 shows a frequency distribution of change in shell

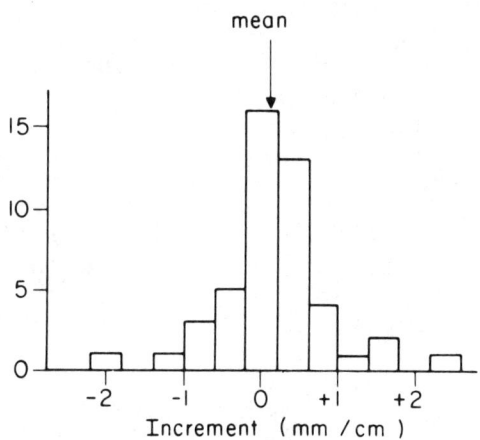

Figure 2. Histogram of change in mean diameter per centimeter of section for *Kosmoceras* data shown in Figure 1. Mean change is +0.005 mm/cm.

diameter per centimeter of stratigraphic section. The question of whether the upward and downward probabilities of change are equal need not be answered in the affirmative for a random-walk analysis to be applicable. Although it would be interesting if the probabilities were equal - suggesting evolution by genetic drift - unequal probabilities are just as reasonable mathematically and biologically. If it could be established, for example, that the probability of an increase in shell diameter in *Kosmoceras* was higher than that of a decrease we would have a measure of the pressure of natural selection favoring a size increase.

In the *Kosmoceras* example, the mean increment per centimeter is slightly positive (see Fig. 2). This explains, of course, why there was a net increase in shell diameter in this particular sequence. But the mean increment per centimeter is not different statistically from zero and thus, we cannot in this situation reject the null hypothesis of genetic drift. If directional selection was operating, it was not significant statistically for the 1200-cm sequence.

The importance of treating Figure 1 as a Markov chain can be illustrated by what happens if conventional independent events statistics are used. Suppose one were to ignore the Markovian aspects of this situation and ask: "Is shell diameter significantly correlated with position in the stratigraphic sequence?" or "Is the increase in shell diameter from bottom to top of the section statistically significant?" To answer the first question one *could* compute a correlation coefficient (r) for all samples, with one variable being shell diameter and the other being centimeters above the base of the section. The r turns out to be 0.74 (N=48) which is different significantly from zero. The second question can be answered by comparing means and variance of the bottom and top samples. A conventional t-test indicates that these two means, in fact, are different. But both exercises are invalid because they ignore the Markovian aspects of the problem.

The bottom line here is that random walks that seem to show significant trends may not. The problem is not limited to evolutionary series of morphologic characters. Any time series in historical geology, such as a sealevel curve or paleotemperature trend is prone to the same difficulty.

COMPUTER ASPECTS OF RANDOM WALKS

The accumulation and processing of data as well as some of the statistical analysis of time series can be carried out most appropriately by computer. But far more important is the use of Monte-Carlo simulations to improve the intuition and generally raise the consciousness of those working with time series. This is particularly important in view of the fact that most people see time series as if they were *non*Markovian, independent-events processes. For these people, the results of random-walk analysis are counter-intuitive.

Figure 3 shows several classic random walks generated by a random-number generator. Each is 50 time units long and the probability of upward and downward change is everywhere constant and equal to 0.5. For each pair of random walks, conventional correlation coefficients (r) have been computed for each series against time and for each series against the other. Using this (invalid) approach, most r values are statistically significant - matching one's intuitive impressions of the graphs. But

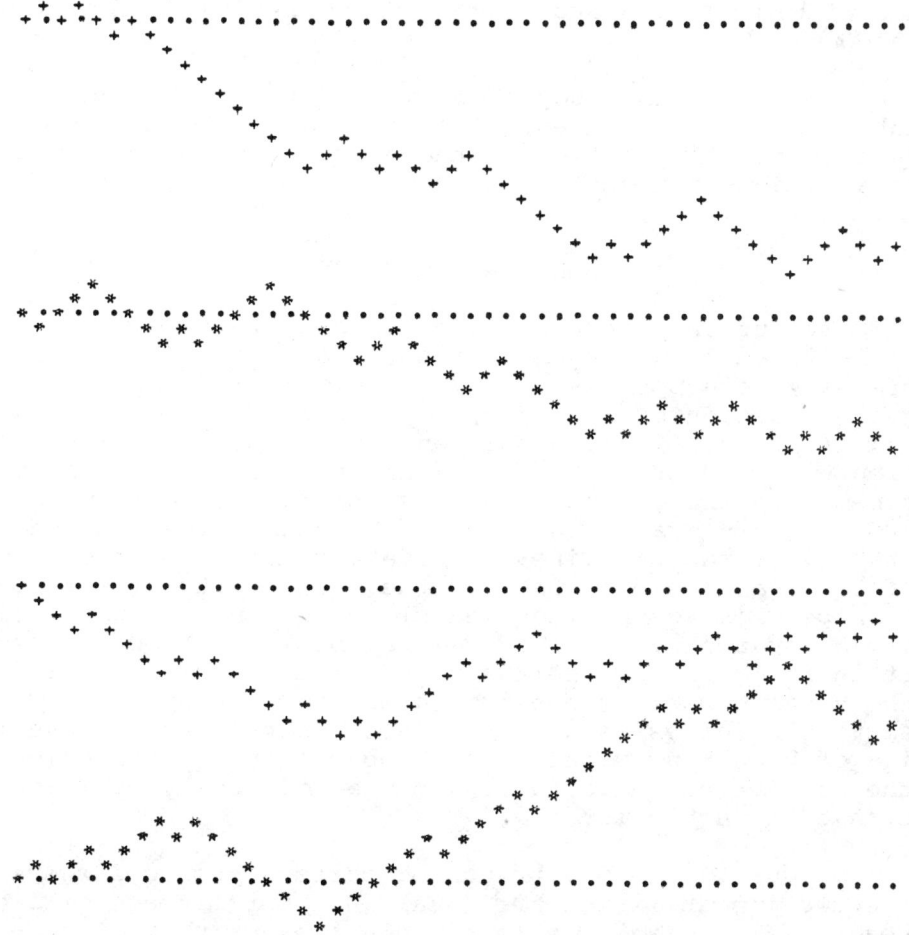

Figure 3. Four computer-generated random walks. Time goes from left to right (50 arbitrary time units). In each time series, upward and downward movements had equal probabilities. As is typical of all time series, path followed by random walk gives appearance of being correlated with time; also, members of pair of random walks may appear correlated with each other. This exemplifies Yule's "nonsense correlation."

the correlation analysis, of course, is invalid because none of the several walks show a significant departure from chance expectations. The end-points of all walks are within the 95 percent confidence limits of the classic random walk.

Only through study of many random walks known to be unbiased can the student or professional researcher gain an intuitive impression of the normal range of outcomes in a random-walk situation.

BRANCHING PROCESSES

Evolution at or above the species level generally is depicted as a branching tree. Each segment of the tree may be a species, genus, or higher monophyletic group. Segments (lineages) may originate through branching of preexisting lineages and may terminate (extinction). The type of branching process termed time homogeneous is the one most readily applicable to actual evolutionary branching patterns. In the time homogeneous branching process, lineages have a constant probability of termination (expressed as the probability of extinction per lineage million years) and a constant probability of branching. The system may or may not be balanced. If it is balanced, the two probabilities are equal and the total number of coexisting lineages (standing diversity) fluctuates as a random walk. But if the probability of branching exceeds the probability of extinction, the statistical expectation is that diversity will increase through time.

Figure 4 shows a branching pattern produced where the two probabilities are equal (0.1 per lineage million years). The result is reasonable subjectively if compared with real evolutionary trees from the fossil record but the applicability of the time homogeneous branching process to evolution is yet to be established fully. In the real world, there are situations where the frequencies of branching or extinction change through time such that assuming temporal constancy of probabilities would be an oversimplification and lead to error. The most obvious such situations are those where extinction probability suddenly increases or branching probability suddenly decreases to produce what is known as a mass extinction. Also, it is clear that some biologic groups differ consistently from others in the values of the two probabilities. For example, Jablonski (1978) showed that molluscan lineages without a free-swimming larval stage have higher probabilities of extinction and branching than those with a free-swimming larval stage. Even in

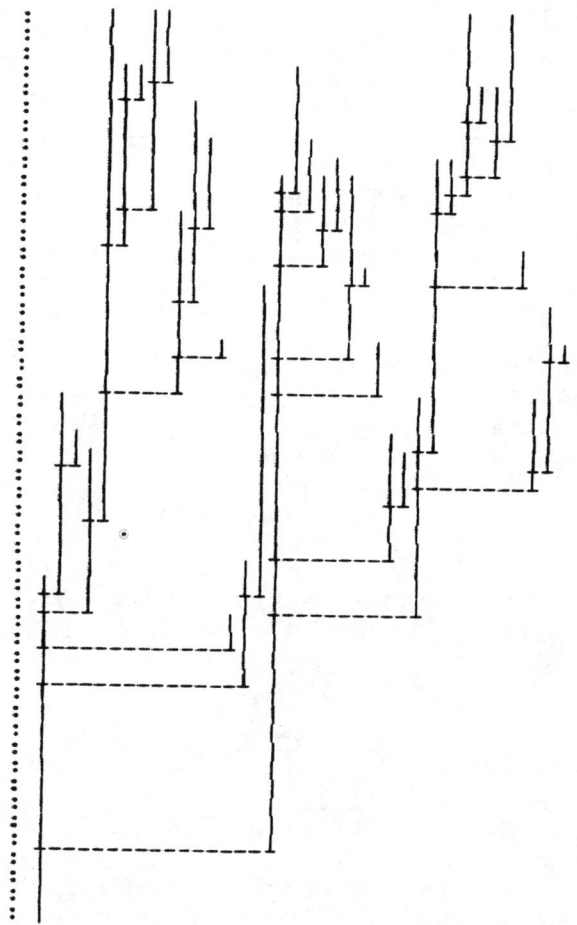

Figure 4. Tree-like branching pattern. Process starts with single lineage at time=0 (lower left) and proceeds through chance branching and termination. Thus, fact that more branching events than termination events occurred was matter of chance.

these situations, the mathematical framework of the Markovian branching process provides a sensible null hypothesis against which to test real world data.

COMPUTER ASPECTS OF BRANCHING PROCESSES

The mathematical analysis of the simpler branching processes has been developed in other fields. Equations have been derived by which one can predict the mean expectation (and its variance) for various aspects of the branching pattern. These aspects include standing diversity as a function of time, total progeny produced in a branching system, probability of extinction of the whole system, and many others. In short, if the type of branching process has been defined and if the probabilities of branching and extinction are known or can be postulated, it is possible to describe with rigor the probable anatomy of the resulting branching pattern. Some of the predictions lend themselves to computer processing but as before, the main value of the computer is to raise the consciousness of the researcher by Monte-Carlo simulation of branching patterns. Several examples are shown in Figure 5. All use precisely the same model and the same probabilities as in Figure 4, but the results differ greatly. Some branching systems abort early whereas others undergo what would be termed an adaptive radiation (even through we know there is nothing adaptive about the radiation).

CONCLUSION

I will close by discussing a Markov situation which probably has applicability to the geologic and paleontologic records and which illustrates better than any other the counter-intuitive nature of Markov processes. This is inspired by the excellent and intriguing paper by Cohen (1976) on the so-called Polya's Urn problem. Suppose we have an urn which contains one white ball and one black ball. Chooose a ball at random and note its color; then add a ball of the same color to the urn, thus making a total of three balls (two of one color and one of the other). If this process is repeated many times, the urn gradually fills up. The ratio of colors starts at 1:1; after the first iteration, the ratio is inevitably 2:1 or or 1:2 depending on which color was chosen; after the next iteration, the ratio may revert to 1:1 or move to 3:1, and so on. The problem is to predict statistically the temporal change in the ratio of colors.

The Polya's Urn problem has many obvious biological analogs. Suppose, for example, that the initial balls represent two species of an organism landing as waifs on

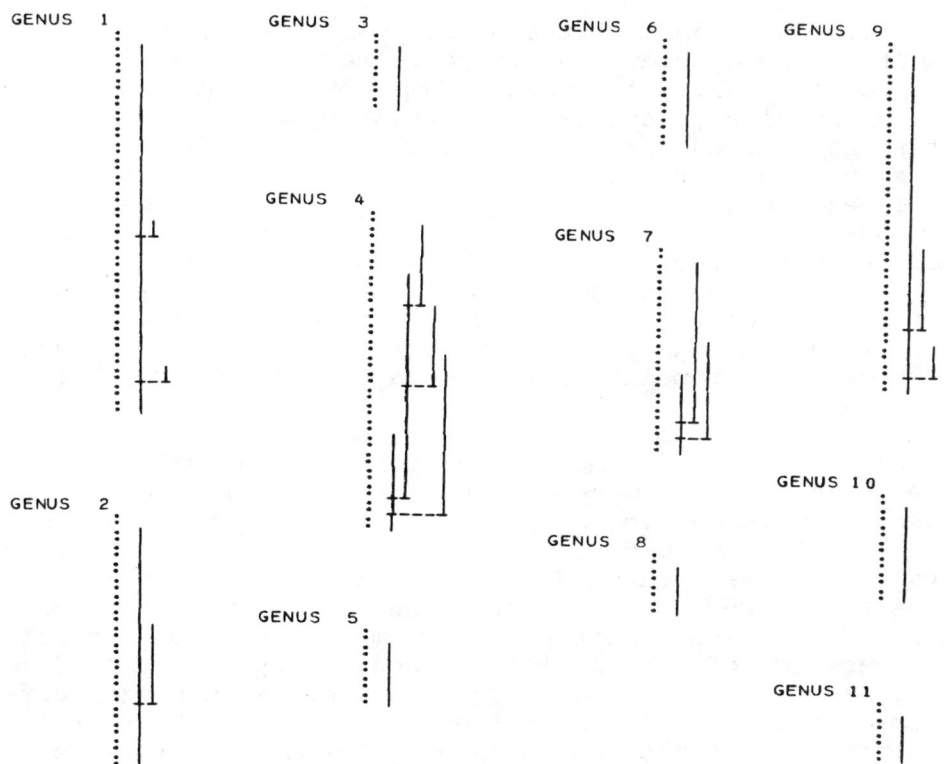

Figure 5. Eleven branching trees made with same program that produced Figure 4. Probabilities of branching and termination were same as in Figure 4 but all eleven "genera" went extinct after few time iterations and few new branches were formed.

an uninhabited island and suppose that the species have the same probability of population growth through reproduction. How will the ratio of population sizes in the two species behave over time? As Cohen has indicated, the answer to the Polya's Urn problem is counter-intuitive and is missed by most otherwise competent ecologists and population biologists. Usual scenarios proposed are (a) that the proportion of one color will fluctuate about 0.5 indefinitely, (b) that one color will dominate the population quickly (tending toward a frequency of 1.0) with the dominant color being a matter of chance in the early iterations, and (c) that frequency will behave as a random walk through the range from 0.0 to 1.0.

In fact, none of the scenarios are obtained. If Monte-Carlo simulations are performed, the frequency of one color usually fluctuates for a few iterations but then invariably settles at a frequency which is constant for the remainder of the simulation and which is effectively permanent. The interesting aspect is that all frequencies between zero and one have the same probability of being chosen as the constant, final frequency. One can predict, therefore, that a steady state will be achieved but one cannot predict anything about that constant frequency. In practice, Polya's Urn achieves steady state in remarkably few iterations: usually a few hundred with a total population size of a few thousand individuals.

In the biological example of colonization of an island, the species composition of the island will soon reach what seems to be an equilibrium and if such an example were monitored in the real world, the ecologist could not be blamed for postulating that the equilibrium value is a reflection of the competitive interaction between the two species. But, thanks to knowledge of Polya's Urn, we know that equilibrium is inevitable even in the absence of interaction between the species. In fact, the result is the same if the two species colonized different islands and were not even aware of each other's existence. It also can be shown that steady state is achieved with equal rapidity in situations where there are several species. How many fossil communities which seem to have fixed species composition were formed by a Polya's Urn process?

The Polya's Urn case illustrates again the importance of revising one's intuitive and mathematical approaches when faced with historical processes which are Markovian and, as noted earlier, historical geology and paleontology are primarily Markovian. Computerized simulation techniques are the obvious vehicle for teaching and exploring the Markov process.

ACKNOWLEDGMENT

This work was supported in part by the Systematic Biology Program, National Science Foundation, NSF Grant DEB 78-22568.

REFERENCES

Brinkmann, R., 1929, Statistische-biostratigraphische Untersuchungen an mitteljurassischen Ammoniten ueber Artbegriff und Stammesentwicklung: Abh. Ges. Wiss. Goettingen, Math-Phys Kl., N.F., v. 13, no. 3, p. 1-249.

Cohen, J.E., 1976, Irreproducible results and the breeding of pigs: Bioscience, v. 26, p. 391-394.

Jablonski, D., 1978, Transgressions, regressions, and endemism in Gulf Coast Cretaceous molluscs (abst.): Geol. Soc. America Abstr. v. 10, no. 7, p. 427-428.

Jones, B., 1979, Data storage and retrieval for the palaeontology collections, University of Alberta: Palaeont. Assoc., Sp. Papers in Palaeontology, v. 22, p. 175-187.

Krumbein, W.C., 1969, The computer in geological perspective, *in* International Symposium of Computer Applications in the Earth Sciences: Plenum Press, New York, p. 251-275.

Raup, D.M., and Gould, S.J., 1974, Stochastic simulation and evolution of morphology -- towards a nomothetic paleontology: Syst. Zoology, v. 23, no. 3, p. 305-322.

THE COMPUTER IN PALEOECOLOGY

R.A. Reyment

Uppsala Universitet

ABSTRACT

Access to electronic computers is necessary for modern quantitative paleoecology, as almost all problems are highly multivariate. Diversity studies are becoming increasingly abundant in computer-based paleoecology. Multivariate statistical applications are the most important area of computerized paleoecology, involving both standard multivariate methods as well as special adaptations. A useful concept in paleoecological reconstructions based on borehole data is the ecolog, a union of physical and biological values.

INTRODUCTION

Paleoecology is clearly a complicated subject. It involves not only the whole range of complications proper to ecology but also geologic problems of a specific nature. When quantified, the subject easily becomes highly multivariate. It therefore is obvious that the computer has a self-appointed role and it is no exaggeration to say that there would not be much advanced quantitative paleoecology in research if there were no computers. This can be seen in another way, notably, by reference to the literature. There is little significant work in multivariate ecology of any type before 1960 and most has been done since 1970. The theoretical background was available, to a large extent, but not the means for practically testing it.

For most purposes, it is convenient to think of a quantitative paleoecologic analysis at two levels:

(i) The most adequate and relevant of the available models or methods of statistical ecology applicable to the problem.
(ii) Special methods of geostatistics designed to give the approximate results obtained under (i) more substance and geologic meaning.

There can be no simple approach of the "problem-solution" type.

DIVERSITY STUDIES

Diversity is currently a popular subject among statistical ecologists and therefore it is appropriate to note some applications from the sphere of paleoecology.

Paleoecologists seem to use techniques developed mostly by numerical taxonomists which all rely heavily on computers. Some examples are cited here.

Birks and Deacon (1973) used lists of species of Recent and fossil vascular plants in 12 geographic regions in Britain for paleoecologic purposes. Using 4 similarity indices (the coefficients of Jaccard, Dice, Simpson, and Braun-Blanquet), converted to dissimilarity coefficients, and nonmetric scaling, two-dimensional dispositions of points representing the regions were determined for each time interval. A marked north-south gradient was demonstrated.

Cheetham and Hazel (1969) studied the performances of various similarity coefficients, 23 in all, for analyzing associations of microfossils. Henderson and Herron (1977), in their paper on a probabilistic method of paleobiogeographic analysis, concluded that diversity studies in paleontology tend to suffer from a number of troublesome defects, most of which are obvious to the quantitative worker. Further references to diversity work are Hazel (1970), Kaesler (1969), and Kaesler and Mulvany (1976). Another paper that can be mentioned here is by Forester (1977), who considered the problem of measuring the relative abundance of microorganisms - in this connection he uses the Poisson parameter for producing an abundance coefficient.

G.P. Patel and his co-workers as a matter of fact have shown that there is really only one similarity index. Most if not all the variants proposed in the past by numerous workers reduce algebraically to the same formula.

ORIENTATION ANALYSIS

An important area of statistical paleoecology is that of the quantitative study of the orientations of fossils. Two main problems belong here, (1) the orientions of fossils in situ and (2) the orientations of transported fossils. The analysis of the former may yield significant information on the mode of life of the organism or organisms involved. The latter can provide details concerning the water currents prevailing during the time at which the biological flotsam and jetsam was stranded.

In regard to the statistical theory for these analyses, we are in a strong position today thanks to the developments of the mathematical analysis of geomagnetism. Thus, the entire field of directional statistics can be taken over without modification (cf. Mardia, 1972).

Neoecology does not have the same interest in analysis of directions, although I can conceive of several situations in which this type of approach could be put to better use than hitherto has been done. The most rewarding area of research in paleoecology has been that of the dispersal of cephalopod shells, after the death of the organism. The essentially trivariate nature of the majority of orientational problems in paleoecology remains to be exploited fully and almost all studies made are based on the circular distribution.

POPULATION DYNAMICS

Population dynamics is one of the main fields of activity in the neoecologist. A large part of the books on statistical ecology by Pielou (1974, 1977) are concerned with this aspect of the subject. For understandable reasons, population dynamics cannot be given the same prominence in statistical paleocology.

Nonetheless, in favorable situations, it is possible to develop an approximate analysis using a classical population dynamic approach. Micropaleontology offers opportunities, as complete growth sequences of ostracods may occur in sediment that has not been reworked. In fact, the preparation of a life table (Reyment, 1971, p. 112) for a sample of ostracods may be used with great effect to judge whether a deposit has been reworked, a secondary outcome of the study. For example, life tables have been prepared for fossil pelecypods, Pleistocene bears, and other vertebrates, although, perhaps, not always correctly.

For more than one species, the number of analyzable paleoecological situations is rare, being limited, for all practical purposes, to the predator-prey relationship and semiquantitative inferences on competition between species (c.f. Reyment, 1971, fig. 24). The predation relationship can be given only adequate statistical study in paleontology for situations where the predator has left an observable trace on the shell of the prey. The best example of this is provided by drilling gastropods, (c.f. Reyment, 1971, p. 130-150). For marine invertebrate paleontologists, at least, predation by drills is of considerable significance and therefore it is a rewarding subject for paleoecological research. Fossil ostracods, pelecypods, and gastropods may be drilled by naticids, less frequently by muricids, and inasmuch as the first-mentioned group is an abundant component of borehole samples, sufficient material can usually be obtained to permit a satisfactory statistical study.

It should be noted here that the analysis of a predator-prey relationship in paleoecology is much of a gamble in that the observed prey and predator frequencies cannot be claimed with certainty to represent the actual maxima attained by them. Not only are the sampling fluctuations dependant on factors outside the normal limits of statistics, but there is the added vexation of the unknown extent of migration as well as post-mortem transport of the drilled shells. A statistical analysis therefore must be preceded by a detailed qualitative study of the material.

SPATIAL PALEOECOLOGY

Within certain limits, it is possible to carry out useful studies on spatial paleoecology (Reyment, 1971, chapter 6). The confines for such studies of necessity are narrow, and may verge on paleobiogeography. I have had occasion to discuss morphometric variations in Paleocene ostracods occurring throughout the Early Paleocene epicontinental transgression across West and North Africa (Reyment and Reyment, 1980). The morphometric difference identified could be related to the possible existence of a a climatic gradient. This example certainly is not referable to the main concept of spatial ecology (cf. Pielou, 1974). Only sessile organisms, such as corals and bryozoans, are liable to leave sufficiently good traces of their erstwhile spatial relationships to permit a usual type of spatial analysis such as developed by Matern (1960), although the study of Pleistocene plant associations would seem to offer certain possibilities.

ECOLOGICAL DIVERSITY

Ecological diversity for fossil species may be analyzed with a fair degree of accuracy and perhaps there are more examples of this category of statistical paleoecologic analysis in the literature than of any other. These may be in the form of semiquantitative comparisons of faunal lists, usually involving percentages. Considerable use also has been made of "indices" of which many variants have been proposed (Reyment, 1971, p. 160, ff).

One of the preferred tools for analyzing ecological diversity is the "Shannon-Wiener Index" (or "Shannon-Weaver Index"), which uses relative abundances (Pielou, 1974, p. 290) and which is gaining some vogue of late in statistical paleoecology thanks to its desirable mathematical properties (see also the discussion on Diversity Studies).

ANALYSIS OF SPECIES FREQUENCIES

Seventeen species of ostracods of Early Paleocene age were analyzed by Joreskog's maximum likelihood model of factor analysis (Joreskog, Klovan, and Reyment, 1976). These data were treated originally in Reyment (1963). Here, it was determined that the relative frequencies of different species in samples may be interpreted in terms of the major environmental factors to which the organisms reacted, that is five unspecified major environmental components. It also was concluded here that although the factor analysis of fossil species associations seldom can be expected to disclose whether a species is stenohaline or euryhaline, and stenothermal, or eurythermal, it can indicate whether it is stenooic or euryoic. In connection with the more detailed analysis of the material given in Reyment (1966) it was thought possible to identify one factor as bathyal, extrapolating from our knowledge of the depth distribution of living ostracods.

The Shannon-Weaver index, previously mentioned, was introduced into geology by Pelto (1954) and modified by Miller and Kahn (1962) as a method of studying multispecies systems [in part, analogous to the examples of Pielou (1977)]. Pelto's (1954) suggestion was to use the function

$$H = -\sum_i p_i \log_e p_i$$

for studying multicomponent systems. Here, p_i is the percentage of the i-th component, and $\sum_i p_i = 100\%$. Pelto made use of the concept of relative entropy, Hr, which is defined as the ratio of the actual entropy to the maximum entropy, Hm, for the number of components under consideration:

$$100 \, Hr = \frac{-100 \sum_{i=1}^{N} p_i \log_e p_i}{Hm}.$$

Here, p_i is the proportion of the i-th component in an N-component system and Hm is

$$Hm = -\sum \frac{1}{N} \log_e \frac{1}{N} = \log_e N.$$

In the application devised by Miller and Kahn (1962), it is required that the species be divided into "biofacies". For my study of the Lower Paleocene ostracods, I accepted the seven factors as representing paleobiofacies based on the 17 most abundant species. It was found that the entropy approach yields valuable additional information, particularly for the identification of environmental components that may have been overlooked in the earlier analysis. Thus, in addition to the bathyal component outlined in the factor analysis, calcareous and pelitic components were isolated (Reyment, 1966, p. 48).

SECULAR FLUCTUATIONS IN THE ABUNDANCE OF SPECIES

The study of secular fluctuations in the relative frequencies of species is one that has a specific paleoecologic flavor. A classical, early study is that of Chaney (1924), reanalyzed in Reyment (1971, p. 174 ff). Chaney was concerned with attempting to identify shifts in relative abundances of plants in the Bridge Creek flora of Late Oligocene age in Oregon, U.S.A.

THE SPECIES x LEVELS MATRIX

The simplest sequential representation of chronological variations in a set of species can be produced in terms of the categories + (= an increase in average size), - (= a decrease in average size), 0 (= no change). The species by stratigraphic levels matrix of these observations is a useful indicator for picking out a sustained ecologic trend in a multicomponent set of observation, that is a common mode of reaction to the totality of environmental fluctuations. An example is given in Table 1. This representation shows that in the earlier levels of the sequence, most species follow the same pattern of variation, presumably ecologically controlled. Further aspects of the interpretation of this material are given in Reyment (1966, p. 90-93).

THE ECOLOG

I shall now consider an example in which the correlations between frequencies of organisms, on the one hand and geochemical components of the host sediment, on the other, are used to produce what can be referred to as an *ecolog*, that is a log in which diagnostic chemical elements are related to fluctuations in the frequencies of species through time. The example is taken from Reyment (1976).

Samples from 26 levels in a Nigerian borehole in sediments of Late Campanian (Cretaceous) age were analyzed with respect to the 14 elements Si, Fe, Mg, Ca, Na, K, Ti, P, Mn, V, Mo, Sr, Pd, and Zn. The frequencies of the foraminifers *Afrobolivina afra* Reyment, *Gabonella elongata* de Klasz & Meijer, and *Valvulineria* sp. were recorded for those levels. The ostracods, being relatively rare, were pooled for the purposes of the analysis.

Table 1. Size-directional changes in a series of seven species of Nigerian Paleocene ostracods (based on fluctuations in the length of the carapace).

Species	Direction of change upwards in borehole
Cytherella sylvesterbradleyi	− 0 − 0 0 − + 0 +
Ovocytheridea pulchra	− + + 0 − 0 0 − −
Leguminocythereis lagaghiroboensis	− + + 0 − − − − −
Trachyleberis teiskotensis	0 + + 0 + − + + +
Buntonia beninensis	− 0 + + 0 − + + 0
Buntonia bopaensis	0 + + 0 − − − 0 0
Buntonia livida	0 + + 0 − − − 0 0

The aim of the study was to facilitate the graphical expression of a difficult paleoecologic and biostratigraphic problem. In one direction, it was thought to be of interest to show how all variables considered in the one connection change through time. In another direction, interest was concentrated on tracing temporal covariation in frequencies and geochemical variables.

The method of canonical correlations (Blackith and Reyment, 1971) was used for studying the relationships between sets. Canonical correlations have been little used in ecology owing to certain problems of interpretation, not the least of which is that a high canonical correlation is not associated necessarily with the greatest part of the information in the material. A biological example of the application of canonical correlation to an ecological problem is given by Reyment (1975) for ostracods in the Niger Delta. In this study, pH, Eh, bathymetry, phosphorus, and sulfur formed the predictor set; the response set was composed of total organic substance, $\Sigma\ CaCO_3$, and the total frequencies of ostracods. The most significant results of the analysis are (1) a significant canonical correlation with Ss weighed against ostracods (a thanatocoenetic relationship), and (2) the distribution of the ostracod species is controlled by depth in a negative association with phosphorus. Canonical-correlation analysis has an added useful side, that is the graphical presentation of the transformed partitioned observational vectors. In this situation, it was ascertained that the samples rich in ostracods form a well-defined cluster. The example reviewed here however is more complex, as direct observations on known ecologic components could not be obtained. Si is correlated significantly and positively with Fe, Mn, and V, and negatively and significantly with Mg, Ca, Na, P, and Sr. The variable Fe is correlated significantly and positively with Mn, V and Mo, and significantly negatively with Na, P, and Sr, whereas Na is correlated significantly and positively with K, P, and Sr. Further significant correlations are as follows: Ti is correlated positively with V, Mo, Pb, and Zn, and P is correlated positively with Sr. Mn is correlated positively with V and Mo. Mo is correlated positively with Pb and Zn and negatively with Sr.

For the microfossils, the following significant relationships between sets occur. There is a negative correlation between ostracod frequencies and Zn, whereas *Afrobolivina afra* is not correlated significantly with any of the geochemical variables. The frequencies for the *Valvulineria* are correlated positively with Mn and Fe and negatively with Ca, whereas *Gabonella elongata* is correlated positively with Fe and Mo, and negatively with Ca and Sr.

In the following, the vector variable z_1 contains the chemovariables and the vector variable z_2, the frequencies of the microfossils. The roots of the determinantal equation for the two sets (the R_{ij} are submatrices of the correlation matrix R)

$$|R_{22}^{-1}R_{21}R_{11}^{-1}R_{12} - \lambda_j I| = 0 \qquad (1)$$

are $\lambda_1 = 0.849$, $\lambda_2 = 0.752$, $\lambda_3 = 0.414$, and $\lambda_4 = 0.182$. The first two of these roots are statistically significant. These roots are the squares of the canonical correlations, that is $R_{c1} = 0.922$ and $R_{c2} = 0.867$ which are the maximum correlations between two linear functions of the two sets of variables.

The structure coefficients for two canonical factors for all 18 variables are given in Table 2. The main steps involved are as follows (extracted from Cooley and Lohnes, 1971). Having determined the roots of equation (1), the vector d is obtained from

Table 2. Structure coefficients (Cooley and Lohnes, 1971) for two canonical factors of the geochemical and species-frequencies data (after Reyment, 1976).

Variable	Geochemical set		Set of species frequencies		
	Factor 1	Factor 2	Variable	Factor 1	Factor 2
Si	0.31	-0.22	ostracods	-0.39	-0.66
Fe	0.36	-0.23	Afrobolivina	0.12	0.14
Mg	-0.26	-0.08	Valvulineria	0.67	-0.74
Ca	-0.38	0.24	Gabonella	0.62	-0.20
Na	-0.08	0.31			
K	0.21	0.23			
Ti	0.12	0.39	Factor redundancy	0.217	0.197
P	0.05	0.37	Total redundancy	0.565	
Mn	0.49	-0.21			
V	0.26	0.33			
Mo	0.67	0.23			
Sr	-0.27	0.24			
Pb	0.31	0.67			
Zn	0.51	0.36			
Factor redundancy	0.102	0.078			
Total redundancy	0.218				

$$(R_{22}^{-1}R_{21}R_{11}^{-1}R_{12} - \lambda_j I)d_j = 0, \qquad (2)$$

with the constraint that $d_j R_{22} d_j = 1$. The d_j are the weights for the j-th canonical factor of z_2. The corresponding weights for the j-th canonical factor of z_i are obtained from the relationship

$$c_j = \frac{(R_{11}^{-1}R_{12}d_j)}{\sqrt{\lambda_j}}.$$

Up to now, these steps are the usual ones of canonical correlation. The expansion of the method into a "redundancy analysis" Cooley and Lohnes, 1971) is done by extracting the variance by the canonical variables, $s_1 s_1 / p_1$, where p_1 denotes the number of variables in z_1

(here, this comprises the 14 chemical elements) and $s_2's_2/p_2$, where p_2 denotes the number of variables in vector z_2 (in this example, this is 4). We also have $s_1 = R_{11}c$ and $s_2 = R_{22}d$. The redundancy of set 1 in the presence of set 2 (set 1 contains the chemovariables, set 2 contains the frequencies of the organisms) is defined as

$$R_{d_x} = s_1's_1 R_{c1}^2/p_1,$$

where R_{c1} denotes the canonical correlation for the first pair of canonical variates and the subscript x labels the canonical factor x. The reverse relationship is expressed by the formula

$$R_{d_y} = s_2's_2 R_{c1}^2/p_2.$$

The first canonical factors are $x_1 = c_1'z_1$ and $y_1 = d_1'z_2$. Likewise, the second canonical factors are $x_2 = c_2'z_1$ and $y_2 = d_2'z_2$.

The first canonical factor (Table 2) for the left set of variables comprises significant loadings for most of them. Only Na, Ti, and P are so low as to indicate nonsignificant correlation. The first canonical factor for the right set contains significant loadings for all frequencies except that of *Afrobolivina afra*. The left-hand canonical variate is correlated positively with Si, Fe, K, Mn, V, Mo, Pb, and Zn, and negatively correlated with Mg, Ca, and Sr. This canonical variate seems to be explainable as a dipolar relationship between sediment richer in carbonates and clastic sedimentary components. The right-hand canonical variate is correlated positively with the frequencies of *Valvulineria* sp. and *Gabonella elongata*, and negatively with ostracods.

Ecolog from the Canonical Correlations

The plot of the scores obtained by substituting the partitioned mean vectors into the first pair of linear relationships can be used to produce a paleoecologic log in which the fluctuations in the frequencies of the organisms are weighted against variations in the chemical components of the host sediment. As to be expected from the rather high corresponding canonical correlation, the oscillations follow the same general trends, although there are numerous deviations in the middle and upper thirds of the plots. These deviations are small but might mark periods during which the chemical influences were overprinted by other environmental factors. The lower third of the figure might be an indication of a phase in development during which the chemical components of the environment dominated, such as arises during periods of pronouncedly chemical sedimentation, as in the formation of a marl.

Ecolog by Principal Coordinates

Using Pythagorean distances between individuals, all 18 variables of the foregoing analysis were grouped into a single principal-coordinates analysis (Gower, 1966). The plot of the first set of coordinates against location in the borehole, illustrated in Figure 1, shows the existence of trend in the points. This could indicate that there was a largely unidirectional ecological trend in the paleoenvironment through the time-interval covered by the samples. An interesting property of this ecolog is that the youngest samples seem to be in a state of ecological equilibrium (levels 17-26 inclusive), whereas the older samples may reflect an ecologically perturbed system. The system could have been in the process of becoming stabilized in some manner or other, not necessarily optimal, for the proliferation of benthic microorganisms. In fact, the youngest samples are characterized by the predominance of *Afrobolivina afra*.

STABILIZED CANONICAL VARIATES IN PALEOECOLOGICAL ANALYSIS

Canonical-variate analysis of living and fossil organisms, based on morphological characters, can be distorted, from the aspect of the biological interpretation of the coefficients of the eigenvectors forming the canonical variates, through the inclusion of redundant within-group directions. Instability is associated with the smallest eigenvalues, particularly if these do not differ greatly from zero. In a study of borehole samples of *Afrobolivina afra* Reyment from Nigeria, stability of the canonical-variate coefficients was attained by removal of a near-redundant direction of within-group variation. This leads to improved interpretability of

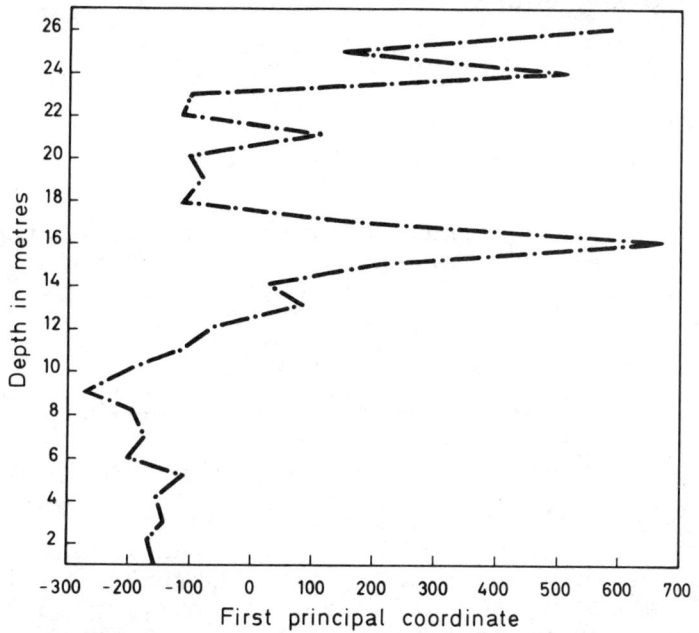

Figure 1. Ecolog of fluctuations in frequencies of microorganisms constructed from first set of principal coordinates.

the morphometric relationships in this species (Campbell and Reyment, 1978). The characters measured on *Afrobolivina afra* are: (1) = length of the test, (2) = width, (3) = width of final chamber, (4) = height of final chamber, (5) = height of second last chamber, (6) = diameter of proloculus, (7) = breadth (measured at right angles to variable 2), (8) = width of aperture, (9) = location of aperture on second last chamber. The main computational steps are set out here.

The within-groups sums of squares and cross-products matrix W on n_W degrees of freedom, and the between-groups sums of squares and cross-products matrix B are computed in the usual manner of canonical-variate analysis, together with the matrix of sample means. It then is recommended that the matrix W then be standardized to correlation form, with similar scaling for B. The standardization is obtained by pre- and post-multiplying by the inverse of diagonal matrix S, the diagonal elements of which are the square roots of the diagonal elements of W. Consequently,

$$W^* = S^{-1}WS^{-1},$$

and

$$B^* = S^{-1}BS^{-1}.$$

The eigenvalues e_i and eigenvectors u_i of W^* then are computed; the corresponding orthogonalized variables are the principal components.

With

$$E = \text{diag}(e_i, \ldots, e_v) \text{ and } U = (u_1, \ldots, u_v),$$
$$W^* = UEU^T.$$

Usually, the eigenvectors now are scaled by the square root of the eigenvalue; this is a transformation for producing within-groups sphericity. Shrunken estimators are formed by adding shrinking constants k_i to the eigenvalue e_i before scaling the eigenvectors. Write

$$K = \text{diag}(k_1, \ldots, k_v) \text{ and define}$$

$$U^* = U(E + K)^{-\frac{1}{2}} = U^*(k_1, \ldots, k_v).$$

Next form the between-groups matrix in the within-groups principal-component space, that is,

$$G_{(k_1, \ldots, k_v)} = U^{*T}_{(k_1, \ldots, k_v)} B^* U^*_{(k_1, \ldots, k_v)}$$

and set d_i equal to the i-th diagonal element of G. The i-th diagonal element d_i is the between-groups sums of squares for the i-th principal component.

An eigen-analysis of the matrix $G_{(0,\ldots,0)}$ yields the usual canonical roots f and canonical vectors for the principal components, a^u. The usual canonical vectors c^u are given by

$$c^u = U^*_{(0,\ldots,0)} a^u.$$

Generalized shrunken (or generalized ridge-) estimators are determined directly from the eigenvectors a^s of

$G_{(k_1,\ldots,k_v)}$, with $c^S = U^*_{(k_1,\ldots,k_v)} a^S$. A generalized-inverse solution results when $k_i = 0$ for $i \leq r$ and $k_i = \infty$ for $i > r$. This gives $a_i^{GI} = a_i^u$ for $i \leq r$ and $a_i^{GI} = 0$ for $i > r$. The generalized inverse solution results from forming $G_{(0,\ldots,0,\infty,\ldots,\infty)} = U_r^{*T} B * U_r^*$, where U_r^* corresponds to the first r columns of $U^*_{(0,\ldots,0)}$. The generalized canonical vectors $c^{GI} = c^S_{(0,\ldots,0,\infty,\ldots,\infty)}$ are given by $c^{GI} = U_r^* a^{GI}$, where a^{GI}, of length r, corresponds to the first r elements of a^u. In practice, marked instability is associated with a small value of e_v and a correspondingly small diagonal element d_v of G. A generalized inverse solution with $r = v-1$ frequently provides stable estimates and usually is simpler conceptually than using shrinking constants.

An easy rule to use is to examine the contribution of d_v to the total group separation, trace $(W^{-1}B)$; the latter is merely trace $(G_{(0,\ldots,0)})$ or $\sum_{i=1}^{v} d_i$. In situations where one or two canonical variates describe much of the between-groups variation, it may be better to examine the relative magnitudes of the first one or two canonical roots derived from $G_{(0,\ldots,0)}$ and $G_{(0,\ldots,0,\infty)}$ rather than a composite measure. Either way, if $d_v/\Sigma d_i$, or the corresponding ratio of canonical roots, is small (say, less than 0.05), then little loss of discrimination will result from excluding the smallest eigenvalue-eigenvector combination ($k_v = \infty$) or, equivalently, from eliminating the last principal component.

The eigenvalues and eigenvectors for all nine variables are listed in Table 3. The smallest eigenvalue accounts for only 1.8 percent of the variation within groups. The eigenvector corresponding to the smallest eigenvalue (hereinafter referred to as the smallest eigenvector) reflects a contrast between variables 2 and

Table 3. Eigenvalues and eigenvectors of within-groups correlation matrix W* for all nine variables; between-groups sums of squares for each principal component (after Campbell and Reyment, 1978).

Eigenvalues (e_i)		1	2	3	4	5	6	7	8	9
		4.31	1.25	1.00	0.69	0.48	0.43	0.38	0.29	0.16
Eigenvectors		v1	v2	v3	v4	v5	v6	v7	v8	v9
	u_1	0.35	0.43	0.42	0.37	0.37	0.15	0.35	0.21	0.23
	u_2	0.33	0.09	0.20	0.19	0.21	-0.30	-0.31	-0.55	-0.53
	u_3	0.28	-0.15	-0.03	-0.10	-0.07	-0.85	0.23	0.31	0.09
	u_4	-0.08	-0.06	0.03	0.09	0.05	0.13	0.01	0.64	-0.74
	u_5	-0.44	-0.05	-0.18	0.44	0.51	-0.25	-0.42	0.17	0.24
	u_6	-0.01	-0.10	0.03	0.74	-0.66	-0.02	0.01	-0.03	0.03
	u_7	0.27	0.26	0.24	-0.20	-0.28	0.03	-0.73	0.33	0.20
	u_8	0.64	-0.48	-0.44	0.14	0.19	0.29	-0.11	0.08	0.10
	u_9	-0.10	-0.69	0.70	-0.05	0.08	0.06	-0.04	0.00	0.10
		1	2	3	4	5	6	7	8	9
diag$\{G_{(0,\ldots,0)}\}$		1.01	0.25	0.23	0.27	0.26	0.07	1.57	0.18	0.21

trace $\{G_{(0,\ldots,0)}\}$ = 4.05.

3, to wit the width of the test, and the width of the last chamber.

These loadings are large and , it may be suspected that if the corresponding between-groups sum of squares is small, as is the situation in our example, instability in the corresponding canonical variate coefficients may result.

The between-groups sums of squares for all principal components shows that 39 percent of the between-groups variation is associated with the seventh principal component and 17 percent with the first principal component. The variation for the seventh principal component results from a contrast between variable 7 and most of the other variables; the first principal component is a "size component" (cf. Blackith and Reyment, 1971).

The canonical variate analysis can be carried out in terms of the principal components (the coefficients for the original variables are estimated by projecting back to the space of the original variables). The coefficients for the first canonical variate (a_i^u in Table 4) highlight the contribution from the seventh

Table 4. Standardized canonical vectors for nine variables, including shrunken estimates (from Campbell and Reyment, 1978).

	v1	v2	v3	v4	v5	v6	v7	v8	v9	Canonical roots
$\underset{\sim}{a}_1^u$	0.54	-0.11	-0.01	0.11	0.06	-0.07	0.79	-0.18	-0.13	
$\underset{\sim}{a}_2^u$	-0.64	-0.32	0.14	-0.13	-0.44	-0.11	0.39	0.05	-0.32	
$\underset{\sim}{c}_1^u$	0.00	-0.59	-0.09	0.43	0.44	0.07	0.99	-0.51	-0.10	2.28
$\underset{\sim}{c}_2^u$	0.65	0.81	-0.27	-0.44	-0.40	0.15	0.04	0.28	0.25	0.76
$\underset{\sim}{c}_{1\,(0,\ldots,\infty)}^{GI}$	0.04	-0.37	-0.31	0.43	0.41	0.06	1.01	-0.51	-0.12	2.25
$\underset{\sim}{c}_{2\,(0,\ldots,\infty)}^{GI}$	0.58	0.29	0.31	-0.53	-0.36	0.25	-0.03	0.37	0.32	0.70
$\underset{\sim}{c}_{1\,(0,..,\infty,\infty)}^{GI}$	-0.17	-0.21	-0.17	0.32	0.42	-0.04	1.06	-0.53	-0.16	2.17
$\underset{\sim}{c}_{2\,(0,..,\infty,\infty)}^{GI}$	0.53	0.32	0.35	-0.40	-0.50	0.22	-0.03	0.36	0.33	0.68

principal component. The coefficients for the original variables are determined explicitly from the principal-component canonical vector; any inflation in these latter coefficients results in inflated coefficients for those among the original variables contributing to the eigenvector from which the principal component is derived. Note that the first principal component contributes most to the second canonical variate (see a_2^u in Table 4).

The first canonical variate amounts to 56 percent of the between-group variation. The first two canonical variates account for 75 percent of the variation between groups. The coefficients for the standardized original variables for the first two canonical variates are shown in Table 4, namely c_1^u and c_2^u.

SHRUNKEN ESTIMATES

The effect of shrinking the contribution of the smallest eigenvector (and associated eigenvalue), namely, the ninth principal component, is shown in Table 4 (here, $k_9 = \infty$ implies the elimination of the ninth principal component from the analysis).

The two sets of coefficients for the original varbles (c_i^u) and $k_9 = \infty$ ($c_{i(0,\ldots,\infty)}^{GI}$) are similar, except

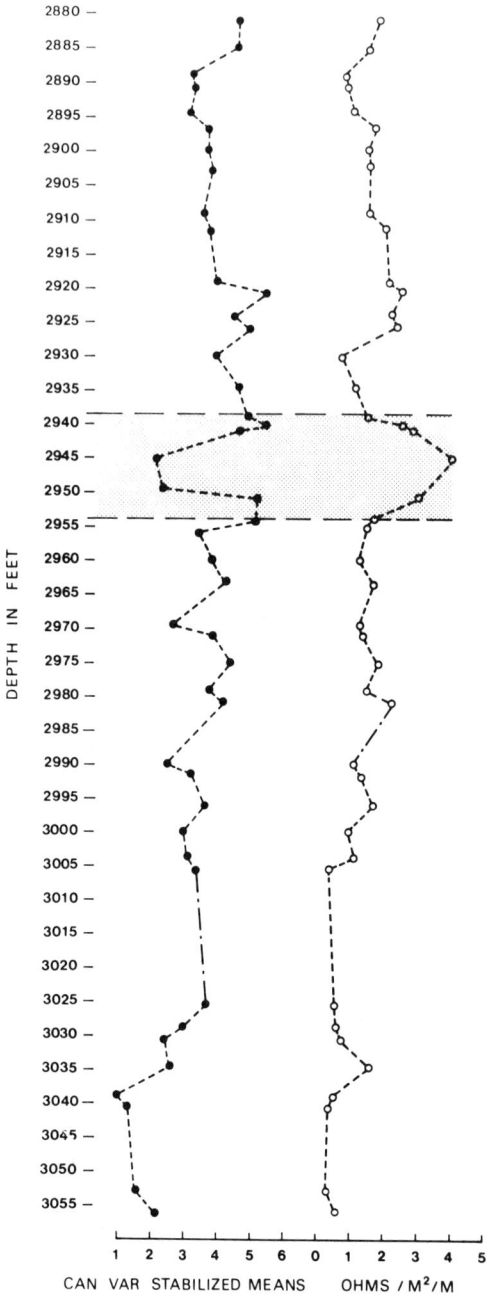

Figure 2. Biolog formed from growth-reduced principal coordinates of 10 samples of *Afrobolivina afra*. Right-hand curve is short-normal electrical resistivity log.

for variables 2 and 3. The decrease in the magnitude of the coefficient for the second variable and the change in sign of the coefficient for variable 3 are apparent. The sum of the coefficients for these two variables is relatively stable and the canonical roots are little affected by the elimination of the smallest eigenvector.

A plot of the shrunken estimates for all nine variables of the canonical variate means for the first canonical variate against location indicates that there is a general drift over time in the morphology of the species, manifested here as a trend to the right. This shift seems to be due to a long-term environmental effect (Fig. 2).

The improved interpretability of paleoecological data brought about by the foregoing approach to canonical variate analysis offers attractive prospects for the future. From the aspect of computing, the procedure, of necessity, is an interactive one.

REFERENCES

Blackith, R.E., and Reyment, R.A., 1971, Multivariate morphometrics: Academic Press, London and New York, 412 p.

Burnaby, T.P., 1966, Growth-invariant discriminant functions: Biometrics, v. 22, no. 1, p. 96-110.

Campbell, N.C., and Reyment, R.A., 1978, Discriminant analysis of a Cretaceous foraminifer using shrunken estimators: Jour. Math. Geology. v. 10, no. 4, p. 347-359.

Chaney, R.W., 1924, Quantitative studies of the Bridge Creek flora: Am. Jour. Sci., v. 8, no. 44, p. 127-144.

Cooley, W.W. and Lohnes, P., 1971, Multivariate data analysis: John Wiley & Sons, New York, 364 p.

Gower, J.C., 1966, A Q-technique for the calculation of canonical variates: Biometrika, v. 53, pts. 3/4, p. 588-590.

Gower, J.C., 1976, Growth-free canonical variates and generalized inverses: Geol. Inst. Univ. Uppsala Bull. (new ser.), v. 7, p. 1-10.

Joreskog, K.G., Klovan, J.E., and Reyment, R.A., 1976, Geological factor analysis: Elsevier, Amsterdam, 178 p.

Mardia, K., 1972, Statistics of directional data: Academic Press, London and New York, 357 p.

Miller, R.L., and Kahn, J.S., 1962, Statistical analysis in the geological sciences: John Wiley & Sons, New York, 357 p.

Pelto, C.R., 1954, Mapping of multicomponent systems: Jour. Geology, v. 62, no. 5, p. 501-511.

Pielou, E.C., 1974, Population and community ecology: Gordon & Breach, New York, 424 p.

Pielou, E.C., 1977, Mathematical ecology: John Wiley & Sons, New York, 385 p.

Reyment, R.A., 1963, Multivariate analytical treatment of of quantitative species associations: an example from palaeoecology: Jour. Anim. Ecology, v. 32, p. 535-547.

Reyment, R.A., 1966, Studies on Nigerian Upper Cretaceous and Lower Tertiary Ostracoda. III, Stratigraphical, palaeoecological and biometrical conclusion: Stockh. Contr. Geol., v. 14, 151 p.

Reyment, R.A., 1971, Introduction to quantitative paleoecology: Elsevier, Amsterdam, 226 p.

Reyment, R.A., 1975, Canonical correlation analysis of hemicytherinid and trachyleberinid ostracodes in the Niger Delta: Bull. Am. Paleontologist, v. 65, no. 282, p. 141-145.

Reyment, R.A., 1976, Chemical components of the environment and Late Campanian microfossil frequencies: Geol. Foren. Stockh. Forh., v. 98, p. 322-328.

Reyment, R.A., and Banfield, C., 1976, Growth-free canonical variates applied to fossil foraminifers: Geol. Inst. Univ. Uppsala Bull. (new ser.), v. 7, p. 11-21.

Reyment, R.A., and Reyment, E.R., 1980, The Paleocene
    trans-Saharan transgression and its ostracod fauna:
    Proc. 2nd Conf. on Geology of Libya, Tripoli,
    in press.

THE FUTURE OF INFORMATION SYSTEMS IN THE EARTH SCIENCES

P.G. Sutterlin

Canada Centre for Mineral

and Energy Technology

ABSTRACT

Much of the effort in modern information systems has been concentrated in two aspects. The first is in the design and generation of various types of data bases. The second is in the design and development of methods to retrieve selectively relevant information from these data bases. These activities have had the effect of forcing the geoscientist to reexamine the methods in which earth-science data has been collected, recorded, and stored. As a result some fundamental progress has been made in geoscience information systems apart from the fields of computer and information sciences. The next phase already seems to be in progress. The use of distributed processing techniques, the availability of relatively inexpensive microcomputers and an emerging communications technology will significantly affect the manner in which geoscientists will gather, store, and access information.

INTRODUCTION

The rate of change in technology of computer-aided data processing and electronic data communications is mind-boggling. It is difficult to anticipate future developments in geoscience information systems which undoubtedly will be affected by these new techniques. Perhaps the best that can be done is to review the progress of the past couple of decades and, in the light of technology already in place, attempt to gauge the impact this may have on the geoscience community in the coming decade.

The word SYSTEM is being used more and more frequently as a tool to market a whole host of products. There are stereophonic sound systems, home rug-cleaning systems and, most recently, word-processing systems. These systems probably would be described as respectively: one or more electronic components designed to reproduce music from a plastic disk or magnetic tape; a machine with a set of fancy attachments designed to apply a cleaning agent to wall-to-wall carpeting; and a stand-alone mini- or microcomputer which allows direct entry, dynamic revision, and composition of textual material. The emphasis would be focussed most likely on the physical "things" which serve to process music, clean rugs, and manipulate alphanumeric characters. In the same vein, an information system has come to indicate in most instances nothing more than a set of computer programs or a "software package" which permits the processing of various types of information. This, however, is a much too restricted point of view.

In a recent report to the Provincial Ministers of Mines in Canada, C.F. Burk (1979) of the Canada Centre for Geoscience Data defined an information system as follows:

> "Things, people, procedures and information organized and managed to achieve an objective."

In this context, perhaps the two most common information systems are a library and a room full of filing cabinets, each with their respective staffs who acquire information and data in order to:

(A) Develop and maintain an information resource, and
(B) Apply procedures and management techniques to insure efficient and effective utilization of the information resource.

To determine the impact of the application of computer-aided techniques on development of information systems for the earth sciences, it is useful to analyze in a bit more detail the functions involved in information-resource development and information-resource utilization, using the library and the file room as models.

INFORMATION-RESOURCE DEVELOPMENT

The development of an information resource involves, in general terms:

- (A) Information Resource Acquisition
    1. Information selection
    2. Information collection
- (B) Information Processing
    1. Information classification
    2. Information indexing and abstracting
    3. Information cataloging
- (C) Information Storage
    1. Information storage medium design
    2. Information organization.

INFORMATION-RESOURCE ACQUISITION

The manner in which the basic information resource of any information system is acquired has been affected only marginally by computer technology. Only where information and data are recorded directly in analog or digital form onto a permanent storage medium has the methodology of selection and collection of information changed, and then only in response to a change in the information-generation methods. In all other situations, collection and selection of information to comprise the resource remains a manual operation. This, however, is one area in which the emerging communications technology may have a marked impact in the near future.

INFORMATION PROCESSING

The processing of information as part of the resource-development function becomes necessary when the number of individual items in an information resource reaches the point where time constraints preclude item-by-item examination of all information sources pertinent to a particular problem. Information is classified, indexed and, in some situations, cataloged to produce subject and author indexes in the form of card catalogs in libraries, or various indexes to the contents of a series of filing cabinets. These indexes constitute "secondary" or "derived" information resources designed to facilitate identification and selective retrieval of those items of information (the books, documents, and file folders) which are most likely to contain pertinent information. In fact, they are organized or "structured" compilations of information or data. In this sense, these indexes can be regarded as data bases. It thus would seem that one of the major objectives in the processing of an information resource in its development phase is the generation of data bases. The manner in which those data bases are to be accessed has a definite influence on their design. Therefore, the design of the data bases has been, and is yet, the objective of much research and study.

DATA BASES

There is an almost continuous gradation from a data base whose records contain only numerical measurement information (presumably unsullied by any subjective interpretation or regurgitation) to one whose records contain only references to the sources of information (with or without keywords or abstracts). It has been popular to contend that the former "numeric" data base comprises "real" data whereas the latter "bibliographic" data base is composed of information. It has been contended further that failure to recognize the difference between data and information, particularly as it pertains to information-systems design, will lead only to disaster. However, the point at which objectivity ceases and subjectivity begins, particularly in the earth sciences, is not easy to determine. Hence, many of the more frequently used geoscience data bases contain within their records some elements of both these "information end members (Fig. 1). It would seem, then, that the design of a geoscience information data base - and consequently the design of the information system to process the data base - is more sensitive to the "mix" of information types which form the information resource than to some arbitrary distinction between data and information. Therefore it is convenient to regard "data" as a part of the total spectrum of information, and a data base as structured information. In that manner it is possible to use the terms data and information interchangeably.

DATA BASE ASPECT	END MEMBER 1		END MEMBER 2
Type of Information	Secondary (Information)	Mixed (Infordata)	Primary (Data)
Relative volume of Information	Large	Small	Large
Information Structure	Simple	Complex	Simple
Information Contents	Bibliographic citations	Measurements Observations	Digitized log data

Figure 1. Range of data bases in geoscience.

INFORMATION STORAGE

There is not much point in processing information and creating data bases to identify the resources unless they are arranged and stored in such a manner as to relate to the data bases and can be retrieved readily. Generally, in manual information systems, this involves some sort of sequential ordering within specified classes, and the whole arranged physically to optimize access time. Again, we are reminded of the file room and the library shelves. As more and more information became available, most storage facilities became inadequate physically, and many methods have been devised which will enable the storage of more and more information in less per unit space. Computer technology first made significant inroads on the information resource development function with respect to the storage of information.

INFORMATION-RESOURCE UTILIZATION

The utilization of an information resource involves, in general terms:

    (A) Information Resource Utilization
            1. Information Processing
                   I. Information problem definition
                  II. Information resource identification
                 III. Information search strategy definition
                  IV. Information selective retrieval
    (B)   2. Information Acquisiton
                   I. Information transmission
                  II. Information (document) delivery.

The conventional methodology of information retrieval presupposes a relative physical location for each item of information within the information resource. Given a well designed set of data bases and an orderly storage algorithm, items of information can be identified, selected and retrieved directly. The search strategy will depend on the number of data bases or indexes available to be used to narrow down the search. The definition of the information required and the search strategy employed yet remains a problem. On the one hand, not many librarians or file clerks are sufficiently subject-oriented. On the other hand, not many users are sufficiently familiar with information systems. This point will be elaborated next.

THE ROLE OF THE COMPUTER INFORMATION SYSTEM

During the past four decades, the rate of increase in the amount of information has been staggering. All predictions point to the fact that this rate of increase will be at least maintained (if not accelerated) in the next twenty years. There have been essentially two stimuli in the application of computer technology to information systems.

1. The indexes to most of the major information-resource depositories have themselves become sufficiently large, in terms of the number of records they contain, that item-by-item manual searching of these data bases has become impractical.
2. The volume of material encompassing the information on any one subject - or even one subdiscipline - has become sufficiently great that storage of this information at a single location is becoming impractical. This has resulted in the development of indexes which cover more than one information resource. Depository access to these composite indexes, some of which by now have assumed gargantuan proportions, is impractical using solely manual techniques.

No single geoscience library is capable of selecting, acquiring, and storing all of the earth-science information now available and which will be generated in the future. Similarly, although a single exploration company may be able to organize and store its own field data and information, but is unlikely to have the manpower, facilities, or inclination to store centrally all exploration data available.

The electronic computer therefore has been used extensively for the last fifteen years as a tool in those parts of information systems which have to do with information storage - and consequently data-base design - and with the retrieval of information from these data bases. The application of computer technology to geoscience information then should have been relatively straightforward. After all, card catalogs and file folders have been in existence for a long time. What more could have been required than a simple transposition of information and data from one storage medium to another? Is there something unique about geoscience information which seemingly has required the development of specialized data bases and information systems? Or perhaps is it the nature of the science itself, or the methodology of its practitioners which has produced the host of data bases and information systems which now are being used by geoscientists? The answer is - probably a little bit of both (Hubaux, 1973).

BIBLIOGRAPHIC DATA-BASE DESIGN

It is relatively easy on a library catalog card to determine which combination of alphanumeric characters refer to the author (or authors), the title, the citation, the abstract (if present), and the keywords. The information in each of these "records" does not have to be ordered or "structured" precisely. The record contents are decipherable regardless of their internal consistency. However, a computer, not being able to make the same type of value judgments, requires internal record consistency. This immediately raises the question not only of what information should be included in each record of such a data base but how this information should be structured. Although this question has not been resolved entirely among the indexers, abstractors, and catalogers of bibliographic data, progress is being made. In any situation, each document or book can be described in terms of a record containing one or more author names, a title, a series of keywords which summarize the essence of the document's contents, a citation or reference, and an abstract. In one way or another, each of these items of information can be assigned a field which must be "flagged" in such a manner as to identify the contents of the fields. Although there may be many authors, a host of appropriate keywords, a lengthy citation, and a comprehensive abstract making up each record, the information in each field applies equally to the entire record (Fig. 2). This permits the structure of the record to remain essentially linear because each field is linked to a common root, giving rise to a simple tree. Moreover, as long as each individual item of information within the record which might be used as a search key is assigned a separate field, the fields can be readily rearranged - that is to say that their sequence can be revised easily and altered. This indicates that many, if indeed not most, bibliographic data bases can be adapted for processing by a number of generalized "data-base systems" or computer programs designed to accommodate simple data structures. This is not to say that, from a systems analyst's point of view, these "systems" are simple or unsophisticated - to the contrary. It does indicate, however, that the data-base design need not be governed directly by the nature of the computer programs which may be used eventually to process the information.

Because much of the information collected by geoscientists is not yet "quantifiable", the definition of the language used to communicate data and concepts perhaps has not been as precise in the geosciences as it has in the other physical and natural sciences. Much reliance has been placed on the context in which certain words and phrases are used. However, when bibliographic data bases were being converted to those to be processed by computer, it became apparent that many so-called "descriptors" were meaningless, especially (as is usually the situation) when they were used as keywords out of context. This has provided the need for the development of a number of thesauri to control the vocabulary of computer-processable geoscience bibliographic data bases. Ultimately, the use of these thesauri in conjunction with the general availability of the data bases should serve to help "tighten up" geological terminology to the point where communication of information will be less ambiguous and more complete and effective. This is a benefit which has been the direct result of the application of computer concepts to bibliographic data-base design, and which may not have taken place otherwise. Certainly, without the stimulus of computer-processable data bases, it is unlikely that the development of a multilingual geoscience thesaurus would have been initiated. This joint undertaking of COGEODATA, IUGS, and the ICSU Abstracting

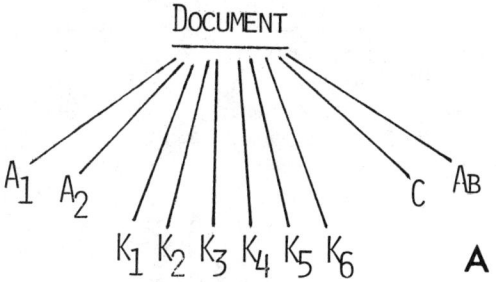

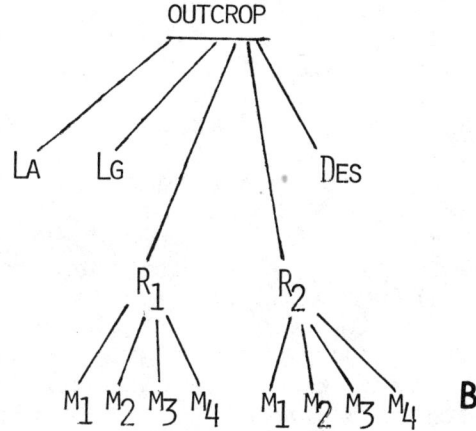

Figure 2. Distinction between structure of secondary (A) and primary (B) information.

Board, under the Chairmanship of Dr. Harm Glasshof of West Germany will have a major international impact on geoscience information systems apart from the aspects of computer processing which prompted this project.

NONBIBLIOGRAPHIC DATA-BASE DESIGN

Now, let's turn to the other information-system model. When looking at the geological information and data on the sheets of paper within a single file folder — which is a record analagous to a bibliographic citation — the items of information contained by implication are related to a geological concept or object and therefore also to one another. However, because geological objects and concepts are in themselves the product of the interaction of many physical, chemical, and biological phenomena, the interrelations among the items of information required to describe them adequately tend to be more complex than when the record is a bibliographic reference. Let's take the example of an outcrop file in which each record contains information about the outcrop itself as well as information relating to a number of rock types within that outcrop. Presumably, within the file folder, there would be one sheet of paper containing general information relating to the location and description of the outcrop itself, and a series of sheets of paper each containing information relating to a separate rock type within the outcrop. We see immediately that there are within this outcrop record a number of subrecords, the fields of which do not necessarily relate to the outcrop as a whole (Fig. 2). In manual systems, the nature of the recording and storage techniques used usually makes it unnecessary to spell out every single link between information items. However, the computer again is unable to make such implied correlations among a set of complexly interrelated information items. The example I have given is a relatively simple one. It should be easy to imagine, however, that complex tree, hierarchy, and network information structures can result easily when working with geological objects, concepts, and phenomena. Much care then must be exercised when designing data bases which contain purely numerical or a combination of numerical and observational information in order that the item-to-item relationships are preserved. This problem has led to the development of a wide array of data bases. The data-base design to a large degree has come to be dictated by the perceived availability of a set of computer programs to process a specific data set. This, however, has not always been the situation.

When geoscientists first began employing computers to store and retrieve structured information, the generalized software packages designed to process data were able only to accommodate the most simple information structures (Sutterlin and May, 1978). These had been developed to respond to the need for computer-aided financial and bookkeeping functions which had to accommodate relatively uncomplicated data structures. Therefore, there evolved a conviction among geoscientists that it was necessary to develop a specific software package to accommodate each of the data structures which were being encountered. Unfortunately, this belief yet exists to a large degree. However, the fact that the content of a set of items of information may be unique scientifically is no longer justification for this specific systems design approach. It is possible now to accommodate the structure of almost any set of geological information using one of the generalized "data-base management systems" available today. However, unlike the example with bibliographic data bases, the nature of the computer programs selected to process a particular data set can have a direct influence on the design of the data base. This is because it is not always a simple matter to rearrange, without loss of information, the data as they become structured more complexly.

It is precisely these constraints and considerations which can serve to benefit geoscientists. One of the more significant outcomes of designing geoscience data bases for computer processing has been the discovery (or affirmation, if you like) that geoscientists with few exceptions have a rather cavalier attitude towards "primary" information (what could be termed data). The prevailing point of view seems to have been, perhaps unconsciously, that as long as no one other than the generator or collector of information was likely to use a field book or a "personal" file, it was of little consequence how information was recorded or stored as long as the records were decipherable by the original observer. In other situations, the volumes of information have been too great and this has precluded attempts to communicate "primary" information. Therefore, geoscientists have become accustomed to communicating in terms of concepts and conclusions, which are synthesized or "secondary" information (Hubaux, 1973). These two factors have served to foster the notion that the communication of "primary" geoscience information is at best unnecessary and at worst potentially incriminating.

Computer technology, however, has made it possible to communicate economically and rapidly hitherto unprecedented volumes of "primary" information. So, it is slowly becoming unacceptable to describe and record the location of an outcrop as being "about a quarter of a mile northeast of the upper end of Lake Skaneateles". This description might lead the original observer back to the spot, but anyone else would be unlikely to find it. In this respect, the application of the computer has forced some reevaluation of information collection and recording methods.

BIBLIOGRAPHIC INFORMATION RETRIEVAL SOFTWARE

There are many geoscience bibliographic data bases in existence, and almost as much corresponding computer software designed to retrieve information from these data bases. However, it seems that in the last two years or so, there is beginning to be some recognition that perhaps a few large bibliographic data bases which are accessable widely are in the best interests of the geoscience community. As a result, the AGI Geo.Ref data base and Geosystem's GeoArchives data base have been implemented using respectively Lockheed's DIALOG software and SDC's ORBIT software. These data-base vendors have made their data bases widely available through the TYMSHARE, TYMNET, and DATAPAC communications networks. A software package named DIANE is available through the European Economic Commission and uses the EURONET communications network. Retrieval of bibliographic information from these data bases is fast, efficient, and relatively inexpensive. Records are stored randomly on magnetic disk so that access is direct. Comprehensive indexes to the data bases allow a great amount of flexibility in the search strategy used and the form of output desired. Experience has shown that the average retrieval cost is one dollar per minute of connect time.

As long as care is taken in the design of a bibliographic data base, computer conversion of such a data base for processing by a generalized software package is relatively straightforward on a one-to-one basis. Indexes can be generated automatically for each separate field in the record, so that only the indexes are searched in order to identify the storage location of these records which satisfy the search criteria. This not only greatly reduces retrieval time, but also permits a certain amount of Boolean logic to be applied if the search criteria involve more than a single descriptor. Most of the commerically used software packages permit searching of text where the record contains an abstract. However, it should be noted that this latter capability should not be taken as indication that the control of vocabulary, and thus thesaurus development, is no longer important, even though from a strictly computing point of view, it may no longer be a requirement to effect comprehensive search searches of bibliographic records. The benefits of thesaurus construction as an aid to indexers and as an important adjunct to "sharpen up" communication effectiveness.

Other refinements of software design continually are emerging. One example is the QUIKLAW System developed and operated as a data-base vending service by QL Systems in Kingston, Canada. This software computes a ranking for each of the records retrieved which fulfill the search criteria on the basis of the frequency of occurrence of all the search terms (including those in the abstract) used in the retrieval request. The user then may preselect the number of records which he may wish to see displayed or printed on his terminal based on this statistical ranking. Search strategy commands in DIALOGUE and ORBIT may be preserved and subsequently modified, obviating the need to "start from scratch" every time a similar search or modification of an analogous search is desired. The command language required to use these systems is being simplified constantly so that even the uninitiated user with one day's instruction can master the basics of search techniques. In fact, it is becoming almost easier now to formulate the search commands than it is to master the commands required to insure that one is identified correctly by the system as a legitimate user.

NONBIBLIOGRAPHIC INFORMATION RETRIEVAL SOFTWARE

As has been mentioned before, in the late 1960's and the 1970's many data-specific as well as generalized software packages were developed to process what essentially were nonbibliographic or "primary information" data bases (Fig. 3). This has not been an attempt on the part of geoscientists to seek greener fields in the areas of information and computer science, although many geoscience grant committees seem convinced that this continues to be the sole motification. The fact remains that, until the middle 70's, when System 2000 and Honeywell's IDS II were developed, computer scientists who were developing data-processing software packages were not being faced with information which is structured as complexly as it is in the geosciences. Information which arranges itself in trees, hierarchies, and networks is not as abundant in other fields and where it has occurred, the practical solution was to divide the information into less complexly structured subsets.

As a result of this dilemma, geoscientists developed, in conjunction with (and sometimes over the protests of) the computer professional, a whole host of *data specific* as well as a few *generalized* software packages, some of which are displayed in Figure 3. These represent an

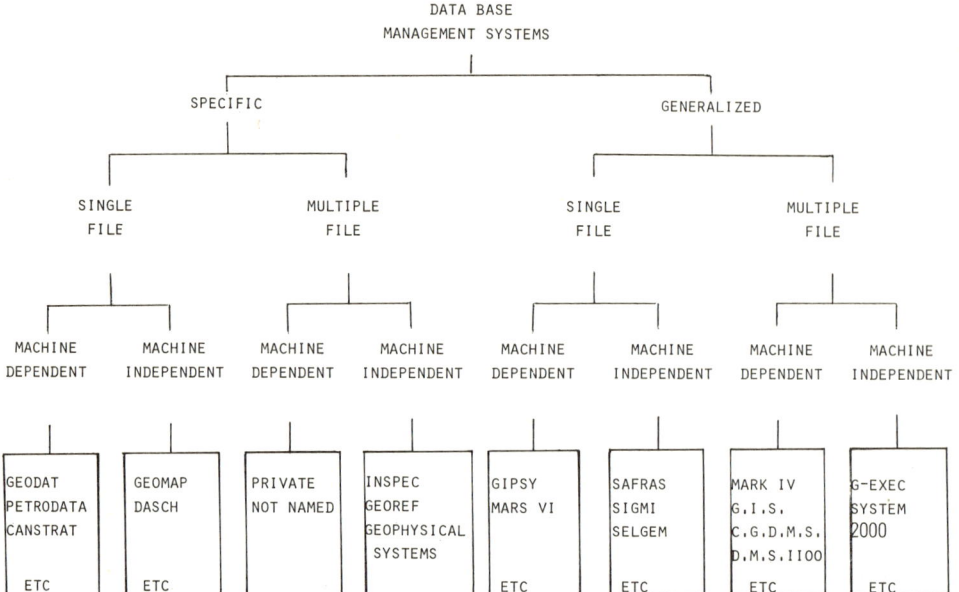

Figure 3. "Primary information" data bases.

almost total spectrum with respect to the degree of programming sophistication incorporated in them. The primary aim, of course, has been to design and develop "The Software Package" which would accommodate readily any conceivable geoscience information structure, and then to convince the geoscientific community to use this system. The analogy to the DIALOGUE, ORBIT, and DIANE systems is clear. Communication, using automated techniques, then would become a reality because a common system would force adherence to standards of both information content as well as basic information structures.

Unfortunately, the conversion of primary information data bases from one form to another is not at all straightforward. When information structures become more and more complex, it becomes desirable from a practical point of view to use the structure embodied in the programming language being used to program the software package in order to define some of the interrelationships among the items of information in each record. Then, unlike the situation with the bibliographic data base, conversion from one form to another involves more than a mere rearrangement of fields within the records - in the worst situation it may involve considerable redesign and manual regeneration of the data base itself.

Even conversion of a data base to another "format" by developing a computer program to do so is not always easy, and becomes more difficult as the information structures become more complex. It can be done, but in many situations is not justifiable if the costs are taken into account. The FILEMATCH concept, which was enunciated three years ago, was an attempt at solving this problem (Sutterlin, Jeffery, and Gill, 1977). This concept suggested that each item of information in a data base have appended to it, in one way or another, enough structural information to define explicitly its relationship to all of the other items of information in the record, apart from the structural information embodied in high-level programming languages. Even though the implementation of FILEMATCH may have been somewhat unsophisticated, it may be time to revive the idea in the light of modern distributed processing concepts. Until something like this is done, communication of large amounts of primary information in machine-processable form will lag behind the significant progress in this direction which are evidenced with the bibliographic information systems.

FUTURE TRENDS

Future trends in information systems, including those which will be used in the earth sciences, will be a product of the interrelated developments in the areas of microcomputer design, distributed processing techniques, and communications technology.

Information processing may come to revolve around the types of functions which mini- and microcomputers are able to perform (Wigington, 1979). Present costs of the these machines is about $15,000 per megabyte of storage which includes an input-output console, a printer, and operating system as well as applications software. For double this amount, such equipment can be configured with a hard disk capable of storing up to 12 megabytes of information and a number of input-output terminals. The software is sophisticated sufficiently to permit on-line entry and editing of information as well as some data-base search capabilities independent of a large mainframe computer. The potential of these machines to generate small data bases at the point where the data originate and are used is becoming obvious. The cost benefits of file generation using mini- and microcomputers in terms of the efficient use of personnel and physical resources is becoming more and more attractive as the power of these machines increases in relation to their cost.

It is possible, using primarily hardware features included with most of these mini- and microcomputers to transfer files of information from their storage media, usually floppy disks, to the storage media of mainframe computers. The generation of data bases using one mainframe computer and several microcomputers can form a powerful and effective distributed processing network when the primary information resource (as may be the situation) is deposited in several physically separated locations.

Most microcomputers have available some sort of database management file software which permits generation of nonbibliographic data bases and selective retrieval, in random access mode, of information from these data bases. Presumably, the time is not far off when some of these software packages will be able to accommodate fairly complex information structures. It would be ideal then to incorporate these data bases into larger data bases using distributed processing methods. However, much of the structural information will continue to be part of the high-level programming languages which are used in developing the software packages, making communication of the generated files to another computer for incorporation in a large data base difficult without first converting the data bases to the format compatible with that of the software package implemented on the other machine.

Now, given the potential benefits of distributed data processing in the generation of primary information data bases, it might be appropriate to reexamine the FILEMATCH concept. It may be advantageous to study the feasibility of establishing some sort of protocol whereby structural information, hopefully automatically generated by the software as the data base is being generated, can be appended to every item of information in the records of a data base. At present, data-base management systems which produce random access files do generate as the file is built (in the form of indexes) information as to the precise location of records in the storage medium. Could this technique not be applied in the manner suggested?

The one area where there has been technologically the least progress, has been in the delivery of the information to the user. It is not justifiable economically to store the entire contents of a document or notebook in a data base for reproduction and transmission to a cathode-ray terminal or a high-speed printer. However, there are times when a user requires the information more quickly than is possible using interlibrary loan and other similar transmission and delivery facilities. The answer, already tried in Great Britain, seems to be to utilize radio transmission rather than telephone lines for packet switching of information. If cable television companies were to make available one frequency or channel for radio packet switching, about 55,000 users at one time could be accommodated. The cost of data transmission, as opposed to the present telecommunications networks, could be reduced dramatically. This technology, coupled with a videodisk capable of storing 54,000 frames of information, could make the problem of document delivery and data transmission a thing of the past. It would be prudent for geoscientists to closely monitor these developments.

There is one aspect of information systems which purposely has been left to the last, but not because it it is the least important. This discussion began by defining an information system as:

"Things, people, procedures and information organized and managed to achieve an objective".

Much has been said about things, procedures, and information, but little about people. Of course, many people including the information generators, indexers, abstractors, data-base designers, software developers, and users all have an impact on information systems. But, in research centers, industry, and government, the task of making efficient and effective use of the new information technology has been treated rather lightly. The trained librarian generally is given these responsibilities. However, the library staff may be ill-equipped to handle these responsibilities. The person charged with this task must be a subject-matter specialist. Then, knowing where the information is available, where to obtain it, and how to prepare it for use assumes more than a superficial understanding of the technological changes which are affecting the information-systems functions. This person is one of a new breed - the information specialist.

The modern geoscience information professional must be truly interdisciplinary - so much so that he has in the past and continues to suffer from a lack of identity.

There is an almost urgent need for people trained to merge the geoscience, information science and computer science disciplines to help in the solution of complex information problems. It is not too drastic to predict that, if such people are not produced, the discipline of geology ultimately will be assimilated by those other scientific disciplines which better understand the vital part that information is playing, and will continue to play, in the evolution of research and development in their fields.

However, it must be remembered that geoscience data bases at present in reality are generated by those interested in geoscience information primarily for their own use. One of the goals which must be kept in mind is that, through research and further development, information technology should be simplified to the point where the tools can be put into the hands of the end user.

REFERENCES

Burk, C.F., 1979, National system for geological information: Rept. Provincial Ministers of Mines, Canada Centre for Geoscience Data, Energy, Mines and Resources Canada, Ottawa.

Hubaux, A., 1973, A new geological tool - the data: Earth Science Reviews, v. 9, no. 2, p. 159-196.

Sutterlin, P.G., Jeffery, K.G., and Gill, E.M., 1977, FILMATCH: A format for the interchange of computer-based files of structured data: Computers & Geosciences, v. 3, no. 3, p. 429-442.

Sutterlin, P.G., and May, R.W., 1978, Geology, *in* Encyclopedia of computer science and technolgoy: v. 9, Marcel Dekker Inc., New York, p. 27-56.

Wigington, R.L., 1979, Minis and micros in perspective, unpublished paper presented at Spring Meeting, Assoc., Information and Dissemination Centres (ASIDIC), Ottawa, Ontario.

TRENDS IN COMPUTER APPLICATIONS IN STRUCTURAL GEOLOGY:

1969-1979

E.H. Timothy Whitten

Northwestern University

ABSTRACT

A brief review of the journal literature since 1969 (supplementing an earlier review covering 1959-1969) indicates the manner in which computer-based techniques have permeated most fields of structural geology. The biggest new thrust has been in the application of finite-element methods, which seem destined to play an increasingly important role in structural-geology research.

INTRODUCTION

Ten years ago, at the beginning of a paper with the same principal title, it was asserted that "Relatively few attempts have been made to exploit the potential usefulness of computers to structural geology" (Whitten, 1969, p. 223). At that time, computers generally had been available only to academic geologists for a little more than a decade, and producing a complete review of the use of computers in structural geology was a relatively simple task. Several goals and objectives could be identified clearly. Now, after a further decade of research publication, the situation is radically different. The computer and quantitative thinking have permeated most domains of the geological sciences. The availability of computers and the widespread ability of students to program and use computers on a routine basis allows these tools to be used without particular comment in a large proportion of modern structural-geology research. "Computer applications" is ceasing to be a distinctive subdiscipline of structural geology; rather the use of computers is tending to be just one of the useful techniques that qualified structural geologists currently use.

As ten years ago, "structural geology" is limited deliberately here so as to exclude structural work primarily based on geophysical data, rock mechanics, and theoretical research on strain and stress, although, in such fields, computers continue to play significant roles. In this context, in 1969, I used "experimental work" instead of "rock mechanics," but recently an important class of computer experiments has emerged that is based on finite-element methods.

Before proceeding further, it is appropriate to summarize the situation that was reported in 1969. Whitten (1969, p. 223-224) was able to categorize published work on the use of computers in structural geology under four headings:

(1) storage and collation of structural observations,
(2) calculation of generalized structural properties from arrays of observations
(3) description of the morphology of structural features, and
(4) mapping and analyses of the spatial variability of structural elements.

Obvious future potentialities were seen to include the development of computer programs to permit:

(i) identification of the multiplicity of geological factors that control the nature and spatial variability of the geometry of actual folds,
(ii) objective evaluation of the relative importance of the identified controlling geological factors, and
(iii) erection of process-response models to permit simulation of fold systems and objective (quantitative and qualitative) comparison of such simulations with the nature of actual fold systems.

These categories remain useful in classifying papers published in the past decade, although papers in categories (1), (2) and (3) have declined in relative numbers as the methods became well known and the topics did not warrant new journal articles. As new techniques for spatial analysis were developed within the past decade, many of them have been applied to structural geology; hence, there is a considerable volume of new work under category (4). So far as future developments anticipated in 1969 are concerned, items (i) and (ii) have received considerable attention, but most of the new research does not fall within the subject matter of this review. For example, there has been significant quantitative and qualitative work on the genesis of foliation and whole folds, but it has not relied primarily on computer technology. By contrast, the finite-element method, made possible by the use of large computers, permitted the development of process-response models (item iii) for a wide variety of structural problems; this is a fast-growing and powerful new field of research.

The principal purpose of this article is to review the actual published work in these several fields. However, a major problem is that the total volume of literature has expanded so greatly within the past 10 years that this review (a) cannot claim to be complete - many important papers from the world literature may well have been missed inadvertantly, and (b) can make only cursory allusion to most of the publications cited. The bibliography at the end of this paper is long, and the references cited are worldwide and frequently from little-known sources. However, a principal contribution of this paper is the bibliography, which serves to (a) place individual contributions in a global perspective, and (b) provide a brief annotated reference to all computer applications in structural geology within the past decade. This paper is concerned with only work published since Whitten (1969), although reference is made to a few articles that had not been seen when that review was written.

STORAGE AND COLLATION OF STRUCTURAL OBSERVATIONS

In the previous review (Whitten, 1969), the largest number of articles was attributed to Prof. R.E. Adler, who has continued to champion the use of computer methods for the collation, storage, and interpretation of structural data. Abundant papers by Adler and his coworkers early in the decade reviewed speak for themselves (Adler, 1968a, 1968b, 1968c, 1968d; 1969a, 1969b, 1969c, 1969d, 1969e; 1970a, 1970b; 1971a, 1971b, 1971c; 1972a, 1972b;

Adler and Bodechtel, 1970; Adler, Fenchel, and Pilger, 1965; Adler and others, 1969, 1970; Adler and Paffrath, 1972; Kruckeberg, 1968). The initial steps, begun before 1969, were development of punch cards (Adler, 1968d), field-data sheets (Haugh, Brisbin, and Turek, 1967; Berner and others, 1972; Pferd, 1975a, 1975b), coding devices (Adler, 1972b), and storage of the quantitative data (Adler, 1969d; Burns, 1973). Numerous computer-based systems for the processing, reporting, and storage of such geological data have been developed (e.g., in Australia, Burns and Dunnet, 1973; in the USSR, Vaytekunas and Yankunayte, 1975), although most systems have been set up for general geological information, rather than specifically for structural geology. Allen and Herriott's (1976) system was for tiltmeter, strainmeter, and magnetometer data from southern California.

Electronic-data processing was applied specifically to tectonic analysis of the Weyebusch-Siegerland area of Germany (Eckert, 1970), and to the Grenville Province of Canada (Wynne-Edwards and others, 1970; Laurin and others, 1972).

CALCULATION OF GENERALIZED STRUCTURAL PROPERTIES FROM ARRAYS OF OBSERVED DATA

Whitten (1969, p. 225-227) cited numerous papers in which computer methods were used to plot and analyze β- and π- diagrams. More recently, Kruhl (1974) and Nuttall and Cooper (1978) reviewed the use of computers for the statistical processing of structural measurements (cf. Shatagin and Dergachev, 1978). Numerous computer programs have been published since 1969 for the preparation and contouring of Schmidt stereographic projections. Adler, Fenchel, and Pilger (1965) and Krausse (1970) dealt with Schmidt nets, and La Fountain (1970), Franssen and Kummert (1971), and Behrens and Siehl (1975) provided computer programs; Lam (1969) and Arpat (1970) also published FORTRAN programs for β- diagrams. Tocher (1979) showed how continuous flowing curved-line contours on such diagrams can be computer generated. General purpose computer programs for stereographic structural analyses are included in Braun (1969), Bonyun and Stevens (1971), Brandle and Aparicio (1973), Bouchez and Mercier (1974), Shatagin and Sandomirskiy (1974), and Cubitt and Celenk (1976). Ramsden and Cruden (1979a, 1979b) recently analyzed in detail the problem of estimating densities on contoured orientation diagrams.

Dumitrescu (1974) briefly described a useful method for computer preparation of block diagrams, and Srivastava and Merriam (1975, 1976) described a technique for computer construction of optical-rose diagrams.

There has been considerable recent interest in statistical analyses of linear and vectoral data in structural geology. For example, Venkitasubramanyan (1971) discussed ambiguities in evaluating conical folds on the basis of unit-vectoral data, Mark (1973) showed that, in use of computer time, eigenvalue methods are more efficient for analyzing axial-orientation data than is the rotational-vector approach, and Allredge, Mahtab, and Panek (1974) provided a method for estimating the confidence region for the mean of certain axial data. Venter and Spang (1974) considered numerical analyses of multimodal directed (e.g., cross beds) and nondirected (e.g., joint poles) orientation data, Bailey (1975) used cluster analysis for polymodal structural-orientation data, and Dudley, Perkins, and Gine M (1975) in a detailed critical review, showed the inadequacy of existing tests of preferred orientation in fabric diagrams, and because valid tests involve such lengthy computations, they advocated use of a convenient manual test. Ramsden (1975) wrote his thesis on numerical methods in fabric analysis, whereas Ramsden and Cruden (1979b) used computer simulations for estimating densities on orientation diagrams. Published computer programs for the analysis of directional data are less abundant, but they include those of Lutzner and Maaz (1969) and Schuenemeyer, Koch, and Link (1972); Winchell (1967) alluded in an abstract to another.

In petrofabric studies, computer-based methods have been used occasionally. For example, Siemes (1967) used a FORTRAN program to evaluate fabric data measured by X-ray diffractometer. Burger (1972) described a computer technique that involved the solution of a series of spherical triangles for universal-stage data in order to calculate calcite compression and tension axes. Starkey (1974) described three computer programs for quantitative analyses of orientation data obtained by X-ray fabric analyses; he used quartz-orientation information for Bergsdalen quartzites, Norway, for illustrations. Groshong (1974) compared observed calcite petrofabric data with those predicted by a computer-based model, whereas Spang (1974) and Spang and van der Lee (1975) showed that detailed numerical dynamic analyses of quartz, calcite, and dolomite deformation lamellae gave excellent agreement with the results of conventional dynamic petrofabric techniques. Lister, Paterson, and Hobbs (1978) used the Taylor-Bishop-Hill model for polycrystalline deformation as the basis for a computer program to simulate preferred crystallographic orientations in deforming quartzites.

DESCRIPTION OF THE MORPHOLOGY OF STRUCTURAL FEATURES - FOLDS

There has been relatively little computer-based work on the shape of folds, although reference should be made to Stabler's (1968) Fourier analysis of a wide range of fold shapes. In the USSR, Goncharov (1971) extended his earlier research with a new mathematical model of fold structure, and Vikhert (1967, 1973) and Vikhert and Goncharov (1969) continued their development of deterministic and probabilistic models for the description and classification of folds. Shcherbakov (1973) used automatic-data processing to enhance interpretation of the structure of the sedimentary rocks of western Latvia. Edel'shteyn (1972) used quantitative methods to study folds of platform areas and Kulyndyshev (1972) also used "logical analysis" to describe elementary structural surfaces. Hudleston (1973) showed the value of harmonic analysis for describing fold shapes. In an abstract, George and Sowers (1975) alluded to comparison of folds computed for elastic, viscous, viscoelastic, and plastic models, and concluded that the latter may be the dominant rheological response in relatively shallow geological folding.

Wray (1973) provided a FORTRAN computer program to construct cross sections of curved and faulted structural surfaces; his illustrative example from Butte, Montana, showed the geometry prior to three stages of faulting. Nidd and Ambrose (1971) discussed computerized solutions for some geometric problems of conical folds, whereas more recently, Cheshire, Spang, and Stockmal (1978) briefly outlined an empirical nonlinear computer technique for evaluating conical-fold structural elements. Cruden and Charlesworth (1972) used an APL program to generate cylindrical and conical folds and then tested for optimum methods of calculating their fold axes; Charlesworth, Langenberg, and Ramsden (1975, 1976) subsequently described computer methods for determining the axes, axial planes, and sections of four folds from the Canadian Rocky Mountains. In an extension of this technique, Langenburg, Rondeel, and Charlesworth (1977) used computer methods to produce structural sections through five homogeneous domains within a 50 km^2 area of the Belgian Ardennes.

For the Kriging method of spatial analysis (see the section on Mapping and Analysis of Spatial Variability..), semivariograms with respect to a selection of azimuths are computed first to determine the weighting factors to be used for the moving-average calculations; the preparation of a semivariogram involves calculating the spectrum of variance (along a particular azimuth across the map area) of the dependent variable at all spacing intervals between samples. Whitten (1977a) determined that such semivariograms provide a novel method of describing quantitative attitudes of fold shape, and, because the variance of the elevation on a bed is a minimum parallel to the axis and a maximum normal thereto, semivariograms have proved to be useful tools in identifying the fold axes in subsurface terrain. Whitten (1977a, 1979) used this technique to locate fold-axial trends in the Michigan Basin.

In petroleum exploration, computer applications to structural problems have become widespread, although for proprietary reasons they rarely are alluded to in the literature. A few unusual examples are cited here. Kontorovich and others (1971) used computer methods to predict productivity of structures in the western Siberian Platform on the basis of recognition-pattern algorithms. Petrov, Ellanskiy, and Zverev (1972) discussed general problems of utilizing mathematical methods in petroleum geology, whereas Davis, Doveton, and Hambleton (1975) briefly alluded to a case history of using probabilistic analysis for oil exploration. Interactive graphics (*Interactive Exploration*) as a tool in petroleum search permitting direct control of large central, but remote, computers by field-based explorationists was referred to in abstracts by Branisa and Johnson (1975) and Jones, Johnson, and Herron (1976). Juneman (1971) earlier had used interactive computer graphics for analyzing three-dimensional geological data sets.

Burns, Marshall, and Gee (1969) and Borrmann (1972a) were responsible for some of the early exercises in computer-assisted geological structural mapping. Of course, cartography now is an extensively computerized discipline, but no attempt is made here to assay the tremendous advances in this field and in computer graphics generally, despite their immediate and direct usefulness to structural geology.

The analysis of strain in folded rocks has received considerable attention in the past decade, although most studies did not use computer computations. However, the

detailed analyses of strain in folded layered rocks by
Ramberg (1970) and Hobbs (1971) involved considerable computation. Matthews, Bond, and van den Berg (1974) used a
FORTRAN IV program in their algebraic analysis of strain
using elliptical markers. Flinn's (1978) computation of
three-dimensional progressive deformations also is relevant here. Finally, Kruhl (1978) published a useful computer program for unravelling the structural complexity
of highly strained current-bedding in Scottish Precambrian
quartzites.

During the past decade there has been a steady growth
of interest in using computer technology and satellite
imagery for geological structural interpretation. This
is a computer application that is going to expand considerably in the next few years. Representative studies
include the papers by Short (1973a, 1974a, 1974b) and
Kronberg (1975) on geological structures seen from the
Earth Resources Technology Satellite (ERTS-1) imagery.
In analyzing structural geology, ERTS imagery also was
used by Bodechtel and Lammerer (1973) for the Alps and
Appenines, by Bodechtel and Nithack (1974) for northern
and central Italy (also using SKYLAB data), by Lambert
(1975) for the northwestern part of the Massif Central
of France, by Burkhardt and Endlicher (1977) for the
eastern Bavarian basement, and by Offield and others
(1975) for the southern copper region of Brazil. More
recently, Cassinis (1977) illustrated the value of satellite imagery in interpreting the structural geology of
Italy. In the USSR, satellite imagery also has been used
extensively to elucidate tectonic and major structural
features; for example, one may cite the work of Bush and
Kats (1978) on the Sredizemnomorsk Alpine Region, of Kapustin, Przhiyalgovskiy, and Trofimov (1978) on the tectonic
framework of the Caspian Depression, of Grishkyan, Parfenov, and Ufimtsev (1977) on the Baikal Rift Region, of
Solov'yeva (1978) on central Asia, and of Volchegurskiy
and Pronin (1978) on the major structural forms within
the Russian and Turanian plates.

DESCRIPTION OF THE MORPHOLOGY OF STRUCTURAL
FEATURES - FAULTS, LINEAMENTS, ETC.

The rate of publication on faults and faulting has
increased enormously within the past 10 years, and in
attempting to obtain objective quantitative answers to
problems, computer syntheses and analyses have been used.
The distribution of faulting in space and time (e.g.,

Cogbill, 1979) has important implications in earthquake-prediction studies.

Numerous descriptive studies of fissures, fractures, and faults have been made. For example, Jeran and Mashey (1970) used a FORTRAN program for stereographic analysis of coal fractures and cleats, Altinli and Eroskay (1971) used automatic data-processing techniques to analyze the fractures of the Ikizdere Granitic Complex, and Ghez and Janot (1974) described a method for statistically computing the matrix-block volume of a fissured reservoir. The relationship between joint spacing (frequencies and orientation) and bed thickness was analyzed by Bock (1971). Borrmann (1972b) described a computer program for the statistical analysis of fracture and joint orientation data, and Vychev (1976) a program for fracture-density determination in rocks from field observations. Podwysocki (1973) developed computer methods for analyzing fracture patterns identified from remote-sensing imagery.

Whitten (1969) referred to James' (1968) earlier work on faulted surfaces using least-squares methods with discontinuous functions. James (1970) applied his method to subsurface data for the Ramsey Oil Pool, Payne County, Oklahoma, where there was independent evidence of high-angle faulting. More recently, Attoh and Whitten (1979) published a FORTRAN program that extends this method and permits estimates of the throw of one or more faults to be computed on the basis of subsurface data. Peikert (1970) also described an interactive computer-graphics system for elucidating subsurface faulted structures. As mentioned, most geophysical studies are excluded from this review, but Sharma and Geldart's (1968) use of Fourier transforms in the analysis of gravity anomalies associated with two-dimensional faults such as the Canadian Logan Fault, and Szumilas' (1977) brief report on a computer time-to-depth conversion and structure mapping system designed for evaluating complexly faulted areas are relevant here.

Serra (1973) described a computer program for calculating and plotting the stress distribution and possible fault trajectories under specified crustal boundary conditions. Dabovski (1975) provided a mathematical model for stress and displacements around faults (and magma chambers), whereas Alarcon Guzman, Fernandez, and Gonzalez-Rubio (1975) published a computer program for stress analysis of structures.

A major new development in the computer-based study of faults and faulting involves simulation with finite-element methods. This topic is discussed in the section on Finite-Element Methods..., where reference is made to Dieterich's (1969) work on the mechanical properties of seismically active fault zones, Kosloff's (1976) model of creep zones of finite width (associated with faults) resembling the Palmdale Uplift, California, Miyatake's (1977) simulation of dynamic faulting processes, Rodgers and others (1977) thrust-fault mechanisms, Stein and Wickham's (1978) study of drape-fold development associated with faulting, and the analyses of Stein and Wickham (1979) on fault-zone propagation, and of Wickham, Stein, and Reddy (1979) on various faulting models.

Stearns (1978) developed an analog model to produce simulations of fault patterns during the deformation of thick, homogeneous, isotropic continuous rock masses; analogies were drawn with structures in the Rocky Mountain Foreland Province.

Terrestrial lineaments recently have received considerable attention because of the general availability of air photos and satellite imagery. Crain (1972, 1973) reviewed in detail lineaments at different levels of geotectonic analysis, and he analyzed Monte-Carlo simulations of such structures. Short (1973b) briefly alluded to mapping joints, faults, and contacts from ERTS imagery. Various aspects of computer-assisted analysis of lineaments from air photographs have been discussed by many authors; for example, Maffi and Marchesini (1964), Hanley (1975) for the fracture pattern of Catawba Mountain, Virginia, Huntington (1975) for fracture systems of the Arabian Shield, the Karroo Basin, and some Paleozoic folded rocks of northern Wales, Podwysocki, Moik and Shoup (1975), Burns, Huntington, and Green (1977), Arakelyan and Karakhanyan (1978), and Yeromenko and Katterfel'd (1978). Rice, Davis, and Johnson (1976) specifically alluded to inferences about subsurface geological structure based on computer analyses of lineaments.

The acquisition of satellite imagery involves extensive use of computers. Kowalik (1975) compared lineaments read from SKYLAB and LANDSAT images with actual joint orientations in north-central Pennsylvania, whereas Correa and Lyon (1976) analyzed Californian linear features on the basis of optical Fourier analyses of ERTS-1 imagery. Tricart (1976) presented evidence for lineaments in the French Vosges Mountains read from LANDSAT-1 imagery. Burns, Shepherd, and Berman (1976) investigated the reproducibility of lineaments read from satellite imagery. Cardamone and others (1977) used LANDSAT-2 images to study lineaments in the Friuli earthquake area, and Masson (1978) also used LANDSAT data to analyze structures of the Levantine Rift. Artamonov, Vostokov, and Sheremet (1978) combined satellite and geologic-geophysical data in a study of the fracture tectonics of the Peri-Baltic Syneclise.

MAPPING AND ANALYSIS OF SPATIAL VARIABILITY OF STRUCTURAL ELEMENTS - FILTERING SYSTEMS

Whitten (1969, p. 229-233) gave a detailed review of the use of trend-surface and other filtering analyses for subsurface structural mapping. Although work in this area has expanded considerably, several problems identified in 1969 remain today (e.g., most mathematical models require data on a regular grid, whereas actual data are almost always irregularly spaced).

The basic purpose of trend-surface analysis is to separate three components of the total variability: (a) regional effect or trend, (b) localized features of geological significance, and (c) error and other random components (including significant geological variability produced by features smaller than the data-point spacing). A detailed guide to the practical use of trend-surface analysis methods was published by Whitten (1975); perhaps the most important advance since 1969 was the introduction of orthogonal-polynomial trend-surface analysis for irregularly spaced data (Whitten, 1970), which eliminates the troublesome task of inverting large matrices and permits the Z^2-array and the contribution of each coefficient to be identified; a FORTRAN computer program is available (Whitten, 1974). Trend-surface analyses of folded subsurface strata now have become relatively routine and tend not to be noted specially in English-language publications; we may note the use of mathematical trend analysis by Lokhmatov and Alayev (1967) and Lokhmatov, Alayev, and Yevdokimova (1968) for Lower Cambrian strata of the upper Angara River Region, Sibera, by Belonin and Zhukov (1970) for the surface of the Aleksevvku Uplift in the Kuibyshev District, and by Kumar (1977) for structural analysis of the Miocene rocks in part of southern Louisiana. In the Peoples' Republic of China, Xu and Sun (1966) described trend-surface analyses of stratigraphic data, and Zhang (1977) used polynomial and Fourier trend-surface analyses for evaluating subsurface structure on the basis of 71 irregularly spaced data points in 682 km^2. Myasnikova and Shpil'man (1973) used trend analyses to investigate the interrelationships between sediment accumulation and structural evolution.

Recently, numerous additional techniques for the analysis of subsurface structure have been developed that approximate surfaces to the actual observed data. Such artifices, which, unlike trend-surface analyses, make no attempt to separate component (c) from (a) and (b), necessarily make the implicit assumption that all observed data are accurate (error free) and deserve equal weight being placed on them. Although these factors can be a disadvantage in some spatial analyses, stratum-elevation data used in structural work usually are determined with virtually no error. However, the fact that irregularly spaced data can be used directly in orthogonal-polynomial trend-surface analyses gives a powerful advantage over most techniques which construct surfaces that pass precisely through the data-point values, because computation of almost all of the latter requires gridded data. Approximating gridded from irregularly spaced data can introduce significant unacceptable errors and biases (Whitten and Koelling, 1973).

Experiments with numerous slightly different techniques have been made, but perhaps the most important in the past decade have been:

(1) a method of spatial filtering in which two-dimensional Fourier transforms are used with gridded raw data,
(2) bicubic-spline surfaces, and
(3) surfaces developed by a variety of Kriging techniques.

(1) The early work on spatial filtering with two-dimensional Fourier transforms developed by Robinson, Charlesworth, and coworkers in Alberta was reviewed by Whitten (1969, p. 232-233). This useful technique for identifying and enhancing subsurface structural patterns was elaborated and further described in a series of papers (Robinson, 1969, 1970; Robinson and Charlesworth, 1969a, 1969b, 1975; Robinson and Merriam, 1972). Recently, Eschner, Robinson, and Merriam (1979) discussed the comparison of spatially filtered geological maps. Unfortunately, gridded data (or data interpolated on a grid) are a prerequisite of this method.

(2) Spline surfaces are used widely by engineers for continuous smooth surfaces passing precisely through specified points. In principle, spline surfaces should be useful in modeling subsurface structure (cf., Davis, 1975). Koelling and Whitten (1973) published a FORTRAN program for computing bicubic-spline surfaces for gridded data, but as they (Whitten and Koelling, 1973) pointed out, the necessity of interpolating values on a grid introduces unacceptable errors in subsurface structural analyses where only irregularly spaced raw data are available. It seems that it should be possible to develop a system of equations to permit construction of bicubic-spline surfaces for irregularly distributed data points. Whitten and Koelling (1975, 1978) described a set of equations and directly computed a spline surface on the basis of irregularly spaced data for the top of the Devonian Dundee Limestone in Michigan; although, in this example, problems did not arise, their set of 15 equations to solve for 16 coefficients should not always yield a complete solution.

(3) Prof. G. Matheron expanded the weighted moving-average methods developed by Dr. D.G. Krige (for gold and uranium assay data) to a sophisticated series of techniques generally referred to as Kriging. Although initially developed in France by Matheron and coworkers for mine assay and development, Kriging is valuable in many other domains in which spatial variability is important. In a general article on stochastic models in geology, Whitten (1977b) drew attention to the extensive background literature on Kriging. Several authors have used Kriging to construct structural stratum-contour maps. Agterberg (1970) exposed the problems of estimating autocorrelation functions from irregularly spaced data; he approximated autocorrelation functions for top of the Arbuckle Group (Kansas) and used Kriging and polynomial trend surfaces to portray its subsurface structure. Sampson (1975a, 1975b) illustrated contoured maps (and maps of the expected errors) for the top of the subsurface Pennsylvanian Lansing Group in Stafford and Graham Counties, Kansas, based on universal Kriging of orthogonal data interpolated from the original irregularly spaced observations. Olea (1975) illustrated similar maps for the top of the Tobifera Series, Magellan Basin, Chile, which also were based on universal Kriging of orthogonalized data. Whitten (1977a, 1979) also used Kriging for a structure map of the Dundee Limestone of part of the Michigan Basin. Two computer programs for computing such maps by Kriging are available (David, 1977; Sampson, 1975b); unfortunately,

both of these programs require gridded-data input, so that the questionable interpolation of values at the nodes of a rectangular grid is necessary with virtually all real subsurface data. In addition, these programs use simple Kriging, rather than universal Kriging. The latter provides superior results when a significant regional gradient (trend) is present, but computer programs for universal Kriging are only available in proprietary sources.

Trend surfaces and these three additional types of surface approximation all make the assumption of continuous variation across the study area, whereas, in practice, faults may be a complicating factor. As mentioned previously, in 1968 James introduced a simple nonlinear model that permitted faulted surfaces to be mapped, by using what in fact are faulted polynomial trend surfaces. James (1970) applied the method to faulted subsurface structures in Oklahoma. Attoh and Whitten (1979) extended James' technique and their FORTRAN program permits estimation of subsurface-fault throws. Bolondi, Rocca, and Zanoletti (1975) and Dahlberg, Deland, and Creed (1975) provided brief abstracts concerned with mapping faulted surfaces; to augment customary methods of inferring subsurface faults, the latter used computer statistical methods to correlate stratigraphic features with possible faults.

In conclusion, a few additional studies must be cited. Arabadzhi, Vasil'yev, and Mil'nichuk (1967) described the use of correlation analysis in the compilation of structural maps of the Caspian Basin area, and Berlyand (1971) investigated the use of autocorrelation analysis in studying the structure of a gravity field. Demirmen (1972, 1973) published details of a numerical-description technique for folded surfaces. Srivastava and Merriam (1974) and Srivastava (1975) described quantitative optical- and digital-processing methods which they applied to numerous subsurface horizons in Kansas. Jones and Jordan (1975) briefly alluded to an automated procedure for structural mapping with data containing points that do not intersect the horizon of particular interest. Whitten (1977a, 1979) published surfaces (prepared by Kriging) depicting areal differential-subsidence rates of certain Cretaceous units in the southeastern USA; this technique should have value in analyzing the rate of warping and folding during sedimentation (Whitten, 1976).

FINITE-ELEMENT METHODS, SIMULATION, AND RELATED TECHNIQUES

Without doubt, the most dramatic advances in computer applications in structural geology in the past decade have occurred in this domain. Finite-element methods were explained simply in Zienkiewicz's (1971) text, which gives a computer program that can be adapted readily to many structural problems. Zienkiewicz (1974, 1976), a civil engineer, recently contributed some geological papers on flow and geophysical topics.

An impressive array of papers has been based on finite-element methods to analyze various aspects of folds. In the compass of this review, one can do little apart from catalog the types of problems that have been assayed. The following list of about 40 papers and abstracts indicates the chronological development of analyses of folding and folds that used finite-element and related types of computer methods:

Date	Author/s	Subject
1964 1968	Chapple	Solution of finite-difference equations by computer in study of finite-amplitude folding.
1968	Fumagalli	Simulation of rock-mechanics problems.
1969	Chapple	Fold shape through time that results from folding of a viscous-plastic layer.
1969	Dieterich and Carter; Dieterich and Onat	Finite-element determination of the stress field in two-dimensional, large-amplitude folding of a viscous layer and other solids.
1970	Voight and Dahl	Analysis of nonlinear rock deformation by numerical continuum approaches.
1970	Dieterich	Computer simulation of time-dependent deformation: mechanics of finite-amplitude folding.
1970a 1970b	Chapple	Growth of folds from shape perturbations in layered rocks and instability during growth of folds.

1971	Stephansson and Berner	Finite-element methods for analyzing folds, boudinage, and isostatic adjustments.
1972	Fujii	Finite-element methods in structural analysis.
1973	Bridwell	Finite-element applications for mechanical problems in structural geology.
1973	Parrish	Nonlinear finite-element fold model involving steady-state flow of quartzite and marble.
1973	Stephansson	Finite-element solution of various structural-geology problems.
1973	Heinze and Goetze	Computer simulation of stresses and strains in heterogeneous polycrystalline solids.
1973	Hudleston and Stephansson	Experiment, analog modeling, and finite-element study of single-layer buckling.
1974	Ikeda and Shimamoto	Numerical experiment on viscous bending folds.
1974	Chapple and Spang	19 samples of Greenport Center syncline, NY, used for comparison with layer-parallel slip folding simulated by finite-element methods.
1974	Bridwell	Stress-strain finite-element model of nonlinear bending.
1974	Bridwell and Swolfs	Stability analysis (using finite-elements) of experimentally deformed Indiana limestone.
1974	Hardy and others	Computer simulation of inelastic-beam behavior.
1974	Heinze and Goetze	Numerical simulation of stress concentrations.
1974	Owens	Mathematical modeling of magnetic anisotropy of deformed rocks.
1975	Bird, Toksoz, and Sleep	Finite difference models of continent-continent convergences.
1975 1976	Parrish, Krivz, and Carter	Finite-element folds of similar geometry.

1976	Cosgrove	Finite-element study of growth of buckling instabilities into finite fold structures.
1976	Stephansson	Finite-element analysis of folds.
1976	Robinson	Computer simulation of folds.
1976	Wickham and Anthony	Compares actual Appalachian structures in carbonates to finite-element models.
1976	DeBremaecker and Becker	Numerical simulation of folding taking incompressibility into account.
1977	Cobbold	Finite-element analysis of fold propagation.
1978	DeBremaecker and Becker	Finite-element models of folding.
1978	Anthony and Wickham	Fold shape and strain studied in finite-element simulation of asymmetric folding.
1978	Lewis and Williams	Finite-element study of fold propagation in a viscous layer.
1978	Williams, Lewis, and Zienkiewicz	Finite-element analysis of the significant role of initial perturbations in folding.
1978	Stein and Wickham	Drape folds and related faulting.
1978	Neugebauer and Spohn	Finite-element modeling of flexure of US Atlantic continental margin.
1979	Cobbold	Removal (by finite-elements) of finite deformation, using strain trajectories.
1979	Reddy and Wickham	Finite-element studies of incompressible flow due to gravity in rocks, etc.
1979	Reches and Johnson	Finite-element analysis of monocline development.

In addition, numerous other structural problems have been assayed with the versatile finite-element method. Again, the list of applications is long. Voight and Samuelson (1969) discussed the application of finite-element methods to stress analysis in the earth sciences.

Douglas (1970) calculated the crustal displacement under sediment loading by finite-element methods, and Sturgul and Grinshpan (1975) modeled possible isostatic rebound of the Grand Canyon by finite elements. By using a three-dimensional finite-element relaxation model, Kosloff (1976) modeled a structure resembling the Californian Palmdale Uplift. Feenstra and Wickham (1976) used finite-element methods to model simple-shear deformation superimposed on symmetrical folds that is relevant to understanding the development of structures similar to those of the Ouachita Mountains. Salt domes and other diapiric structures have been studied by several authors; for salt domes, Howard (1971) used computer-simulation models and Woidt (1978a, 1978b) used finite-element methods, whereas Fletcher (1972) studied mantled gneiss domes with finite elements. Finite-element modeling of surface deformation associated with intrusion of magma in reservoirs within the Kilauea Volcano, Hawaii (Dieterich, 1972; Dieterich and Decker, 1975), Himalayan Orogeny (Bird and Toksoz, 1976), and strain paths and folding in carbonate rocks near the Blue Ridge (central Appalachians) (Wickham and Anthony, 1977) further illustrated the versatility of the method. Forster and Leonhardt (1972) used mathematical simulation to study kinematic stream lines within the Eastern Alps. At a smaller scale, Stromgard (1973) examined the stress distribution during boudinage and pressure-shadow formation with photoelastic and finite-element techniques, and Selkman (1978) used finite elements for displacement analysis of boudinages.

Hattori and Mizutani (1971) described computer simulations of fracturing in layered rocks, whereas finite-element studies of the responses of jointed rock under quasistatic loading (Baligh, 1972), and of preexisting fractures on the scale of laboratory experiments (Minear, 1972) were followed by Tapp and Wickham's (1978) finite-element study of predicted fracture occurrence during folding, and Tapp, Wickham, and Reddy's (1979) numerical modeling of fracture density in single-layer folds. Cundall (1971) developed a computer model for simulating progressive large-scale movements in blocky rock systems, and, in the field of petroleum geology, du Prey and Boisse-Codreanu (1975) generated numerical simulations of fissured-reservoir production.

Dieterich (1969) modeled by finite-element methods the mechanical properties of a seismically active fault zone. Miyatake (1977) and Wickham, Stein, and Reddy (1979) prepared numerical simulations of the dynamical-faulting process, Stein and Wickham (1979) of fault-zone propagation, and Rogers and others (1977) of thrust-fault mechanisms.

Three additional abstracts are relevant here. Jackson (1972) dealt with numerical simulation of the main and aftershock earthquake sequence of a fault system, whereas finite-element techniques were used by Alewine and Jungels (1972) in a study of the 1964 Alaskan earthquake, and by Clancy, Turcotte, and Kulhawy (1977) in analysis of strain accumulation and release on the San Andreas Fault.

MISCELLANEOUS APPLICATIONS

As mentioned earlier, computer applications have permeated thoroughly virtually all domains of structural geology. Because they have become so commonplace, it is an elusive task to track down all of the interesting and useful developments that are important in modern work. Also, several applications do not fit neatly into the major headings dealt with here. A few examples may be briefly cited.

Nonlinear-regression models were used by Mundry (1972) to determine rheological constants in time-dependent stress-strain measurements. Deist, Salamon, and Georgiadis (1973) described a new digital method for three-dimensional stress analysis in elastic media, whereas Pollard and Holzhausen (1978) provided a FORTRAN program for calculating stress intensity factors, stresses, and displacements associated with a fluid-pressurized fracture.

Kinematics and strain can be tackled at widely dissimilar domain sizes. In geodynamics, innumerable papers are based on computer computation; for example, papers range from those of Coode (1966, 1967) on spherical harmonic analyses of major tectonic features to the least-squares Bullard-type plate-tectonic continent-fit reconstructions of Smith and Hallam (1970) for the southern continents and of Smith (1971) for the Mediterranean area, and to Neugebauer and Spohn's (1978) work on buckling along the trailing edge of a passive plate margin and its effect on sediment accumulation rates. Attempts to model the creep deformation of solids using modified finite-difference equations (Andrews, 1971) and to model the tectonic flow behind island arcs (Andrews and Sleep,

1974) are examples of a further class of computer-based geodynamical structural study. Smith and Kind (1972) developed a least-squares strain-variation picture for southern Nevada on the basis of a regional strain-meter array; at a different size level, Watson and Smith (1975) developed a computer simulation of grain shape during isotropic growth of grains in an aggregate, a study of potential value in numerous strain studies. Using homogeneous-strain transformation in plane-polar coordinates, Hirsinger (1976) estimated total strain by matching the deformed shape of fossils with their initial geometries.

CONCLUDING REMARKS

It seems clear that in the next decade the use of computer techniques will have permeated and revolutionized so thoroughly all facets of structural geology that it probably will be simpler to identify those fields of structural geology in which computers are not used, rather than those in which they have been utilized. Perhaps the most significant new insights will be derived from exploiting (a) the potential of the finite-element techniques for structural simulation and prediction and (b) the interface between geophysics and structural geology.

REFERENCES

Adler, R.E., 1968a, Gelandevermessung mit einfachen Hilfsmitteln: Breithaupt Mitteilungen, Kassel II, p. 1-61.

Adler, R.E., 1968b, Der Einfluss der Tektonik auf maschinell mit der Streckenvortriebsmaschine Wohlmeyer aufgefahrene Flozstrecken: Gluckauf-Forschungshefte, 29 Jahr., no. 3, p. 149-156.

Adler, R.E., 1968c, Zur Tektonik des Polsumer und Hulsdauer Sprunges im nordlichen Ruhr-Karbon: Neues Jahrb. Geol., Palaont., Monat., no. 5, p. 257-276.

Adler, R.E., 1968d, Lochkarte.., ein Hilfsmittel der modernen Tektonik: Clausthaler Tekt. Hefte., v. 8, p. 93-149.

Adler, R.E., 1969a, Electronische Datenverarbeitung in der Tektonik: Z. deutsch. Geol. Ges., Jahr. 1967, v. 119, p. 219-244.

Adler, R.E., 1969b, Kleintektonische Beobachtungen aus dem Ruhrkarbon: Forschunsber. Land. Nordrh.-Westf., no. 2008, p. 5-53.

Adler, R.E., 1969c, Moderne Daten- und Informationsgewinnung und -verarbeitung in der Tektonik: Zbl. Geol. Palaontol., Teil ID, no. 6, p. 1053-1079.

Adler, R.E., 1969d, Instrumentelle Aufnahme und elektronische Auswertung von tektonischen Flachen und Linearen: Geol. Rundsch., v. 59, no. 1, p. 152-162.

Adler, R.E., 1969e, Neue Arbeitsmethoden in der tekonishen Forschung: Geol. Mitt. (Aachen), v. 9, no. 2, p. 97-108.

Adler, R.E., 1970a, Tektonische Geologie, photogrammetrische Datenerhebung und elektronische Datenbearbeitung: Gluckauf-Forschungshefte, 31 Jahr., no. 6, p. 318-332.

Adler, R.E., 1970b, Elektronische Datenverarbeitung in der modernen Tektonik: Clausthaler Tekt. Hefte, v. 10, p. 25-47.

Adler, R.E., 1971a, Mathematische Geologie: Zbl. Geol. Palaontol. Teil I, p. 3-14.

Adler, R.E., 1971b, Praktische Tektonik 2 - Geschichtliche Entwicklung: Zbl. Geol. Palaontol., Teil I for 1970, no. 7-8, p. 915-938.

Adler, R.E., 1971c, Tektonik und Datenverarbeitung: Zbl. Geol. Palaontol., Teil I, no. 1-2, p. 15-28.

Adler, R.E., 1972a, Praktische Tektonik Stellung, Bedeutung, Moglichkeiten: Zbl. Geol. Palaontol. Teil I for 1971, no. 5-6, p. 333-357.

Adler, R.E., 1972b, Ableitung einer Kennzahl zur Charakterisierung tektonischer Deformationen: Neues Jahrb. Geol. Palaont., Monat., no. 5, p. 257-259.

Adler, R.E., Berling, D., Bodechtel, J., and Vieten, W., 1970, Anwendung der Photogrammetrie zur Erfassung tektonischer Daten: Clausthaler Tekt. Hefte, 10, p. 337-358.

Adler, R.E., and Bodechtel, J., 1970, Tektonische Datenerfassung aus terrestrischphotogrammetrischen Aufnahmen sowie deren Weiterverarbeitung in der EDV: Bildmessung und Luftbildwesen, v. 38, no. 5, p. 267-272.

Adler, R.E., Fenchel, W., and Pilger, A., 1965, Statistische Methoden in der Tektonik II. Das Schmidtsche Netz und seine Anwendung im Bereich des makroskopischen Gefuges: Clausthaler Tekt. Hefte, 4, p. 1-111.

Adler, R.E., Kruckeberg, F., Pfisterer, W., Pilger, A., and Schmidt, M.W., 1969, Elektronische Datenverarbeitung in der Tektonik: Clausthaler Tekt. Hefte, 8 for 1968, p. 1-157.

Adler, R.E., and Paffrath, A.A., 1972, Ein neues Schema zur Fixierung geologischtektonischer Informationen: Neues Jahrb. Geol. Palaont., Abh., v. 140, p. 1-32.

Agterberg, F.P., 1970, Autocorrelation functions in geology, in Merriam, D.F., ed., Geostatistics, a colloquium: Plenum Press, New York, p. 113-142.

Alarcon Guzman, J.A., Fernandez T, D.A., and Gonzalez-Rubio, H.E., and others, 1975, Stress analysis of structure by computer with dynamic option: Intern. Inst. Seismol. Earthquake Engineer., Individ. Stud., v. 11, p. 106-146.

Alewine, R.W., and Jungels, P., 1972, A study of the 1964 Alaskan earthquake using finite element and generalized inversion techniques (abst.): Am. Geophys. Union Trans., v. 53, no. 11, p. 1119.

Alldredge, J.R., Mahtab, M.A., and Panek, L.A., 1974, Statistical analysis of axial data: Jour. Geology, v. 82, no. 4, p. 519-524.

Allen, S.S., and Herriot, J.W., 1976, A computer graphics data system for low-frequency geophysical data (abst.): Am. Geophys. Union Trans., v. 57, no. 3, p. 154.

Altinli, I.E., and Eroskay, S.O., 1971, Toplam vektor metodu ile Ikizdere Granit Karmasiginin Yonelim Tahlili: Istanbul Univ. Fen Fak Macm., Ser. B., v. 36, no. 3-4, p. 115-136.

Andrews, D.J., 1971, A numerical method for creep deformation of solids (abst.): Am. Geophys. Union Trans., v. 52, no. 4, p. 347.

Andrews, D.J., and Sleep, N.H., 1974, Numerical modelling of tectonic flow behind island arcs: Roy. Astron. Soc., Geophys. Jour., v. 38, no. 2, p. 237-251.

• Anthony, M., and Wickham, J., 1978, Finite-element simulation of asymmetric folding: Tectonophysics, v. 47, no. 1-2, p. 1-14.

Arabadzhi, M.S., Vasil'yev, Y.M., and Mil'nichuk, V.S., 1967, Postroyeniye strukturnykh kart metodom korrelyatsionnogo analiza na primere Prikaspiyskoy vpadiny: Prikl. Geofiz., v. 50, p. 133-139; (English translation) Exploration Geophysics, 1969, v. 50, p. 107-113.

Arakelyan, R.A., and Karakhanyan, A.S., 1978, Opoznavaniye i deshifrirovaniye razlomov na kosmicheskikh snimkakh razlichnykh urovney yestestvennoy generalizatsii: Vysshoe Uchebnoye Zavedeniye, Izv., Geol. Razved., no. 10, p. 35-39.

Arpat, E., 1970, A computer method for preparing beta diagrams (program in FORTRAN IV-H, using an IBM 360/67 computer): **Mineral** Res. Explor. Inst. Turkey Bull. (foreign edit.), no. 74, p. 34-42.

Artamonov, M.A., Vostokov, Y.N., and Sheremet, O.G., 1978, Razlomnaya tektonika Baltiyskoy sineklizy i prilegayushchikh territoriy po kosmicheskim i geologo-geofizicheskim dannym: Vysshoye Uchebnoye Zavedeniye, Izv. Geol. Razved., no. 10, p. 141-146.

Attoh, K., and Whitten, E.H.T., 1979, Computer program for regression model for discontinuous structural surfaces: Computers & Geosciences, v. 5, no. 1, p. 47-71.

Bailey, A.I., 1975, A method of analyzing polymodal distributions in orientation data: Jour. Math. Geology, v. 7, no. 4, p. 285-293.

Baligh, M.M., 1972, Finite element study of the response of jointed rock (abst.): Am. Geophys. Union Trans., v. 53, no. 11, p. 1118.

Behrens, M., and Siehl, A., 1975, GELI 2 - ein Rechenprogramm zur Gefuge- und Formanalyse: Geol. Rundsch., v. 64, no. 2, p. 301-324.

Belonin, M.D., and Zhukov, I.M., 1970, Geometrical properties of the surface of the Alekseevka uplift in the Kuibyshev district, in Topics in mathematical geology, Romanova, M.A., and Sarmanov, O.V., eds.: Consultants Bureau, New York, p. 186-199.

Berlyand, N.G., 1971, O vozmozhnostyakh avtokorrelyatsionnogo analiza pri izuchenii struktury gravitatsionnogo polya: Fizika Zemli, no. 1, p. 68-78.

Berner, H., Estrom, T., Lilljequist, R., Stephansson, O., and Wikstrom, A., 1972, Geomap - a data system for geological mapping: Proc. 24th Intern. Geol. Congress (Montreal), Sect. 16, p. 3-11.

Bird, P., and Toksoz, M.N., 1976, Himalayan orogeny modelled with finite elements (abst.): Am. Geophys. Union Trans., v. 57, no. 4, p. 334-335.

Bird, P., Toksoz, M.N., and Sleep, N.H., 1975, Thermal and mechanical models of continent-continent convergence zones: Jour. Geophys. Res., v. 80, no. 32, p. 4405-4416.

Bock, H., 1971, Uber die Abhangigkeit von Kluftabstanden und Schichtmachtigkeiten: Neues Jahrb. Geol. Palaont., Monat., no. 9, p. 517-531.

Bodechtel, J., and Lammerer, B., 1973, New aspects on the tectonic of the Alps and the Apennines revealed by ERTS-1 data: NASA Sp. Publ. 327, p. 493-499.

Bodechtel, J., and Nithack, J., 1974, Geologisch-tektonische Auswertung von ERTS-1 und SKYLAB-Aufnahmen von Nordund und Mittelitalien: Geoforum, no. 20, p. 11-24.

Bolondi, G., Rocca, F., and Zanoletti, S., 1975, Contouring faulted surfaces (abst.): Soc. Explor. Geophys. Annual Intern. Meet., no. 45, p. 58.

Bonyun, D., and Stevens, G., 1971, A general purpose computer program to produce geological stereo net diagrams, in Data processing in biology and geology, Cutbill, J.L., ed.: Academic Press, London, p. 165-188.

Borrmann, H.-G., 1972a, Automatische Kurven- und Kartenkonstruktionen (abst.): Dtsch. Gesell. Geol. Wissen. Ber., Reihe A., Geol. Palaont., v. 17, no. 1, p. 13.

Borrmann, H.-G., 1972b, Programm zur statistischen Auswertung von Richtungsmessungen (Kluft und Gefugestatistik) mit Hilfe der EDV (abst.): Dtsch. Gesell. Geol. Wissen. Ber., Reihe A, Geol. Palaont., v. 17, no. 1, p. 14.

Bouchez, J.-L., and Mercier, J.-C., 1974, Construction automatique des diagrammes de densite d'orientation: Sci. Terre, v. 19, no. 1, p. 55-64.

Brandle, J.L., and Aparicio, A., 1973, Un programa basico para diagramas de petrofabricas: Estud. Geol. (Instit. Investig. Geol. 'Lucas Mallada'), v. 29, no. 4, p. 315-317.

Branisa, F., and Johnson, C.R., 1975, Interactive exploration: an application of interactive computer graphics (abst.): Geophysics, v. 40, no. 1, p. 130.

Braun, G., 1969, Computer calculated counting nets for petrofabric and structural analysis: Neues Jahrb. Mineral., Monat., no. 10, p. 469-476.

Bridwell, R.J., 1973, Finite element applications to mechanical problems in structural geology: unpubl. doctoral dissertation, Univ. Utah.

Bridwell, R.J., 1974, Nonlinear in-plane bending of a plane stress-strain finite element model: Jour. Geophys. Res., v. 79, no. 11, p. 1674-1678.

Bridwell, R.J., and Swolfs, H.S., 1974, Stability analysis of an experimentally deformed single layer of Indiana limestone using finite elements: Jour. Geophys. Res., v. 79, no. 11, p. 1679-1686.

Burger, H.R., 1972, Computerized solution for calculating calcite compression and tension axes: Geol. Soc. America Bull., v. 83, no. 8, p. 2439-2442.

Burkhardt, R., and Endlicher, G., 1977, Geologische Interpretation des ERTS-2-Satellitenbildes des Ostbayerischen Grundgebirges und angrenzender Gebiete: Acta Albertina Ratisbonensia, v. 37, p. 91-102.

Burns, K.L., 1973, Structural data files: Tectonics Struct. Newsl., Geol. Soc. Australia, no. 2, p. 21-22.

Burns, K.L., and Dunnet, D., 1973, Structural data processing: Tectonic Struck. Newsl., Geol. Soc. Australia, no. 2, p. 20-21.

Burns, K.L., Huntington, J.F., and Green, A.A., 1977, Computer-assisted photointerpretation of geological lineaments: perception method: Intern. Symp. Applic. Comput. Oper. Res. Mineral Indust., Papers, no. 15, p. 275-285.

Burns, K.L., Marshall, B., and Gee, R.D., 1969, Computer-assisted geological map: Australas. Inst. Mining Metall., Proc., no. 232, p. 41-47.

Burns, K.L., Shepherd, J., and Berman, M., 1976, Reproducibility of geological lineaments and other discrete features interpreted from imagery: measurement by a coefficient of association: Remote Sensing Environ., v. 5, p. 267-301.

Bush, V.A., and Kats, Y.G., 1978, Tektonicheskoye rayonirovaniye Sredizemnomorskogo al'piyskogo poyasa po rezul'tatam deshifrirovaniya kosmicheskikh snimkov: Vysshoye Uchebnoye Zavedeniye, Izv., Geol. Razved., no. 10, p. 74-79.

Cardamone, P., Leche, G.M., Cavallin, A., and others, 1977, Application of conventional and advanced techniques for the interpretation of LANDSAT 2 images for the study of linears in the Friuli earthquake area: Intern. Symp. Remote Sens. Environ., Proc., no. 11, p. 1337-1353.

Cassinis, R., 1977, Applications of satellite studies for structural geology in Italy *in* Remote sensing of the terrestrial environment, Peel, R.F., and others, eds., Butterworths, London, p. 169-181.

Chapple, W.M., 1964, A mathematical study of finite-amplitude rock folding (abst.): Am. Geophys. Union Trans., v. 45, no. 1, p. 104.

Chapple, W.M., 1968, A mathematical theory of finite-amplitude rock-folding: Geol. Soc. America Bull., v. 79, no. 1, p. 47-68.

Chapple, W.M., 1969, Fold shape and rheology: the folding of an isolated viscous-plastic layer: Tectonophysics, v. 7, no. 2, p. 97-116.

Chapple, W.M., 1970a, The initiation and spacing of folds in viscous multilayered media (abst.): Geol. Soc. America, Abs. Prog., v. 2, no. 4, p. 276.

Chapple, W.M., 1970b, The finite-amplitude instability in the folding of layered rocks: Can. Jour. Earth Sci., v. 7, no. 2, pt. 1, p. 457-466.

Chapple, W.M., and Spang, J.H., 1974, Significance of layer-parallel slip during folding of layered sedimentary rocks: Geol. Soc. America Bull., v. 85, no. 10, p. 1523-1534.

Charlesworth, H.A.K., Langenberg, C.W., and Ramsden, J., 1975, Determining axes, axial planes and profiles of macroscopic folds using computer-based methods (abst.): Geol. Soc. America, Abs. Prog., v. 7, no. 6, p. 734.

Charlesworth, H.A.K., Langenberg, C.W., and Ramsden, J., 1976, Determining axes, axial planes, and section of macroscopic folds using computer-based methods: Can. Jour. Earth Sci., v. 13, no. 1, p. 54-65.

Chesire, S.G., Spang, J.H., and Stockmal, G.S., 1978, Am empirical non-linear technique for the analysis of conical folds (abst.): Geol. Soc. America, Abs. Prog., v. 10, no. 5, p. 212-213.

Clancy, R.T., Turcotte, D.L., and Kulhawy, F.H., 1977, Finite element studies of strain accumulation and release of the San Andreas fault (abst.): Am. Geophys. Union Trans., v. 58, no. 12, p. 1227.

Cobbold, P.R., 1977, Finite-element analysis of fold propagation - a problematic application?: Tectonophysics, v. 38, no. 3-4, p. 339-353.

Cobbold, P.R., 1977, Removal of finite deformation using strain trajectories: Jour. Struc. Geol., v. 1, no. 1, p. 67-72.

Cogbill, A.H., 1979, The relationship between crustal structure and seismicity in the western Great Basin: unpubl. doctoral dissertation, Northwestern Univ.

Coode, A.M., 1966, An analysis of major tectonic features: Roy. Astron. Soc., Geophys. Jour., v. 12, no. 1, p. 55-66.

Coode, A.M., 1967, The spherical harmonic analysis of major tectonic features, *in* Mantles of the Earth and terrestrial planets, Runcorn, S.K., ed.: Interscience Publ. London, p. 489-498.

Correa, A.C., and Lyon, R.J.P., 1976, Application of optical Fourier analysis to the study of geological linear features in ERTS-1 imagery of California: Utah Geol. Assoc., Publ. no. 5, p. 462-479.

Cosgrove, J.W., 1976, The formation of crenulation cleavage: Jour. Geol. Soc., London, v. 132, no. 2, p. 155-178.

Crain, I.K., 1972, Statistical methods for geotectonic analysis (abst.): 24th Intern. Geol. Congress (Montreal), Abst., p. 70-71.

Crain, I.K., 1973, A statistical approach to the analysis of geotectonic elements: unpubl. doctoral dissertation, Australian National Univ.

Cruden, D.M., and Charlesworth, H.A.K., 1972, Some observations on the numerical determination of the axes of cylindrical and conical folds (abst.): Geol. Soc. America, Abs. Prog., v. 4, no. 6, p. 372-373.

Cubitt, J.M., and Celenk, O., 1976, FORTRAN program for producing stereograms in geology: Computers & Geosciences, v. 1, no. 3, p. 207-211.

Cundall, P.A., 1971, A computer model for simulating progressive, large-scale movements in blocky rock systems, *in* Rock fracture: Intern. Soc. Rock. Mech., unpaginated.

Dabovski, K., 1975, Matematicheskaya model' napryazheniy i peremeshcheniy okolo magmaticheskikh kamer i razlomov: Geotektonika, Tektonofiz., Geodinamika, v. 3, p. 17-30.

Dahlberg, E.C., Deland, C.R., and Creed, R.M., 1975, Probability mapping of subsurface faults from stratigraphic data (abst.): Geol. Soc. America, Abs. Prog., v. 7, no. 1, p. 45.

David, M., 1977, Geostatistical ore reserve estimation: Elsevier Scientific Publ. Co., New York, 364 p.

Davis, H.T., 1975, Multivariate prediction and spline functions, *in* The search for oil: some statistical methods and techniques, Owen, D.B., ed.: v. 13, Marcel Dekker, Inc., New York, p. 33-40.

Davis, J.C., Doveton, J.H., and Hambleton, W.W., 1975, Oil exploration by probabilistic analysis: a case history (abst.): Am. Assoc. Petroleum Geologists Soc. Econ. Pal. Miner., Ann. Mtg., v. 2, p. 18.

De Bremaecker, J.-C., and Becker, E.B., 1976, Numerical simulation of folding (abst.): Am. Geophys. Union Trans., v. 57, no. 4, p. 321.

De Bremaecker, J.-C., and Becker, E.B., 1978, Finite element models of folding: Tectonophysics, v. 50, no. 2-3, p. 349-367.

Deist, F.H., Salamon, M.D.G., and Georgiadis, E., 1973, A new digital method for three-dimensional stress analysis in elastic media: Rock Mechan. (Vienna), v. 5, no. 4, p. 189-202.

Demirmen, F., 1972, Mathematical procedures and FORTRAN IV program for description of three-dimensional surface configurations: Kansas Geol. Survey, KOX Tech. Rept., 131 p.

Demirmen, F., 1973, Numerical description of folded surfaces depicted by contour maps: Jour. Geology, v. 81, no. 5, p. 599-620.

Dieterich, J.H., 1969, Mathematical modeling of fault tectonic and seismicity (abst.): Geol. Soc. America, Abs. Prog., pt. 7, p. 47-48.

Dieterich, J.H., 1970, Computer experiments on mechanics of finite amplitude folds: Can. Jour. Earth Sci., v. 7, no. 2, p. 467-476.

Dieterich, J.H., 1972, Numerical modeling of deformations associated with volcanism (abst.): Geol. Soc. America, Abs. Prog., v. 4, no. 3, p. 146.

Dieterich, J.H., and Carter, N.L., 1969, Stress-history of folding: Am. Jour. Sci., v. 267, no. 2, p. 129-154.

Dieterich, J.H., and Decker, R.W., 1975, Finite element modeling of surface deformation associated with volcanism: Jour. Geophys. Res., v. 80, no. 29, p. 4094-4102.

Dieterich, J.H., and Onat, E.T., 1969, Slow finite deformations of viscous solids: Jour. Geophys. Res., v. 74, no. 8, p. 2081-2088.

Douglas, A., 1970, Finite elements for geologic modelling: Nature, v. 226, no. 5246, p. 630-631.

Du Prey, E.J.L., and Bossie-Codreanu, D.N., 1974, Simulation numerique de l'exploitation des reservoirs fissures: World Petrol. Congr., Proc. IX, v. 4, p. 233-246

Dudley, R.M., Perkins, P.C., and Gine M, E., 1975, Statistical tests for preferred orientation: Jour. Geology, v. 83, no. 6, p. 685-705.

Dumitrescu, V., 1974, Blocdiagrame realizabile la computer: Stud. Cercet. Geol., Geofiz., Geogr., Ser. Geogr., v. 21, no. 1, p. 117-119.

Eckert, H.U., 1970, Einsatz der elektronischen datenverarbeitung bei kleintektonischen Untersuchungen auf Blatt Weyerbusch/Siegerland; 1:25000: Clausthaler Tekt. Hefte, 10, p. 191-228.

Edel'shteyn, A.Y., 1972, O kolichestvennyhk metodakh izucheniya platformennoy skladchatosti: Geol. Geofiz (Akad. Nauk. SSSR, Sib. Otd.), v. 11, p. 120-124.

Eschner, T.R., Robinson, J.E., and Merriam, D.F., 1979, Comparison of spatially filtered geologic maps: summary: Geol. Soc. America Bull., pt. I, v. 90, no. 1, p. 6-7; pt. II, p. 104-134.

Feenstra, R., and Wickham, J.S., 1976, Computer models of simple shear deformation superposed on symmetric folds applied to deformation in the Ouachita Mountains (abst.): Geol. Soc. America, Abs. Prog., v. 8, no. 1, p. 20.

Fletcher, R.C., 1972, Application of a mathematical model to the emplacement of mantled gneiss domes: Am. Jour. Sci., v. 272, no. 3, p. 197-216.

Flinn, D., 1978, Construction and computation of three-dimensional progressive deformations: Geol. Soc. London Jour., v. 135, pt. 3, p. 291-305.

Forster, H., and Leonhardt, J., 1972, Mathematische Simulation ptygmatischer Strukturen: Geol. Rundsch., v. 61, no. 3, p. 883-896.

Franssen, L., and Kummert, P., 1971, Presentation d'un programme de traitement des donnees en geologie structurale: Soc. Geol. Belgique, Annal., v. 94, no. 1, p. 39-43.

Fujii, K., 1972, Structural analysis by the finite element method, *in* Prof. Jun-ichi Iwai Memorial Volume, Hatai, K., Asano, K., and others, eds.: Tohoku Univ., Inst. Geol. Paleont., Japan, p. 471-480.

Fumagalli, E., 1968, Model simulation of rock mechanics problems, *in* Rock mechanics in engineering practice, Stagg, K.G., and Zienkiewicz, O.C., eds.: John Wiley & Sons, London, p. 353-384.

George, L., and Sowers, G.M., 1975, Comparison of computed folds in multilayered media: elastic, viscous, viscoelastic, and plastic cases (abst.): Geol. Soc. America, Abs. Prog., v. 7, no. 2, p. 167-168.

Ghez, F., and Janot, P., 1974, Calcul statistique du volume des blocs matriciels d'un gisement fissure: Inst. Francais Petrole Rev., v. 29, no. 3, p. 375-386.

Goncharov, M.A., 1971, Matematicheskaya model'skladchatoy struktury: Geol. Geofiz. (Akad. Nauk SSSR, Sib. Otd.), no. 4, p. 117-123.

Grishkyan, R.I., Parfenov, L.M., and Ufimtsev, G.V., 1977, Kosmicheskiye izobrazheniya Baykal'skoy riftovoy oblasti i yeye vozmozhnaya kinematicheskaya model', *in* Rol' riftogeneza v geologicheskoy istorii Zemli, Florensov, N.A., ed.: Izd. Nauka, Novosibirsk, p. 104-108.

Groshong, R.H., Jr., 1974, Experimental test of least-squares strain gage calculation using twinned calcite: Geol. Soc. America Bull., v. 85, no. 12, p. 1855-1863.

Hanley, J.T., 1975, Numerical techniques for areal fracture analysis (abst.): Geol. Soc. America, Abst. Prog., v. 7, no. 1, p. 70.

Hardy, M.P., Crouch, S.L., Fairhurst, C., and others, 1974, A hybrid computer system simulating inelastic seam behavior: Intern. Soc. Rock. Mech., Congr. Proc., III, v. 2B, p. 1015-1021.

Hattori, I., and Mizutani, S., 1971, Computer simulation of fracturing of layered rock: Engineer. Geology, v. 5, no. 4, p. 253-269.

Haugh, I., Brisbin, W.C., and Turek, A., 1967, A computer-oriented field sheet for structural data: Can. Jour. Earth Sci., v. 4, no. 4, p. 657-662.

Heinze, W.D., and Goetze, C., 1973, Computer simulation of fracturing of layered rock (abst.): Am. Geophys. Union Trans., v. 54, no. 4, p. 450.

Heinze, W.D., and Goetze, C., 1974, Numerical simulation of stress concentrations in rocks: Intern. Jour. Rock Mech. Min. Sci., v. 11, no. 4, p. 151-155.

Hirsinger, V., 1976, Numerical strain analysis using polar coordinate transformations: Jour. Math. Geology, v. 8, no. 2, p. 183-202.

Hobbs, B.E., 1971, The analysis of strain in folded layers: Tectonophysics, v. 11, no. 4, p. 329-375.

Howard, J.C., 1971, Computer simulation models of salt domes: Am. Assoc. Petroleum Geologists Bull., v. 55, no. 3, p. 495-513.

Hudleston, P.J., 1973, Fold morphology and some geometrical implications of theories of fold development: Tectonophysics, v. 16, no. 1-2, p. 1-46.

Hudleston, P.J., and Stephansson, O., 1973, Layer shortening and fold-shape development in the buckling of single layers: Tectonophysics, v. 17, no. 4, p. 299-321.

Huntington, F., 1975, A photogeological study of fracture trace patterns using data-processing techniques (abst.): Inst. Min. Metall. Trans., v. 84B, no. 828, p. 156.

Ikeda, Y., and Shimamoto, T., 1974, Numerical experiments on the viscous bending folds (in Japanese): Geol. Soc. Japan Jour., v. 80, no. 2, p. 65-74.

Jackson, D.D., 1972, Numerical simulation of main shock and aftershock sequence (abst.): Am. Geophys. Union Trans., v. 53, no. 11, p. 1047.

James, W.R., 1968, Least-squares surface fitting with discontinuous functions: Tech. Rept. 8, ONR Task 389-150, Contract Nonr-1228(36), 51 p.

James, W.R., 1970, Regression models for faulted structural surfaces: Am. Assoc. Petroleum Geologists Bull., v. 54, no. 4, p. 638-646.

Jeran, P.W., and Mashey, J.R., 1970, A computer program for the stereographic analysis of coal fractures and cleats: U.S. Bur. Mines Info. Circ. 8454, 34 p.

Jones, T.A., Johnson, C.R., and Herron, S.R., 1976, An application of interactive computer graphics to petroleum exploration (abst.): Geol. Soc. America, Abs. Prog., v. 8, no. 1, p. 26-27.

Jones, T.A., and Jordan, N.F., 1975, Structural mapping with data containing points that do not intersect the horizon of interest: an automated procedure (abst.): Am. Assoc. Petroleum Geologists Mtg. Abst., v. 2, p. 40-41.

Juneman, P.M., 1971, Geological applications of interactive computer graphics and "graphic analysis of three dimensional data" (G.A.T.D.): Sci. Terre, v. 16, no. 3-4, p. 303-316.

Kapustin, I.N., Przhiyalgovskiy, Y.S., and Trofimov, D.M., 1978, Primeneniye kosmicheskoy informatsii pri sostavlenii tektonicheskoy karty Prikaspiyskoy vpadiny i yeye obramleniya (stat'ya I): Vysshoye Uchebnoye Zavedeniye, Izv., Geol. Razved., no. 10, p. 40-46.

Koelling, M.E.V., and Whitten, E.H.T., 1973, FORTRAN IV program for spline-surface interpolation and contour-map production: Geocom Programs, v. 9, p. 1-12.

Kontorovich, A.E., Fotiadi, E.E., Berilko, V.I., and others, 1971, Prognoz produktivnosti lokal'nykh struktur tsentral'noy i yugo-vostochnoy chastey zapadno-Sibirskoy plity s primeneniyem algoritmov raspoznavaniya obrazov: Geol. Geofiz. (Akad. Nauk SSSR Sib. Otd.), Novosibirsk, no. 7, p. 84-91.

Kosloff, D., 1976, Numerical investigation of a mechanism of formation of the Palmdale Uplift (abst.): Am. Geophys. Union Trans., v. 57, no. 12, p. 898.

Kowalik, W.S., 1975, Application of satellite photographic and MSS data to selected geologic and natural resource problems in Pennsylvania: III, Comparison of Skylab and Landsat lineaments with joint orientations in north central Pennsylvania: NASA Tech. Memo, no. X-58168, p. 958-969.

Krausse, H.F., 1970, Uber eine erste statistische Auswertung von Gefugedaten mit elektronischen Rechenanlagen: Clausthaler Tekt. Hefte, v. 10, p. 49-62.

Kronberg, P., 1975, ERTS-1 shows buried structures *in* Geoscientific studies and the potential of the natural environment, Deutsche UNESCO-Komm., Koln, Verlag Dok. Munchen, p. 122.

Kruckeberg, F., 1968, Eine Programmiersprache fur gefugekundliche Arbeiten: Clausthaler Tekt. Hefte, v. 8, p. 7-53.

Kruhl, J., 1974, Mitteilung uber die Benutzung von Sichtgeraeten bei der statischen Auswertung von Gefugemessungen: Geol. Mitt. (Aachen), v. 12, no. 4, p. 319-326.

Kruhl, J., 1978, Current bedding in the Moinian quartzites at eastern Loch Leven, Scottish Highlands: Neues Jahrb. Mineral., Abh., v. 132, no. 1, p. 52-66.

Kulyndyshev, V.A., 1972, Logicheskiy analiz metodov opisaniya elementarnykh strukturnykh poverkhnostey: Geol. Geofiz. (Akad. Nauk SSSR, Sib. Otd.), Novosibirsk, no. 1, p. 142-144.

Kumar, M.B., 1977, Computer-aided subsurface structural analysis of the Miocene formations of the Bayou Carlin-Lake Sand area, South Louisiana: Louisiana Geol. Surv., Geol. Bull., no. 43, pt. I, 177 p.

LaFountain, L.J., 1970, Plotted and point-counted stereograms by computer X-Y plotter or microfilm devices: Geol. Soc. America Bull., v. 81, no. 4, p. 1267-1271.

Lam, P.W.H., 1969, Discussion: computer program for plotting beta-diagrams: Am. Jour. Sci., v. 267, no. 9, p. 1114-1117.

Lambert, P., 1975, La structure impactitique de Rochechouart (Haute-Vienne) et la structure de la partie nord-ouest du Massif central francais, interpretation de "photographies obtenues par satellite"; image ERTS (abst.): France, Bur. Rech. Geol. Minieres Bull. (Ser. 2), Sect. 2, no. 1, 21 p.

Langenberg, C.W., Rondeel, H.E., and Charlesworth, H.A.K., 1977, A structural study in the Belgian Ardennes with sections constructed using computer-based methods: Geol. Mijnb., v. 56, no. 2, p. 145-154.

Laurin, A.-F., Sharma, K.N.M., Wynne-Edwards, H.R., and Franconi, A., 1972, Application of data processing techniques in the Grenville Province, Quebec, Canada: Proc. 24th Intern. Geol. Congress (Montreal), Sect. 16, p. 22-35.

Lewis, R.W., and Williams, J.R., 1978, A finite-element study of fold propagation in a viscous layer: Tectonophysics, v. 44, no. 1-4, p. 263-283.

Lister, G.S., Paterson, M.S., and Hobbs, B.E., 1978, The simulation of fabric development in plastic deformation and its application to quartzite: the model: Tectonophysics, v. 45, no. 2-3, p. 107-158.

Lokhmatov, G.I., and Alayev, G.T., 1967, Metod razdeleniya strukturnykh i izopakhicheskikh kart na sostavlyayushchiye: Geol. Nefti Gaza, v. 11, no. 4, p. 32-38.

Lakhmatov, G.I., Alayev, G.T., and Yevdokimova, V.N., 1968, Matematicheskiy metod paleotektonicheskogo analiza platformennykh struktur v usloviyakh monoklinal'nogo sklona: Geol. Nefti Gaza, v. 12, no. 5, p. 46-50.

Lutzner, H., and Maaz, H., 1969, Ein Rechenprogramm zur statistischen Beschreibung von Richtungsmessungen mit geringer Streuung und dessen Anwendung: Deutsche Gesell. Geol. Wissen., Ber., Reihe A, Geol. Palaont., v. 14, no. 5, p. 561.

Maffi, G., and Marchesini, E., 1964, Semi-automatic equipment for statistical analysis of airphoto linears: Photogram. Engineer., v. 30, no. 1, p. 139-141.

Mark, D.M., 1973, Analysis of axial orientation data, including till fabrics: Geol. Soc. America Bull., v. 84, no. 4, p. 1369-1374.

Masson, P., 1978, Essai d'analyse structurale du Rift Levantine d'apres les donnees Landsat: Photo interpret., v. 17, no. 1, p. 17-33.

Matthews, P.E., Bond, R.A.B., and Van den Berg, J.J., 1974, An algebraic method of strain analysis using elliptical markers: Tectonophysics, v. 24, no. 1-2, p. 31-67.

Minear, J.W., 1972, Finite-element models of preexisting fractures (abst.): Am. Geophys. Union Trans., v. 53, no. 11, p. 1118.

Miyatake, T., 1977, Numerical simulation of dynamical faulting process (in Japanese): Zisin Seismol. Soc. Japan Jour., v. 30, no. 4, p. 449-461.

Mundry, E., 1972, Nonlinear regression models in geology (abst.): Proc. 24th Intern. Geol. Congress (Montreal) Abst., p. 522-523.

Myasnikova, G.P., and Shpil'man, V.I., 1973, Izucheniye svyazi kharaktera osadkonakopleniya i rosta struktur s primeniniyem trend-analiza: Geol. Nefti Gaza, v. 4, p. 12-16.

Neugebauer, H.J., and Spohn, T., 1978, Late stage development of mature Atlantic-type continental margins: Tectonophysics, v. 50, no. 2-3, p. 275-305.

Nidd, E., and Ambrose, J.W., 1971, Computerized solutions for some problems of fold geometry: Can. Jour. Earth Sci., v. 8, no. 6, p. 688-693.

Nuttall, D.J.H., and Cooper, M.A., 1978, Computer programmes for the analysis and presentation of orientation data: Geol. Soc. London Jour., v. 135, pt. 2, p. 243-244.

Offield, T.W., Abbott, E.A., Gillespie, A.R., and others, 1975, Enhanced ERTS images for mapping of structural control in the southern Brazil copper region (abst.): Soc. Explor. Geophys., Ann. Intern. Mtg., Abst., no. 45, p. 76.

Olea, R.A., 1975, Optimum mapping techniques using regionalized variable theory: Kansas Geol. Survey, Ser. on Spatial Analysis, 137 p.

Owens, W.H., 1974, Mathematical model studies on factors affecting the magnetic anisotrophy of deformed rocks: Tectonophysics, v. 24, no. 1-2, p. 115-131.

Parrish, D.K., 1973, A nonlinear finite element fold model: Am. Jour. Sci., v. 273, no. 4, p. 318-334.

Parrish, D.K., Krivz, A., and Carter, N.L., 1975, Finite element folds of similar geometry (abst.): Geol. Soc. America, Abs. Prog., v. 7, no. 2, p. 223-224.

Parrish, D.K., Krivz, A., and Carter, N.L., 1976, Finite-element folds of similar geometry: Tectonophysics, v. 32, no. 3-4, p. 183-207.

Peikert, E.W., 1970, Interactive computer graphics and the fault problem (abst.) Am. Assoc. Petroleum Geologists Bull., v. 54, no. 3, p. 556-557.

Petrov, A.P., Ellanskiy, M.M., and Zverev, G.N., 1972, Ispol'zovaniye matematicheskikh metodov pri reshenii zadach klassifikatsii geologicheskikh ob'yektov *in* Matematicheskiye metody v gazoneftyanoy geologii i geofizike, Izd. Nedra, Moscow, p. 10-32.

Pferd, J.W., 1975a, A computer-based system for the collection of detailed structural data from metamorphic terrains (abst.): Geol. Soc. America, Abs. Prog., v. 7, no. 1, p. 106.

Pferd, J.W., 1975b, Computer-compatible collection of detailed structural data in metamorphic terrains: published privately, Amherst, Massachusetts, 39 p.

Podwysocki, M.H., 1973, Computer applications in the analysis of fracture patterns (abst.): Geol. Soc. America, Abs. Prog., v. 5, no. 2, p. 207-208.

Podwysocki, M.H., Moik, J.G, and Shoup, W.C., 1975, Quantification of geologic lineaments by manual and machine processing techniques: NASA Tech. Memo., no. X-58168, p. 885-903.

Pollard, D.D., and Holzhausen, G., 1978, FORTRAN computer program for calculation of stress-intensity factors, stresses, and displacements associated with a fluid-pressurized fracture near the Earth's surface: U.S. Geol. Survey open-file rept. 78-160, 26 p.

Ramberg, H., 1970, Folding of laterally compressed multilayers in the field of gravity, II Numerical examples: Phys. Earth Planet. Interiors, v. 4, no. 2, p. 83-120.

Ramsden, J., 1975, Numerical methods in fabric analysis: unpubl. doctoral dissertation, Univ. Alberta.

Ramsden, J., and Cruden, D.M., 1979a, Estimating densities in contoured orientation diagrams: summary: Geol. Soc. America Bull., v. 90, no. 3, pt. I, p. 229-231.

Ramsden, J., and Cruden, D.M., 1979b, Estimating densities in contoured orientation diagrams: Geol. Soc. America Bull., v. 90, pt. II, p. 580-607.

Reches, Z., and Johnson, A.M., 1979, Development of monoclines: Part II Theoretical analysis of monoclines: Geol. Soc. America Mem. 151, p. 273-311.

Reddy, J.N., and Wickham, J.S., 1979, Numerical modeling of geologic structures I. Finite element formulations of viscous, incompressible flows: in press.

Rice, L.F., Davis, J.H., and Johnson, A.C., 1976, Computer analysis of lineaments to infer existence of subsurface geologic structures (abst.): Am. Assoc. Petroleum Geologists Bull., v. 60, no. 4, p. 714.

Robinson, J.E., 1969, Spatial filters for geological data: Oil and Gas Jour., v. 67, no. 37, p. 132-134, 138, and 140.

Robinson, J.E., 1970, Spatial filtering of geological data: Rev. Inst. Intern. Statist., v. 38, no. 1, p. 21-34.

Robinson, J.E., 1976, Computer simulation of real folds (abst.): Geol. Assoc. Canada, Mineral. Assoc. Canada jt. mtg., prog. abs., v. 1, p. 51.

Robinson, J.E., and Charlesworth, H.A.K., 1969a, Spatial filtering illustrates a relationship between tectonic structure and oil occurrences in southern and central Alberta: Can. Inst. Min. Metall., 20th Ann. Mtg., Edmonton, paper 6935, p. 1-8.

Robinson, J.E., and Charlesworth, H.A.K., 1969b, Spatial filtering illustrates relationship between tectonic structure and oil occurrence in southern and central Alberta: Kansas Geol. Survey Computer Contr. 40, p. 13-18.

Robinson, J.E., and Charlesworth, H.A.K., 1975, Relation of topography and structure in south-central Alberta: Jour. Math. Geology, v. 7, no. 1, p. 81-87.

Robinson, J.E., and Merriam, D.F., 1972, Enhancement of patterns in geologic data by spatial filtering: Jour. Geology, v. 80, no. 3, p. 333-345.

Rodgers, D.A., Gallagher, J.J., Jr., Rizer, W.D., and Spang, J.H., 1977, Analysis of thrust fault mechanisms: II. Mathematical models (abst.): Am. Geophys. Union Trans., v. 58, no. 6, p. 507.

Sampsom, R.J., 1975a, The SURFACE II graphics system, *in* Display and analysis of spatial data, Davis, J.C., and McCullagh, M.J., eds.: John Wiley & Sons, New York, p. 244-266.

Sampson, R.J., 1975b, SURFACE II graphics system: Kansas Geol. Survey, Ser. on Spatial Analysis, 240 p.

Schuenemeyer, J.G., Koch, G.S., Jr., and Link, R.F., 1972, Computer program to analyze directional data, based on the methods of Fisher and Watson: Jour. Math. Geology, v. 4, no. 3, p. 177-202.

Selkman, S., 1978, Stress and displacement analysis of boudinages by the finite-element method: Tectonophysics, v. 44, no. 1-4, p. 115-139.

Serra, S., 1973, A computer program for calculation and plotting of stress distribution and faulting: Jour. Math. Geology, v. 5, no. 4, p. 397-407.

Sharma, B., and Geldart, L.P., 1968, Analysis of gravity anomalies of two dimensional faults using Fourier transforms: Geophys. Prospect., v. 16, p. 77-93.

Shatagin, N.N., and Dergachev, A.L., 1978, A modified procedure for the computer preparation of diagrams of the orientations of structural elements: Moscow Univ. Geol. Bull., v. 33, p. 90-91.

Shatagin, N.N., and Sandomirskiy, S.A., 1974, Postroyeniye krugovykh diagramm oriyentirovok na EVM: Akad. Nauk SSSR, Izv. Ser. Geol., no. 9, p. 97-104.

Shcherbakov, V.S., 1973, Rezul'taty prımeneniya nekotorykh matematicheskikh metodov i ETSVM pri izuchenii stroyeniya osadochnogo chekhla zapadnoy chasti Latviyskoy SSR, *in* Problemy regional'noy geologii Pribaltiki i Belorussi: Minist. Geol. SSSR, Vses. Nauchno-Issled. Inst. Morsk. Geol. Geofiz., Riga, p. 189-200.

Short, N.M., 1973a, Mineral resources, geological structure, and landforms surveys: Symp. Signif. Result. Obtain. ERTS-1, v. III (Goddard Space Flight Cent.), p. 30-46.

Short, N.M., 1973b, ERTS: Applications to tectonics, volcanology, mineral resources, and landforms analysis (abst.): Am. Geophys. Union Trans., v. 54, no. 7, p. 700.

Short, N.M., 1974a, Mineral resources, geological structures, and landform surveys: Third ERTS Symp., v. II, NASA spec. publ. no. 356, p. 147-167.

Short, N.M., 1974b, Mineral resources, geological structure, and landform surveys: Third ERTS Symp., v. III, NASA spec. publ. no. SP-357, p. 33-51.

Siemes, H., 1967, Ein Rechenprogramm zur Auswertung von Rontgen-Texturgoniometer-Aufnahmen: Neues Jahr. Mineral., Monat., no. 2-3, p. 49-60.

Smith, A.G., 1971, Alpine deformation and the oceanic areas of the Tethys, Mediterranean, and Atlantic: Geol. Soc. America Bull., v. 82, no. 8, p. 2039-2070.

Smith, A.G., and Hallam, A., 1970, The fit of the southern continents: Nature, v. 225, no. 5228, p. 139-144.

Smith, S.W., and Kind, R., 1972, Observations of regional strain variations: Jour. Geophys. Res., v. 77, no. 26, p. 4976-4980.

Solov'yeva, L.I., 1977, Neotektonika i informativnost' kosmicheskikh snimkov (na primere Sredney Azii): Vysshoye Uchebnoye Zavedeniye, Izv., Geol. Razved., no. 12, p. 47-54 (English translation) Intern. Geol. Rev., 1978, v. 20, no. 11, p. 1281-1286.

Spang, J.H., 1974, Numerical dynamic analysis of calcite twin lamellae in the Greenport Center syncline: Am. Jour. Sci., v. 274, no. 9, p. 1044-1058.

Spang, J.H., and van der Lee, J., 1975, Numerical dynamic analysis of quartz deformation lamellae and calcite and dolomite twin lamellae: Geol. Soc. America Bull., v. 86, no. 9, p. 1266-1272.

Srivastava, G.S., 1975, Optical and digital processing of geological surfaces in Kansas: unpubl. doctoral dissertation, Syracuse Univ., 326 p.

Srivastava, G.S., and Merriam, D.F., 1974, Quantitative comparison of maps using optically derived parameters (abst.): Geol. Soc. America, Abs. Prog., v. 6, no. 7, p. 964-965.

Srivastava, G.S., and Merriam, D.F., 1975, Computer constructed optical-rose diagrams (abst.): Geol. Soc. America, Abs. Prog., v. 7, no. 1, p. 121.

Srivastava, G.S., and Merriam, D.F., 1976, Computer constructed optical-rose diagrams: Computers & Geoscience, v. 1, no. 3, p. 179-186.

Stabler, C.L., 1968, Simplified Fourier analysis of fold shapes: Tectonophysics, v. 6, no. 4, p. 343-350.

Starkey, J., 1974, The quantitative analysis of orientation data obtained by the Starkey method of X-ray fabric analysis: Can. Jour. Earth Sci., v. 11, no. 11, p. 1507-1516.

Stearns, M.T., 1978, The deformation of thick, homogeneous, isotropic continuous rock masses. Part III Analog model studies (abst.): Geol. Soc. America, Abs. Prog., v. 10, no. 1, p. 26.

Stein, R.J., and Wickham, J.S., 1978, Computer models of drape folding and related faulting (abst.): Geol. Soc. America, Abs. Prog., v. 10, no. 7, p. 497.

Stein, R.J., and Wickham, J.S., 1979, Numerical models of fault zone propagation, in press.

Stephansson, O., 1973, The solution of some problems in structural geology by means of the finite-element technique: Geol. Forenin. Forhandl., v. 95, pt. 1, no. 552, p. 51-59.

Stephansson, O., 1976, Finite element analysis of folds: Roy. Soc. London, Philos. Trans., Ser. A, v. 283, no. 1312, p. 153-161.

Stephansson, O., and Berner, H., 1971, The finite element method in tectonic processes: Phys. Earth Planet. Interiors, v. 4, no. 4, p. 301-321.

Stromgard, K., 1973, Stress distribution during formation of boudinage and pressure shadows: Tectonophysics, v. 16, no. 3-4, p. 215-248.

Sturgul, J.R., and Grinshpan, Z., 1975, Finite-element model for possible isostatic rebound in the Grand Canyon: Geology, v. 3, no. 4, p. 169-171.

Szumilas, D., 1977, Using computer for time-to-depth conversion and structure mapping in complexly faulted areas (abst.): Am. Assoc. Petroleum Geologists Bull., v. 61, no. 5, p. 834-835.

Tapp, G., and Wickham, J.S., 1978, Fracture predictions using finite element computer models (abst.): Geol. Soc. America, Abs. Prog., v. 10, no. 1, p. 26.

Tapp, G., Wickham, J.S., and Reddy, J.N., 1979, Numerical modeling of geologic structures III. Fracture density in single layer folds, in press.

Tocher, F.E., 1979, The computer contouring of fabric diagrams: Computers & Geosciences, v. 5, no. 1, p. 73-126.

Tricart, J.L.F., 1976, Evidence offered by LANDSAT-1 imagery of tectonic lineaments in the Vosges Mountains (eastern France), in Rilevanmento spaziale delle risorse terrestri; Tecnologie dei sistemi e componenti spaziali; Rilevamento spaziale dei fenomeni atmosferici; Energia solare e fonti di energia terrestre non convenzionali: Intern. Tech. Sci. Mtg. Space Proc., no. 16, p. 53-59.

Vaytekunas, I.P., and Yankunayte, I.I., 1975, Informatsionno-poiskovaya sisteme po glubokim skvazhinam Yuzhnoy Pribaltiki (abst.): Izd. Litov. Nauch.-Issled. Geologorazved Inst. Vilna, p. 70.

Venkitasubramanyan, C.S., 1971, Least-squares analysis of fabric data: a note on conical, cylindroidal and near-cylindroidal folds: Can. Jour. Earth Sci., v. 8, no. 6, p. 694-697.

Venter, R.H., and Spang, J.H., 1974, Numerical analysis of multimodal orientation data (abst.): Am. Geophys. Union Trans., v. 55, no. 2, p. 73-74.

Vikhert, A.V., 1967, Tipovyye matematicheskiye modeli raspredeleniya vysot (glubin) strukturnykh poverkhnostey, izobrazhennykh na kartakh v izoliniyakh (abst.): Mosk. Obshchest. Ispyt. Prir., Byull., Otd. Geol., v. 42, no. 3, p. 160.

Vikhert, A.V., 1973, O metodike i vozmozhnostyakh postroyeniya morfologicheskoy klassifikatsii skladchatosti v chislennykh statisticheskikh merakh: Mosk. Obshchest. Ispyt. Prir., Byull., Otd. Geol., v. 48, no. 1, p. 148-149.

Vikhert, A.V., and Goncharov, M.A., 1969, O deterministskikh i verayatnostnykh modelyakh strukturnykh poverkhnostey: Geol. Geofiz. (Akad. Nauk SSSR, Sib. Otd.), Novosibirsk, no. 5, p. 66-71: (English translation) Intern. Geol. Rev., 1970, v. 12, no. 11, p. 1310-1313.

Voight, B., and Dahl, H.D., 1970, Numerical continuum approaches to analysis of nonlinear rock deformation: Can. Jour. Earth Sci., v. 7, no. 3, p. 814-830.

Voight, B., and Samuelson, A.C., 1969, On the application of finite-element techniques to problems concerning potential distribution and stress analysis in the earth sciences: Pur. Appl. Geophys., v. 76, p. 40-55.

Volchegurskiy, L.F., and Pronin, V.G., 1977, Vozmozhnosti vyyavleniya krupnykh strukturnykh form v predelakh Russkoy i Turanskoy plit (po materialam deshifrirovaniya televizionnykh kosmicheskikh snimkov): Vysshoye Uchebnoye Zavedeniye, Izv., Geol. Razved, no. 9, p. 14-19 (English translation) Intern. Geol. Rev., 1978, v. 20, no. 11, p. 1276-1280.

Vychev, V., 1976, Programma dlya opredeleniya rustoty treshchin v porodakh po polevym nablyudeniyam: Geotektonika, Tektonofiz., Geodinamika, v. 4, p. 59-64.

Watson, D.F., and Smith, F.G., 1975, A computer program and study of grain shape: Computers & Geosciences, v. 1, no. 1-2, p. 109-111.

Whitten, E.H.T., 1969, Trends in computer applications in structural geology, *in* Computer applications in the earth sciences: an international symposium, Merriam, D.F., ed.: Plenum Press, New York, p. 233-249.

Whitten, E.H.T., 1970, Orthogonal polynomial trend surfaces for irregularly spaced data: Jour. Math. Geology, v. 2, no. 2, p. 141-152.

Whitten, E.H.T., 1974, Orthogonal-polynomial contoured trend-surface maps for irregularly spaced data: Computer Applications, v. 1, no. 3-4, p. 171-192.

Whitten, E.H.T., 1975, The practical use of trend-surface analyses in the geological sciences, *in* Display and analysis of spatial data, Davis, J.C. and McCullagh, M.J., eds.: John Wiley & Sons, England, p. 282-297.

Whitten, E.H.T., 1976, Geodynamic significance of spasmodic, Cretaceous, rapid subsidence rates, continental shelf, USA: Tectonophysics, v. 36, no. 1-3, p. 133-142.

Whitten, E.H.T., 1977a, Kriging in subsurface structural analyses: Pribram Mining Conf. (Czechoslovakia), Proc., v. 3, p. 801-819.

Whitten, E.H.T., 1977b, Stochastic models in geology: Jour. Geology, v. 85, no. 3, p. 321-330.

Whitten, E.H.T., 1979, Semi-variograms and Kriging: possible useful tools in fold description, *in* Future trends in geomathematics, Craig, R., and Labovitz, M., eds.: Pion Press, London, in press.

Whitten, E.H.T., and Koelling, M.E.V., 1973, Spline-surface interpolation, spatial filtering, and trend surfaces for geological mapped variables: Jour. Math. Geology, v. 5, no. 2, p. 111-126.

Whitten, E.H.T., and Koelling, M.E.V., 1975, Computation of bicubic-spline surfaces for irregularly spaced data: Northwestern Univ., ARO-D Grant Rept. 3, 57 p.

Whitten, E.H.T., and Koelling, M.E.V., 1978, O bikubicheskikh splain-poverkhnostiakh dlia neravnomerno raznesennikh dannikh nabliodenii, *in* Issledovania po matematicheskoi geologii, Romanova, M.A., and Sapagov, N.A., eds.: Leningrad, p. 150-162.

Wickham, J.S., and Anthony, J.M., 1976, Incremental strain paths and folding of carbonate rocks near the Blue Ridge, central Appalachians (abst.): Am. Geophys. Union Trans., v. 57, no. 4, p. 321.

Wickham, J.S., and Anthony, J.M., 1977, Strain paths and folding of carbonate rocks near Blue Ridge, central Appalachians: Geol. Soc. America Bull., v. 88, no. 7, p. 920-924.

Wickham, J.S., Stein, R.J., and Reddy, J.N., 1979, Numerical modeling of geologic structures II. Computer models of faulting: in press.

Williams, J.R., Lewis, R.W., and Zienkiewicz, O.C., 1978, A finite-element analysis of the role of intial perturbations in the folding of a single viscous layer: Tectonophysics, v. 45, no. 2-3, p. 187-200.

Winchell, H., 1967, A computer program concerned with statistics of rock fabrics (abst.): Geol. Assoc. Canada, Mineral. Assoc. Can., Intern. Mtg. Abst. Pap., p. 103-104.

Woidt, W.-D., 1978a, Numerical calculations applied to salt dome dynamics (abst.): Am. Geophys. Union Trans., v. 59, no. 4, p. 386.

Woidt., W.-D., 1978b, Finite element calculations applied to salt dome analysis: Tectonophysics, v. 50, no. 2-3, p. 369-386.

Wray, W.B., Jr., 1973, A computer program to construct cross sections of curved surfaces: Jour. Math. Geology, v. 5, no. 2, p. 149-161.

Wynne-Edwards, H.R., Laurin, A.F., Sharma, K.N.M., Nandi, A., Kehlenbeck, M.M., and Franconi, A., 1970, Computerized geological mapping in the Grenville province, Quebec: Can. Jour. Earth Sci., v. 7, no. 6, p. 1357-1373.

Xu, D., and Sun, H., 1966, Trend-surface analysis of stratigraphic data (in Chinese): Stratigraphy Magazine, v. 1, p. 1.

Yeromenko, V.Y., and Katterfel'd, G.N., 1978, Ispol'zovaniye kosmicheskikh snimkov pri izuchenii regional'nykh i global'nykh sistem lineamentov Zemli: Vysshoye Uchebnoye Zavedeniye, Izv., Geol. Razved., no. 10, p. 23-29.

Zhang, Q.R., 1977, An example of trend-surface analysis of gentle folds (in Chinese): Scientia Geologica Sinica, v. 4, p. 377-389.

Zienkiewicz, O.C., 1971, The finite element method in engineering science: McGraw-Hill Publ. Co. Ltd., London, 521 p.

Zienkiewicz, O.C., 1974, Finite element methods in flow problems: an introduction, *in* Finite element methods in flow problems, Oden, J.T., and others, eds.: Univ. Alabama, Huntsville Press, Huntsville, p. 3-4.

Zienkiewicz, O.C., 1976, The finite element method and the solution of some geophysical problems: Roy. Soc. London, Philos. Trans., Ser. A, v. 283, no. 1312, p. 139-151.

A FORECAST FOR USE OF COMPUTERS BY GEOLOGISTS IN THE COMING DECADE OF THE 80s

D.F. Merriam

Syracuse University

ABSTRACT

The first two decades of computer applications in geology have resulted in many advances. The decade of the 80's promises continued rapid development of hardware and software, both areas generally outside the influence of earth scientists. We can expect continued trends in smaller, faster, cheaper, and more accessible computers, especially in the form of micros and minis. Software will be more user oriented and telecommunications will become commonplace. New and improved algorithms for solving geological problems will be developed. Simulation of geological processes and realistic three-dimensional models also will be important in the coming decade. It is possible that a language to express geological concepts and terminology in mathematical terms will be formulated.

INTRODUCTION

Geologists have been involved with computers now for about 20 years. Early computer use was restricted because of limited background of geologists in quantification. Geophysicists, engineering geologists, and hydrologists, however, readily applied computer-oriented techniques in their work as soon as available. Much has been written on the development and application of computers in geology so rather than look back, it is appropriate to see where we are today and then look ahead.

The discovery (conceptual phase) of computers by geologists took place in the 1950's (Fig. 1). The date of computerization can be given as 1958, when W.C. Krumbein and L.L. Sloss published a simple computer program written in SOAP in the Bulletin of the American Association of Petroleum Geologists. This program was designed to calculate three percentages and two ratios. During the 50's most of the papers (there were very few) were of a general nature and contained only suggestions of possible uses (Table 1).

The 1960's brought development and rapid growth in uses. Papers demonstrated uses for different techniques (mostly borrowed from other disciplines such as biology, engineering, statistics, etc.) with test geological data sets. Research and teaching centers of the subject of geomathematics, geostatistics, and computer applications appeared as well as special publication outlets to handle this new aspect of geology.

Many applications were reported during the 1970's and some specific developments to solve geological problems took place. As the use of computers spread through the subdisciplines of geology (and none have been immune) many problems and shortcomings were found not only in the applications but in the content and generation of the data base itself. Now, as we enter the 1980's, computer use will be integrated completely in geologic studies as this aspect of the science matures.

In some respects this introduction of computers into geology can be likened to the introduction of the petrographic microscope and the seismograph. The computer is an extension of the mind as the microscope extended seeing and the seismograph extended hearing. Also parallel to development and integration of the microscope and seismograph into geology, two groups have developed - those interested in the applications as a science unto itself and those who simply want to use the techniques as a tool. Thus, the subdiscipline of geomathematics (which encompasses geostatistics and computer applications) is at a similar cross roads as geophysics and geochemistry were a decade or two ago.

PRESENT STATUS OF COMPUTER USE IN GEOLOGY

Many of the borrowed techniques used to solve geological problems have been successful. Time-trend,

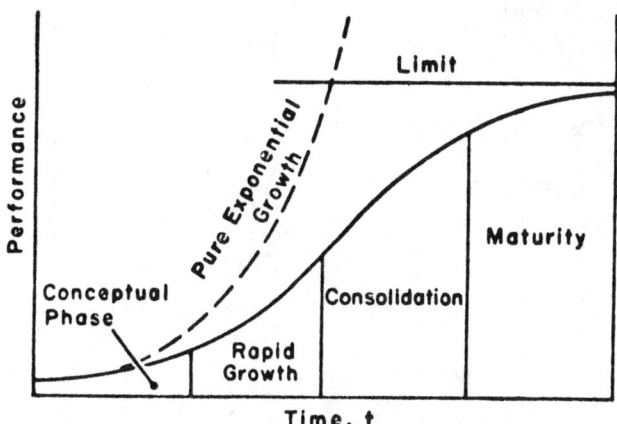

Figure 1. Graphic representation of performance and limit to growth of ideas and applications (from Starr and Rudman, 1973).

cross-correlation and cross-association (which include autocorrelation and autoassociation), and time-series analyses have been used to study sequences. Trend and Fourier analyses and spatial filtering along with numerous enhancement techniques have been used to treat spatial data both two and three dimensions. Multivariate analysis, including cluster, disciminant, principal components, and factor analyses, has been used extensively in a variety of conditions (Merriam, 1980).

Data bases have been assembled and storage and retrieval systems built to use them. Systems such as G-EXEC, GIPSY, SASFRAS, GRASP, and CLAIR are available readily to interrogate these large data bases. Included in this category are the information systems such as Geo.Ref and GeoArchives. Unfortunately, many of the bases are proprietary information.

Graphics have been important especially recently. Geologists are oriented visually and for the most part prefer their data in summary form, for example as maps, cross sections, graphs, etc. Therefore much effort has gone into presenting raw data in some palatable form.

In all of these developments, each step has been more complex (Fig. 2). As geologists became more astute and

Table 1. Historical record of stage of integration of computer concepts or techniques in geology (from Merriam, 1975).

	Publications	Data	Computer Programs	References
Discovery	Papers general with suggestions of possibilities.	None	None	Practically none.
Development	Papers demonstrate use of different techniques.	Madeup	"Borrowed" from other fields intact	Mostly from other disciplines.
Application	Papers acknowledge use of computers and source of programs. Different problems tried.	Sample data sets	Modified and adapted from other fields with some geological bent.	Everything written on the subject in geology.
Assimilation	Completely integrated.	Real data in quantities necessary to solve problems.	Programs written with only parts of "canned" programs used but specific for purpose.	Citation of only those papers of pertinence to work.

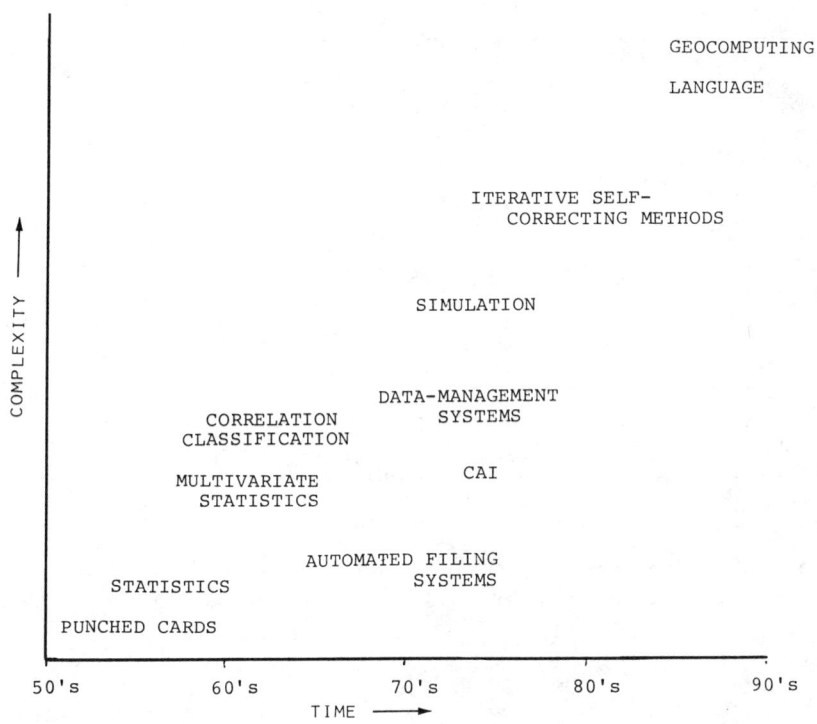

Figure 2. Increasing sophistication of applying computer-oriented techniques to solve geological problems (Merriam, 1980).

competent, more complex analyses were made of more complex problems. Starting with punched cards and sorting machines, we now have advanced to solving highly involved, complex multivariate situations with sophisticated techniques.

As previously stated no subdiscipline of geology has escaped the effect of computers. In the past ten years however, the distribution of interest has shifted. In general it can be seen (Fig. 3) that the use of computers has decreased in relation to other subjects for structure-tectonics, mineralogy-petrology, geochemistry, and stratigraphy-paleontology. Part of the relative decrease has been brought about by an increase in geomathematics (per se), oceanography, geomorphology, and environmental and engineering geology.

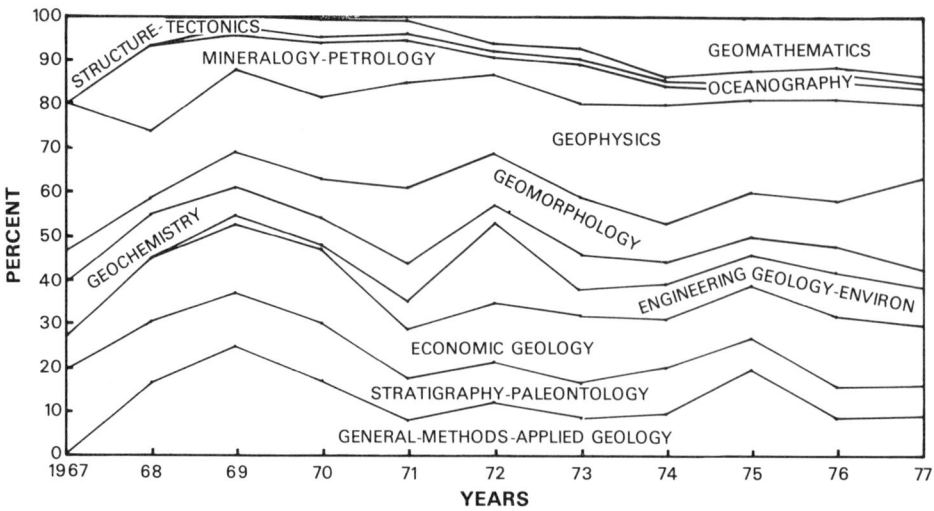

Figure 3. Percentage of papers concerned with geologic subject material for past decade (Merriam, 1980).

PROBLEMS

Along with this computer "revolution" in geology have come many problems with the benefits. These problems, however, have helped emphasize the need for additional studies which in the long run will improve geoloical predictions.

Many of the techniques adapted from other disciplines have been incorrectly used - in other words applied to problems for which they were not intended. In some instances the data did not fit the requirements needed for the analysis, in some applications the technique was simply misused, in others the example was not appropriate or realistic (Sutterlin and May, 1978).

In the construction of data bases it was found that there was a sampling problem, or the data were incomplete or of poor quality. In some situations the desired data, especially proprietary and historical data, were not available. The transcription of visual data bases presented problems in quality control and required decisions on what and how much was to be transcribed (often these decisions were not realistic).

One of the main problems at this point in time is the inability (or unwillingness) of geologists to use what is available. Part of the problem is updating the education of practicing geologists and part is just the knowledge of what is available and where. For example, Geo.Ref, the online bibliographic reference system for geologists, is being utilized only now and its potential is essentially untested. This system alone contains about 500,000 entries and grows at about 4,500 citations per month (incidentally, Australia accounts for only 3 percent of the file in contrast to 42 percent for the US). Coverage unfortunately extends back only to 1961 for North America and to 1967 for areas outside North America. Also 60 percent of the material (from 1967-1978) is in English! And less than 2 percent is concerned with mathematical geology in contrast with 19 percent for structure-geophysics (as determined in 1976-1978 interval of coverage). How many geologists effectively use the system? Probably very few.

CAI (computer-aided instruction) offers tremendous possibilities as a supplemental teaching aid. However, to date little use has been made of it in geology (Merriam, 1976).

WAYS OF THE FUTURE

Now that geology is changing from an observational and historical science to a more quantitative and rigorous one we can look ahead and make some predictions of things to come. Several things should be noted here - first of all we are passing from a stage of "where" to one of "how." In understanding the processes, we seek an understanding of "why" (Merriam, 1975). Second, we have come to appreciate our inherent problems in data distribution and acquisition and now can cope with it. Thirdly, we are dependent on and closely linked with hardware and software developments outside geology. And lastly, it will take an education effort and perhaps a new generation of geologists to develop geologically oriented techniques to solve geological problems.

Hardware

We can expect the continuing development of smaller but bigger capacity, faster machines (Branscomb, 1979b). Recent developments of large-scale integration (LSI and

VSLI), optical fibers, and storage technology in the form of magnetic bubble storage devices have contributed to these trends (Frazer, 1979). They also have contributed to the continued decrease in cost per bit of about 28 percent per year for random access memory (RAM) and 15 percent per year for cost performances for large systems (Branscomb, 1979a). The increase in speed is enormous (Table 2), in fact the difference is much greater than the difference in a person walking 4 miles per hour and the Concord flying at 1200 miles per hour. Size for physical space required for a megabyte of storage has decreased greatly (Table 3) in just 16 years (IBM, 1979). Remember the old first generation machines with their tubes and water-cooling systems? What a spectacular change in such a short time.

There will be improved and easier accessibility of hardware. Minis and micros now are commonplace and the personal computer has arrived. A personal computer consists of a microprocessor (miracle chip), keyboard, memory unit, and TV screen and may be programmed in BASIC for example. They sell for as little as $99 for National Semi-Conductors SC/MP model, $350 for INTEL's SD-80, and $499 for Radio Shack's TRS-80 or $600 for their popular PET. Personal computers accounted for $5 million in sales in 1975 and $55 million three years later in 1978! In a few years every geologist will have one of these personal computers for his work.

Software

Computers will become easier to use; the software will be more user oriented. Eventually synthesizers will be used - that is, machines will be available which hear, speak, and understand (Robinson, 1979). Optical scanners are available already to convert the written word into sound (readers for the blind for example).

Table 2. Computation speeds (Knuth, 1976).

man (pencil and paper)	0.2/sec
man (abacus)	1/sec
mechanical calculator	4/sec
medium-speed computer	200,000/sec
fast computer	200,000,000/sec

Table 3. Cubic feet/1 million characters of storage (IBM, 1979).

1953	400 ft^3
1959	100 ft^3
1970	8 ft^3
1976	0.3 ft^3
1979	0.03 ft^3

Some progress has been made to date in speech recognition for the purpose of automating writing. Imagine being able to dictate your field notes or a manuscript and get a written draft! Many other manual operations will be automated as well so data-acquisition methods will be improved.

Fast algorithms will be developed which in turn will speed up analyses (Kolata, 1978). Already the FFT (fast Fourier transform) has revolutionized some aspects of computing. Now comes the manipulation of polynomials and power series and a system to solve simultaneous linear equations (Davis, David, and Belisley, 1978) which will allow faster computations. Many computations now will become available in practice rather than just in theory.

Also witness the solution of the four-color problem which involved some 1,200 hours of high-speed computing. Does this type of problem solving signal a revolution in the theory of knowledge?

Telecommunications (the marriage of communications equipment and computer technology) will appear commonplace on the geological scene in the not too distant future. Already computer conferencing has been tested in the geoscience community. Many benefits were demonstrated including sharing data bases, joint writing, improving group conferences, maintaining communications on a continuing basis, and ease in dissemination of scientific information (Vallee, Askevold, and Wilson, 1977). Increased useage can be expected with decreasing costs and increased accessibility of hardware and software. The nearest communications network (e.g. TYMSHARE or TYMNET) is only a telephone call away from wherever you are to put you in voice contact with analytical methods, data bases, information systems, other workers in the field.

Geological

In addition to improving the methods already being used in geology new ones will be developed. They will be geologically oriented algorithms designed specifically to solve geological problems. Some of these will be improved simulation methods, some self-correcting iterative methods, and development of a language to express geological concepts and terminology in mathematical terms.

Simulation is important in geology because it allows the condensation of time (an element so important in geological studies). It does not tell you what to do but what happened - an important aspect in unraveling earth history. Models have been simple and straightforward. In the future these models will become more complex (and thus more realistic) as they are tested in three dimensions for real situations. In this respect, CAI also will become more important than now in transferring and demonstrating complex concepts to students.

Iterative self-correcting methods and fast algorithms will be used with increasing efficiency and enhancing results of geological applications. As a result predictions should be improved.

A geocomputing language will be developed specifically to solve geological problems. An example is formalized stratigraphy where stratigraphic concepts and terminology are expressed in terms of mathematical set notation to facilitate computer manipulations (Dienes and Mann, 1977). This will allow stratigraphic problems to be formulated with mathematical definitions and notations and therefore can be solved by more exact methods. To paraphrase Charles Babbage, any branch of mathematics can advance only as far as its notation permits - the same can be said for geology.

SUMMARY

Many exciting changes in geology can be anticipated during the coming decade. The 1980's will show a continued and rapid expansion in the use of computers in the geological profession as this aspect of the science matures. Computers will continue to decrease in size and price as they increase in capacity and efficiency. They will be easier and cheaper to communicate with on a worldwide basis as networks develop and each geologist acquires his own personal equipment. A simple connection will allow him to browse (or if you like, have it read to him) through current abstracts of leading journals, teleconference with other workers on recent developments, access anyone of many data bases for a quick study and call on any one of numerous techniques available for analyses in order to test an idea, check the most recent grants made by any one of several federal, state, or private organizations, or just plain relax and watch a ballgame or play games.

REFERENCES

Branscomb, L.M., 1979a, Computing and communications - a perspective of the evolving environments: IBM Systems Jour., v. 18, no. 2, p. 189-201.

Branscomb, L.M., 1979b, Information: the ultimate frontier: Science, v. 203, no. 4376, p. 143-147.

Davis, M.W.D., David, M., and Belisle, J.-M., 1978, A fast method for the solution of a system of simultaneous linear equations - a method adapted to a particular problem: Jour. Math. Geology, v. 10, no. 4, p. 369-374.

Dienes, I., and Mann, C.J., 1977, Mathematical formalization of stratigraphic terminology: Jour. Math. Geology, v. 9, no. 6, p. 587-603.

Frazer, W.D., 1979, Potential technology implications for computers and telecommunications in the 1980's: IBM Systems Jour., v. 18, no. 2, p. 333-347.

IBM, 1979, Price performance: Data Processor, v. 22, no. 2, p. 1-6.

Knuth, D.E., 1976, Mathematics and computer science: coping with finiteness: Science, v. 194, no. 4271, p. 1235-1242.

Kolata, G.B., 1978, Computer science: surprisingly fast algorithms: Science, v. 204, no. 4370, p. 857-858.

Krumbein, W.C., and Sloss, L.L., 1958, High-speed digital computers in stratigraphic and facies analysis: Am. Assoc. Petroleum Geologists Bull., v. 42, no. 11, p. 2650-2669.

Merriam, D.F., 1975, Computer perspectives in geology, *in* Concepts in geostatistics: Springer-Verlag, New York, p. 138-149.

Merriam, D.F., 1976, CAI in geology: Computers & Geosciences, v. 2, no. 1, p. 3-7.

Merriam, D.F., 1980, Computer applications in geology - two decades of progress: Proc. Geologists' Assoc., v. 91, no. 182, p. 53-58.

Robinson, A.L., 1979, Communicating with computers by voice: Science, v. 203, no. 4382, p. 734-736.

Starr, C., and Rudmann, R., 1973, Parameters of technological growth: Science, v. 182, no. 4110, p. 258-364.

Sutterlin, P.G., and May, R.W., 1978, Data and information management, *in* Encyclopedia of computing science and technology, v. 9: Marcel Dekker, Inc., New York, p. 27-56.

Vallee, J., Askevold, G., and Wilson, T., 1977, Computer conferencing in the geosciences: Inst. for the future, Melo Park, California, 85 p.

INDEX

Alaska, 140, 171
Alps, 340
Analysis, 73
  association, 73
  autocorrelation, 336
  characteristic, 188
  cluster, 73, 120, 327
  correspondence, 73, 223
  discriminant, 203
  dissimilarity, 73
  factor, 73, 203, 245, 287
  gradient, 224
  principal components, 73, 203, 253
  principal coordinates, 73, 294
  redundancy, 292
  sequential, 75, 371
  signal, 128
  spatial, 329
  statistical, 531
  stereographic, 331
  statigraphic, 211
  stress, 339
  systems, 3
  three-dimensional, 225
APCOM, 184
APL, 33, 328
Arabian Shield, 332
archeological seriation, 85
Arkansas, 178
artificial intelligence, 4

Ashby's Law, 4
Atlantic OCS, 171
autocorrelation, 336

BASIC, 376
Bavaria, 330
Belgium, 328
Bernouilli variables, 58
bibliographic data bases, 38
  design, 311
  GeoArchives, 38, 316, 371
  Geo.Ref, 38, 316, 371, 375
Biogenetic Law, 270
Biolog, 300
black box, 8
block diagrams, 327
Blue Ridge, 340
Bolivia, 16
"bow-ties", 151
Bowen's Reaction Series, 15
branching processes, 276
Brazil, 330
bright spots, 157, 216
British Columbia, 16

CAI (computer-aided instruction), 375, 378
California, 38, 176, 332, 340
Canada, 55, 130, 188, 331
Canadian Shield, 55
canonical correlations, 290, 294
characteristic analysis, 188
Chile, 142, 335
China, 333
CIPW norms, 259

CLAIR, 200, 371
cluster analysis, 73, 120, 327
coefficients
  correlation, 72, 274
  difference, 72
  dissimilarity, 284
  distance, 72
  similarity, 70, 72, 284
COGEODATA, 312
communication networks, 316
  DATAPAC, 316
  TYMNET, 316
  TYMSHARE, 316
computation speeds, 376
computer conferencing, 377
computer hardware, 24
computer languages, 26
computer program, 259
contouring, 135
Cope's rules, 270
correlation, 65, 217
  autocorrelation, 336
  crosscorrelation, 218
  geometric, 118
correspondence analysis, 223
cross correlation, 218
cybernetic model, 9

data acquisition, 377
data bases, 129, 200, 308, 371
data-base management systems, 320
  G-EXEC, 371
  GIPSY, 371
  GRASP, 371
  SASFRAS, 371
data-base systems, 38, 220
  Dialog, 38
  Orbit, 38, 316
DATAPAC, 316
data-retrieval software, 316

Deep Sea Drilling Project, (DSDP), 179
Delauney triangles, 136
Delphi method, 189
desk-top computers, 30, 376
deterministic model, 48
Dialog, 38
discriminant analysis, 203

earthquake, 39
ecolog, 289
ecological diversity, 287
ERTS imagery, 330, 332
end-member compositions, 245, 248
England, 87
entropy function, 68
evolutionary sequences, 87
exploration, 183
  geochemistry, 190
  geophysics, 167

factor analysis, 73, 203, 245, 287
fast algorithms, 128, 377
faults, 330, 336
federal leasing policy, 140
FILEMATCH, 319
finite-element method, 325, 332, 337
formalized stratigraphy, 223
FORTRAN, 258, 326, 327, 328, 330, 331, 333, 335, 336, 341
Fourier analaysis, 128, 218, 328, 332
Fourier transformation, 128 153, 215, 334, 377
France, 333, 335
Fresnel diffraction, 148

G-EXEC, 371
games, 4
GeoArchives, 38, 316, 371
Geochautauqua, 39
geodynamics, 341
Geological Survey of Canada, 189

geometric correlations, 118
Geo.Ref, 38, 316, 371, 375
geoscience library, 310
geostatistics, 190
Germany, 179
GIPSY, 371
Goedel Incompleteness, 15
gradient analysis, 224
Grand Canyon, 340
graph theory, 222
graphics, 327, 329, 371
GRASP, 371
Great Britain, 190
Gulf Coast, 69, 140
Gulf of Alaska, 171
Gulf of Thailand, 155

hardware, 320, 375
harmonic analysis, 328
Hawaii, 340

IAMG (International Association for Mathematical Geology, 177
ICSU (International Council of Scientific Unions), 312
IGBA (IGneous BAse) 202
IGC (International Geological Congress), 218
IGCP (International Geological Correlation Program), 66, 202, 218
igneous differentiation trends, 205
image analysis, 52, 106, 107, 330, 332
Information
  processing, 307
  resource, 309
  service, 37
  storage, 309
  systems, 305

independent-events processes, 274
index fossils, 67
interactive graphics, 329
Italy, 330
IUGS (International Union of Geological Sciences) 312

Juggernaut Model, 5

Kansas, 134, 138, 335
Karroo Basin, 332
keywords, 311
Khev-grdzeli section, 10
Kriging, 329

LANDSAT, 107, 332
laws, 9, 16, 270
least-squares fitting, 202
lineament, 106, 332
linear model, 57
Lisbon, 39
logistic model, 58
Louisiana, 140, 178, 333
Lyellian method, 67

map data, 50
Markov models, 10, 220, 271
Mexico, 16, 137
Michigan, 335
mineral-resource evaluation, 17, 43, 188, 191
minicomputer, 132
mining geology, 181
mixing problems, 243
  geochemical, 243
  petrologic mixing, 202, 243
models, 9, 10, 48, 57, 58, 109, 140, 158, 188, 202, 205, 215, 220, 244, 271, 328, 333, 337, 340, 341, 378
Montana, 328
Monte-Carlo simulations, 274, 332
Moon rock, 39
morphology of structural features, 330

multivariate analysis,
    54, 70, 188, 218, 371

Nevada, 342
New York, 73
Nigeria, 289
nonbibliographic data
    bases, 40, 313, 317
nonlinear-regression
    models, 341
Norway, 327
numerical taxonomy, 70
NURE (National Uranium
    Resource Evaluation),
    171

oceanography, 179
Oklahoma, 132, 331
Ontario, 51, 52, 188
optical-rose diagrams,
    327
Orbit, 316
ordination algorithms, 73
ore-reserve estimation,
    191
Oregon, 289
orientation analysis, 285
orientation diagrams, 327
Ouachita Mountains, 340

paleoecology, 222, 283
paleoenvironmental, 222
paleontology, 267
pattern recognition,
    218
petrofabric studies,
    327
petrogenetic system, 11
petroleum geology, 125,
    329, 340
petrological data bases,
    200
    CLAIR, 200, 371
    IGBA, 202
    PETROS, 200
    RKNFSYS, 200
petrological mixing
    models, 202
petrology, 199

petrophysical analysis, 133
PETROS, 200
perception models, 109
personal computer, 30, 376
photointerpretation, 105
point events, 77
Poisson distribution, 49
Poisson regression, 57
Polya's Urn problem, 278
Pomona College, 25
population dynamics, 285
Prestel, 36, 37
principal-components analy-
    sis, 71, 73, 203, 253
principal coordinates, 294
probabilistic approaches,
    218
probability index map, 55
problem-solving, 1
programming languages (also
    see software)
    APL, 33, 328
    BASIC, 376
    FORTRAN, 258, 326, 327,
        328, 330, 331, 333, 335,
        336, 341

Q-mode analyses, 71
quantitative biostratigra-
    phy, 63
Quebec, 55, 56, 188

RBV (relative biostratigra-
    phic value), 69
R-mode analyses, 74
random-process models, 220
random walks, 271
range zones, 81
redundancy analysis, 292
regional assessment, 140
relative entropy, 288
resources
    evaluation, 140, 183
    fuel, 169
    inventories, 175
    mineral, 169
    policy, 170
RKNFSYS, 200
Rocky Mountains, 328, 332

INDEX

salt domes, 340
sampling designs, 189
SASFRAS, 371
satellite imagery, 330
Scotland, 330
seislog, 165
Seismic
  reflection, 145, 157
  stratigraphy, 227
  surveying, 128, 213
seismograms, 158
sequence of events, 76
sequential analysis, 75, 371
seriation, 218
set theory, 222
"Shannon-Wiener Index", 287, 288
shape of folds, 328
shrunken estimators, 296, 299
Siberia, 329, 333
signal analysis, 128
similarity coefficients, 70, 72, 284
simulation, see models
SKYLAB, 330, 332
slotting, 218
software, 320, 376
spatial analysis, 329
spatial information, 106
spatial resolution, 107
spline surfaces, 335
statistical (also see multivariate analysis)
  analysis, 331
  exploration, 45
  geostatistics, 190
  laws, 270
  methods, 202, 336
stereographic analysis, 331
stochastic-process models, 220
stratigraphic analysis, 211, 223
  analysis, 211

  correlation, 217
  data systems, 220
stress analysis, 339
structural geology, 148, 323, 326, 328, 330, 336
subsurface, 334
synthetic seismograms, 215
synthetic sonic logs, 162
systems analysis, 3
systems behavior, 1

telecommunications, 377
teleology, 4
Teletext, 36
  Ceefax, 36
  Oracle, 36
terminals, 30
Texas, 178
text-editing systems, 269
three-dimensional analysis, 225
time series, 215
Treatise on Invertebrate Paleontology, 269
trend surfaces, 225, 333
TYMNET, 377
TYMSHARE, 377

United States, 130
U.S. Bureau of Mines, 189
U.S. Geological Survey, 189
USSR, 179, 330

Viewdata, 37
Virginia, 188, 332
virtual machines, 29
virtual storage, 29

Wales, 332
wavelet processing, 159
well-logging, 132, 221
Western Interior, 88